# EXPANDING
# GEOSPHERES

## ENERGY AND MASS TRANSFERS
## FROM EARTH'S INTERIOR

*C. Warren Hunt*

*Lorence G. Collins*

## THE MAGNIFICENT SCENERY OF THE MOUNT BRUSSILOF MINE
## BRITISH COLUMBIA:

Viewed looking eastward across the Rocky Mountain crest and the Mt. Brussilof magnesite minesite. The scene illustrates the linkage between metal ion transfer from mafic rock to metal ore. The mafic rock, stripped of its metals, becomes granitized, lightened, and rises isostatically, creating mountain scenery in the process.

# ELEVATIONS OF THE GEOID ACROSS THE USA

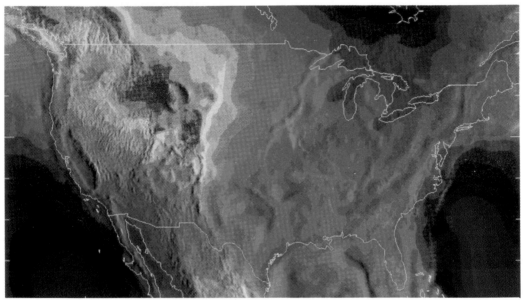

Imagery courtesy of Dr. Dennis G. Milbert, Advanced Geodetic Science Branch, U.S. National Geodetic Survey.

This image is a color-shaded computation of geoid height covering the conterminous United States and environs. A 1 600-km (1,000-mile) diameter welt is the key feature that depicts the aberration from ellipticity of the Earth's level surface. Its center is the Yellowstone volcanic hotspot. The welt is the positive effect produced by excess underlying mass. The variation from the geoid has a net amplitude of 53 m. Arms of the welt extend north to the Canadian Rockies, northwest to the British Columbia Coast Ranges and south to the Colorado Rocky Mountains and into New Mexico. The mid-continent gravity high is also visible as a low-order positive feature.

# EXPANDING GEOSPHERES

## ENERGY AND MASS TRANSFERS FROM EARTH'S INTERIOR

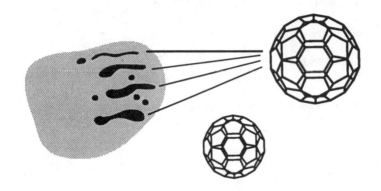

*A Sequel to* **ENVIRONMENT OF VIOLENCE:**

*Editor:*

C. WARREN HUNT

Authors:

C. WARREN HUNT
LORENCE G. COLLINS
E. A. SKOBELIN

ISBN: 0-9694506-1-3
U.S. Library of Congress Catalogue Number: 92-80918

Manufactured in the United States of America
By Conrad Mollath & Co.
San Mateo, California 94403

Published by POLAR PUBLISHING
P.O. Box 4220, Station C
Calgary, Alberta
Canada T2T 5N1

Canadian Cataloguing in Publication Data
Hunt, C. Warren  (Charles Warren), 1924-
    Expanding Geospheres

Includes bibliographical references and index.
ISBN  0-9694506-1-3  (bound)

    1. Earth--International structure. 2. Geophysics.
3. Geology, Structural.  I. Collins, Lorence G.
(Lorence Gene), 1931-  . II. Skobelin, E. A.
III. Title.

QE33.H87 1992       551.1       C92-090159-X

*In Memory of*
**PATRICIA GAYFORD HUNT,**

*Whose unceasing supportive effort
made possible the writing of this book.*

## TABLE OF CONTENTS

\* Initials of principal author: continues to change shown by appearance of different initials.

## CHAPTER V

## Chapter VI

## Chapter VII

## ILLUSTRATIONS

**Frontispiece**: Scenery of Mount Brussilof mine, British Columbia
**Reverse side**: Distortion of the geoid in the western USA

### Chapter III

## Chapter IV

## Chapter V

## Chapter VI

# FOREWORD

To open this book is to take an untrod trail to the fringes of factual knowledge. The reader becomes an explorer in search of new vistas. Awaiting silently are unforeseen insights and the intrinsic thrill of discovery. This is what exploration means. The mission of the editor is achievement of the goal for the widest possible readership.

Tenure of the status quo in geology is a mental stasis akin to the religious yearning for "everlasting peace;" and it is equally unattainable. Nevertheless, stasis is often stoutly buttressed by editors' out-of hand dismissals of innovation and by the arrogations of authority taken by peer reviewers. Only "experts" [so acknowledged by one another] are able to finesse their innovations past these barriers at all, and then only with persistence. Repetitions of conventional wisdom are reentered into the literature for new consideration *ad infinitum*, usually ignoring inconvenient contrary facts. The technical press, thus sanitized, avoids unvetted heresies such as you will read herein.

The curious problem about science that refuses to consider alternative explanations is not what it appears to be, a wish to censor, but rather, a dogged reluctance toward reconsidering fundamentals. In fairness, editors are often chastised by irate "experts," specialists with deeply-held perceptions in their fields. These people leap with fury at a published idea that denies their perceived "truth." Notwithstanding that such "experts" have been proved wrong again and again in science history, they are more numerous today than ever before.

Thus it is that science goes nowhere for long periods. Geological science in particular has been in this state of torpor, since about 1960. That is when geophysicists, needing to explain the newly discovered active incrementation of the submarine seascape, exhumed an old concept, continental drift, from the boneyard of discarded theories. Geologists had dismissed continental drift for insufficient geological support. It was renamed plate tectonics, embellished with the still-scanty geology of the seafloors, bolstered by fanciful mathematical systematics, and hidden behind curtains of newly-minted acronyms. Thus, the paradigm of modern geological science was born.

Some of the resulting adductions of plate tectonics are quite implausible. Convergent masses with nowhere to go express absurdities within the constraints of spherical geometry. Petrology and observational geology are discounted in favor of unseen happenings out of sight under the ocean floors. The concept survives in part because no alternative has been found to counter its preposterous claims and absurdities. CARBIDE-HYDRIDE interactivity comprehensively fills this hiatus. Put forward originally in my 1990 book, ENVIRONMENT OF VIOLENCE (the "EV volume"), this book explores the scientific evidence for the processes of mass and energy emanations from inner Earth and their addition to its outer shells, the "EXPANDING GEOSPHERES."

Thus, what you are about to read is intended to upset the stasis in geology. Wishing to be understood by informed laymen, I am less than happy to have to say that this book has much more technical language in it than the EV volume. Although trying to keep technical language to a minimum, that goal has proven more difficult than anticipated; and some chapters are likely more than a lay person will be able to absorb easily. To try to relieve this problem, supporting technical detail, where possible, has been moved into footnotes. An extremely involved, and lengthy description of the nature and origin of myrmekite, which provides indirect but compelling evidence for silane emission from deep Earth levels, is segregated from the main text into an appendix. In this way the general reader may be assisted by being able to follow the "story" without wholly absorbing the footnotes and appendix.

It has been an exciting journey in science putting this together. The co-authors join me in hoping you find it that way to read.

**C. WARREN HUNT, EDITOR**

# PROLOGUE

### C. WARREN HUNT

**ENVIRONMENT OF VIOLENCE** challenged the Lyellian view of continuous change, the perception that all natural processes are phases of inexorable down-winding. Lyell's view was built on the 18th-century perception of James Hutton, that Earth dissipates heat, which could energize some geological processes. Lyell's winding-down systematics accept that premise by looking only to the dissipation of primordial, endogenic heat or to exogeny for the energy requirements of geology. Modern geologists adhere to this outlook. A recent critic [waggishly] dubbed the attribution of *all* Earth's endogenic energy to primordial endogenic heat "Hutton's perpetual heat machine."(Zysman 1990)

No matter whether we deal with *geological degradation* [such as erosion processes] or *progradation* [as in mountain-building], *all* changes that occur, in Lyellian terms are interpreted as manifestations of the gradual and consequential dissipation of primordial Earth heat energy. This book deals with progradational processes in which heat is an essential component, an interrelationship that must be regarded as a universal truism for growth in general and for geological Earth features in particular. The following quotation sums the conventional outlook:

"Given high initial temperatures, the Earth's thermal inertia is so great that much of the primordial heat is still leaking out. There is no need to postulate new heat sources. Nevertheless, magma generation in the Crust is a problem [regarding] its heat source and volatile content."(Lowman, 1990)[1]

Regarding primordial heat in this way, as the sole significant source for all Earth processes, leads to many problems, however. Especially obvious is the fact that some parts of Earth's Crust are much hotter than others. If everything were winding down, there should be an even flow of heat from the interior, varying only for minor conductivity differences of different rock types. All places on the surface should emit about the same amount of heat. Instead we find extreme, up to 10-fold, variations.

An apparent resolution for the conundrum seemed to be found fifty years ago, when atomic disintegration was recognized as a general heat generation process, and thus, a possible explanation for observed conditions. Unfortunately, quantitative efforts to correlate the known crustal and mantle distribution of radioactive minerals with anomalous regions of high-crustal heat flow have failed completely; and the elusive heat source has remained unidentified. There is no alternative explanation that is consistent with Lyell's constraint.

This brings us back to the dilemma. Mention has already been made of the fact that primordial heat, the converted equivalent of the kinetic energy released by infalling interplanetary debris, must radiate and disperse according to distance from origin, only locally slowed or speeded by minor variations of lithosphere conductivity. "Hot spots," the geologist's term for regions of anomalously high-heat outflow with up to ten times the average, imply a drastically different explanation. The adopted one is that "plumes" of hot Mantle rock rise ("upwell") beneath the lithosphere bringing primordial heat with them.

---

[1] *The idea of heat dissipation as the engine for planetary surface modifications has been elaborately worked over, embellished, and embroidered with cogent observation and geophysical adduction despite the internal inconsistencies one meets in trying to match it with geological reality. If Earth cooling has been continuous over 4.5 billion years, a virtually-perpetual heat machine is, indeed, implied; and heat redistribution into the erratic pattern we find today is impossible without enormous Mantle turnover, convection, as it is euphemistically labelled. The need for convection of the Mantle is used to imply its existence, even a fluidic ease with which it must occur. In the coming pages we will consider this proposition and demonstrate that it is both unnecessary and, in all probability, unreasonable.*

Then, since the upwelling mechanism would produce a surfeit of new crustal rock, the plume, after cooling at the base of the Crust, is imagined [without evidence] to cool, gain density in cooling, and then to plunge downward ("downwell") around the upflow in the manner of a fountain of water. The whole stately mechanism is called a mantle "convection cell." Although little can be proved for or against the idea, and there are many geologist non-believers, no satisfactory alternative has been available.

Here, then, we have two of the linchpins of modern Earth theory: **(1)** the accretionary origin of the Planet, with a heat legacy, and **(2)** convective circulation of the solid Mantle. Neither is demonstrable with direct evidence nor indisputable in view of indirect evidence. It is *not* foregone that the surface energy of our Planet is controlled by mantle plumes that dissipate primordial heat, although one might be forgiven for not realizing as much in view of the reverence of the geoscience community for the idea. These two great "levers of power" of the Earth [to use a political analogy] should stand up to scrutiny. In coming pages the reader will see that with a hard pull on these levers the theories fall apart altogether. In the modern idiom, "Stay tuned!"

If we start at the beginning, that is to say with accretion, there is no denying that Earth history must be a form of winding-down. That means there must have been an initial winding-up. Geoscientific adduction on this point urges upon us the explanation of universal explosion, the cosmic "Big Bang." This concept, a unidirectionally expanding universe, was implied early in our century by Hubble's discovery of the "red shift." The light spectra of stars at greater distances from Earth were shifted toward the red end of the spectrum, suggesting that the stars recede at high velocities from an original central site not far from Earth's galaxy. The initial explosive source was seen to define the center of the Universe. This seemed to fit known facts with a simple solution. It set back the question of a unique origin for the Earth to one of a unique origin for the Universe. Neither was any answer forthcoming on how the energy for the Big Bang was concentrated in the first place.

New data, however, raises doubts as to the efficacy of the "Big Bang" as the explanation for the Universe. One of the first to raise doubts was Halton Arp (1987), who showed that bodies with maximal red shifts, the quasars, may not be so far away as their red shifts suggest but are likely produced in the galaxies adjacent to their astral positions. He lost his tenured professorship, apparently for that heresy. Further doubt was later fuelled by the discovery that "dark" matter represents most of the mass of the Universe and is distributed so unevenly as to discredit the possible interpretation that it represents matter radiated from a single explosive event.

Much like the uneven heat distribution in the Earth, the uneven matter distributed in the Universe does not lend itself to a unitary source. In fact, the Big Bang idea should have been regarded as simplistic and probably implausible from the start. How could we seriously consider the entire Universe to have been energized in one grand cataclysm? How plausible is a long journey of winding-down into ultimate randomness and chaos, if there did not exist some countervailing agency to wind the Universe up in the first place? Where everything else in the Universe appears to occur in multiples both in space and time, is it not presumptive to assign singularity to a perceived initiating process, which then becomes the essential precedent for the opposite, gradualism [geological peace in our time]!?

**Does geologic evidence give any significant support for initial cataclysm followed by 18 billion years of winding down instead of repetitious lesser cataclysms of winding-up interspersed with long intervening periods of winding down? I think not.**

I think it is only too human to contemplate with resignation such perplexing problems. Like the "multitudinous seas incarnate" that overwhelmed Lady Macbeth's psyche, the scientific community, challenged by complexity, seeks a simple *and singular* [and usually wrong] theoretical solution. This inclination toward singularity in a solution uncovers a flaw of human psyche in which singularity becomes a fatal attractor. "Singularity" is a useful mathematical concept to establish and manipulate uniqueness. But that is what it is, a mechanism for manipulating information. Uniqueness, and, hence, singularities, hardly deserve attention as a means to uncover scientific truth because nature abhors them. Singularity should be seen as a psychological trap in observational science, an attractive lure to be avoided on principle.

So where does that leave us? If nature abhors uniqueness, is there any provable case of uniqueness in nature? If there is uniqueness in our Universe, such cannot be demonstrated. Neither can anything be shown to stem from a singularity. To invoke a single cataclysm to explain our Universe defies reason in cosmology just as a unidirectional and gradual slide into chaos is uncharacteristic of observable geology.

Gradational down-winding was shown in the EV volume to be unsatisfactory to explain even degradational processes in geology. In the case of progradational, geological processes this is even more so. The progradational processes create new structure and new Crust on our Planet, while their organic counterpart, the life processes, also contribute to geological construction. All progradational behavior conflicts with the idea of cataclysmic "singularity."

The geologist who treats gradualism according to Lyell's formula, as *the* singularity of geological science, despite good evidence to the contrary, takes a sclerotic and irrational position, as well as one that is ineffective. The uplift and subsidence of mountains, plateaus, and seafloor terranes are known to have happened at times slowly, at other times with sudden motion. Earthquakes, volcanism, "hot spots" in the Crust, all of these are well recognized, and all receive intensive research. But no consistent explanation has emerged, and nothing seems to fit. Many voices object to the status quo, but no alternative has been found, despite all the effort.

A new Earth theory must accommodate the ubiquitous dualism of degradation [destruction] vs progradation [creation], yin vs yang, male vs female, positive vs negative, organic vs inorganic. If no opposites are apparent, we seem to invent them: high road/low road, matter/anti-matter, eternity/Big Bang. Intrinsically, singularity is unnatural, a violation of duality and multiplicity. The observable conditions of duality and multiplicity are essentials in the skein of nature.

My goal, then, in ENVIRONMENT OF VIOLENCE was to give perspective to the essential duality of gradualism and cataclysm, and to explore cataclysmic exogeny by contrast with cataclysmic endogeny in the geological record. Whether I succeeded, readers will decide. The attempt to show a balanced view of the graveyard of cataclysms yielded a surprising bonus. The carbide-hydride hypothesis seeded an entire new theory of the Earth.

The conception of carbide-hydride systematics (CHS) occurred as the EV volume was being set by the printer, and there was no opportunity to explore its profound consequences. This book is intended to carry through on that subject by exploring the behavior of rock in observable situations where possible and, in theory, where not. Systematics is the "in" term I use; but I really like another term better. Hutton might have expressed the emanation of volatile carbide-hydride gases as "humors," a curiously apt term because at once it evokes, appropriately, the images of "essences" and "vapors." Concern that detractors might distort the term "humorous" and aver its author to be a victim of the vapors led me back to the modern term, systematics.

Putting the archaism to bed, I and L. G. Collins will try to demonstrate that the carbide-hydride humors offer nothing less than rationale for resolving two of the greatest conundrums of geology.

**(1)  The energy source of inner Earth, and**

**(2)  The enigmatic mechanism by which vast amounts of silicon and carbon have been relocated from deep to shallow geospheres.**

In prospect of analyzing these problems, it may be instructive to marshal the facts and problems with which we start. They are as follows:

1.  **Continental Crust [felsic rock] has formed and grown through geological time, with a different general composition from oceanic Crust [mafic rock].**

2.  **The origin of granite [felsic rock] is not understood.**

3.  **Basaltic rock is added to the Crust at spreading centers, and in the past at sites of great basalt "floods" [also called "trapp"].**

4.  **Volcanoes normally emit rock fragments that are mainly felsic materials, or, alternatively, they emit lavas, which are mostly basaltic.  In a third category, some explosive volcanism emits only gas without solids [the case in which the volcano is a "gastrobleme"]. The origins of the huge quantities of felsic rock fragments ["ash"] on Earth's surface and the origins of the compressed gas are enigmatic.**

5.  **Block faulting, thrust faulting, development of décollements and nappes, and regional doming and subsidence are geologically prominent occurrences, for**

which provable vertical forces, uplift and gravitation, or hypothetical mantle convection may be [but cannot be proven to be] responsible.

6. The origins of extremely-pure quartz sands and quartzites are enigmatic.

7. Existing theories of the nature of earthquakes have not facilitated prediction of new tremors. Aseismic creep is inexplicable on the basis of existing brittle fracture theory.

8. The provenance of vitrinite coal and its associated hydrocarbons and trace metals [especially germanium] are not understood.

9. The origins of metal ores are not well understood in most cases.

10. The compositional makeup of gas entrapped in crustal rock and ambient in its porosity is inadequately explained in many circumstances. Biogenesis of hydrocarbons and subduction tectonics are unable to explain observed occurrences.

In attempting resolution of these problems our basic perceptions of what is known in geology will have to be examined critically, among these:

A. Earth origin: accretion? - or relict core of a gas giant?

B. Thermal history of Earth: molten? - or solid?

C. Origins of geospheres: gravitational differentiation? - or hydridic differentiation?

D. Separation of continents: plates driven by convecting Mantle? - volatile diapirism with Earth expansion?

This is certainly an enormous spectrum of problems, a challenge to know how to proceed effectively. I resolved to request contributions from others. A paper by Russians, E. A. Skobelin and colleagues (1990) had impressed me. They took a tack entirely new to my thinking. They had asserted that theoretical geology is in disarray and in need of a whole new theory of the Earth. They asked rhetorically:

**"Has plate tectonics resulted in a revolution in geology? ... Plate tectonics is not a theory and, therefore could not result in a scientific revolution. Present-day geology is in a state of deep stagnation, ... [and] mature for a revolution. ... One must struggle against the 'esotericity' (accessibility only to initiated people) of geological science [that is] propagated as plate tectonics. This is a great barrier to progress."**

Nothing so challenging was being said in North America. For example, five noted geologists from American academia in a symposium sponsored prestigiously as a "Nobel Conference" (Carlson, Ed. 1990) expound their beliefs in plate ideology with elegiac prose, without references, and wholly ignoring possible alternatives. A sixth paper in the same symposium volume, by David Ray Griffin, a professor of the philosophy of religion at Claremont, California, stands out in contrast. Griffin, the only non-geoscientist among the authors, is apparently the only one with anything new to say.

Griffin takes a stance of apposition to geologists, wherein he considers *"the internal principle of unrest"* of the Universe. Recognizing Nature's universal capacity for self-organization as contrary to the absolute solutions expected from modern experimental scientific practice, he calls for a new order in which:

**"... the sciences of the simple and the dead will not be considered more scientific and more prestigious than the**

sciences of the more complex and alive and restless. The higher sciences, such as the social psychology of humans, which must deal with many more variables, and the more inclusive sciences, such as geology and cosmology, which must include a large number of the more limited sciences, will be allotted the prestige and support commensurate with their difficulty and importance.

The time is ripe for a major revolution in the worldview adopted by the scientific community."

So, here we have the same call from able minds working in different fields of learning as well as in geographically and politically opposite climates half a world apart. My own background, being much different from that of Skobelin or Griffin, echoes strongly their calls for a worldview unimpeded by prescriptive accommodation of established interpretations of the incomplete data sets with which geoscience works.

How is the scientific community responding to these calls? Not at all, if my experience is any measure. Hostility is rampant to large leaps of theory. Unconventional views are given short shrift, and manuscripts are returned to innovative authors with terse letters of rejection.

Accepted thinking in America, where fads prevail and disputatious behavior is firmly eschewed by disputatious people, non-conformism is staunchly proclaimed to be the universal objective. But let no non-conformist deviate from the rules of non-conformity! **Rule #1:** Small steps only are tolerated [large steps are regarded as arrogance]. **Rule #2:** If one is to be heard, one must have the stature accruing with position in an acceptable hierarchy [e.g. university, larger corporation, or government agency]. The hidden meaning of this attitude is the preservation of influence against the intrinsic challenge to power by one who might rock the boat.

For these reasons geoscience pioneers outside the mainstream innovate more readily than their colleagues, who are subject to the tyrannies of hierarchies. Skobelin and his Soviet colleagues have been isolated by political sequestration and the language barrier. Griffin is immunized by working outside the technical fields altogether. For myself, I decided at an early age to spend my life outside hierarchies, notwithstanding the difficulty of survival without their financial support. Of the contributors, only L. G. Collins has attempted to tread a conventional and western scientific path. The reward for his efforts, apparently because he has digressed from orthodoxy, has been minimal publication.

Before launching into technical items, let the co-authors, E. A. Skobelin and Lorence G. Collins, introduce themselves.

# DISARRAY IN GEOLOGY

### E. A. SKOBELIN

It is most unlikely that a reader would chance to read an introduction to a book by another person whose views were contrary to those of the author. This rare event occurs in this instance; and it implies that there exists a need for "new thinking," for a way out of the impasse in which modern geology finds itself. Perhaps our colleagues should learn from the experience of political leaders, who have managed to work out and implement a new general strategy to bring an end to long-drawn-out political crises on our Planet.

There are two opposing stances on the current state of geological thinking, the "advocates" and the "dissenters," one might call them. The first are, with rare exceptions, orthodox plate tectonocists ("ultra-mobilists"). They take the position that plate tectonics is the only true theory of the Earth, a flawless paradigm, a strong foundation on which science can proceed.

According to this theory, there is no rational possibility for an alternative theory or even an hypothesis, and thus, no basis for even searching for one. They proclaim their excitement upon *"acquiring a paradigm and appearance, in connection with it, of more esoteric (understandable-only-to-initiated-people) scientific research that serves as evidence of maturity in a particular field. Since then, geology and geophysics, based on the new paradigm, have accumulated an abundance of new data and hypotheses; thus scientific research has become increasingly esoteric. This is a very important moment, when it is important to understand modern trends in the Earth sciences."*[2]

---

[2] *Miyashiro 1985; Skobelin 1990*

The dissenter meanwhile, his more focussed and sober outlook easily recognizing the "paean" of excitement by which the advocates celebrate their "grand resolution" of previously-unexplained information and unconnected hypotheses, sees the developments more like a "funeral march" for both the science of geology and for plate tectonics as a theory. He or she will recognize in the esoteric research an element of scientific adventurism, which has remained undetected by the advocates.

**This is certainly a "very important moment ... for understanding modern trends in the Earth sciences."** It must be obvious that no general theory of the Earth exists in modern geology. This state of affairs can be compared with chemistry during the period of alchemy. This is why there is so little rational understanding and, therefore, successful prognosis. The task before us is to create a general theory that will incorporate all that is known and proven in geology, a single internally non-contradictory, unitary theory.[2]

All attempts in this direction should be welcomed. Effort should be made to utilize all the creative potential of geologists in general to reach a solution to this problem. Any incipient monopoly should be quickly terminated as such have been in economics in order to create a free market and to assure competition among ideas, hypotheses, and theories.

This book attempts to deal with a new Earth theory based on the concept of carbides and hydrides at Core level. A closely related proposal was made in the USSR by V. N. Larin (1980).[3]

I have also made an attempt to construct a general geological theory in my book, "Theory of Magmatism and Endogenous Mineralization"[4] in which I have developed completely different ideas in accordance with classical logic - Mr. Hunt is my competitor! But, as the great Voltaire said, "Your opinion is

[3] *Larin, 1980*

[4] *To be published in Russian*

inherently antipathetic to me, but I am ready to sacrifice my life for your right to express it!"

With due respect to Voltaire, and in violation of common logic, I prefer not to be so orthodox. I regard Mr. Hunt not merely as a competitor but also as an ally, since we are united in a mutual goal to save geology from stagnation.

I suggest to our colleagues that they pay close attention while reading this book. They should compare the evidence presented with geological facts known to themselves and then make up their own minds. By so doing, the common aim of finding a solution to our problem will be served with creation of a consensually accepted geological theory.

My appreciation is expressed to Mr. Hunt for inviting me to participate in the introduction to this book and to contribute to its contents.

**E. A. SKOBELIN**
**Krasnoyarskii Rabochii, 159-26**
**Krasnoyarsk, 660093, Russia**

# SYMBIOSIS IN GEOLOGY

**LORENCE G. COLLINS**

In my book, HYDROTHERMAL DIFFERENTIATION AND MYRMEKITE, (Collins 1988a) my theory was clear: that granitic rock, which contained the elusive mineral intergrowth, myrmekite, [Appendix] must have crystallized at temperatures below the melting points of its components and that it must have been derived *in situ* from older mafic rocks. Hydrous fluids, I reasoned, must have brought in silica and potassium from depth to displace the mafic elements, calcium, aluminum, magnesium, iron, and titanium, which must have escaped and been ejected upward as volcanic rock.

Apparently supported by field evidence, this scheme still had deficiencies. If granite formed *in situ* below melting temperatures, where was the heat source for associated volcanism? What source could provide the large amounts of water that would have been essential to catalyze the process and transport the displaced metals away? Whence the large influx of silicon?

Despite the seeming insurmountability of these problems, I favored the *in situ* interpretation of granite formation for many sites I had studied. Thus, I had published my data in the hope that in time the problems would become resolved. My papers (Collins 1988b, 1989) gained the attention of C. Warren Hunt, whose deduction that large volumes of silanes must have emanated and continue to emanate from deep-Earth levels provided the answer to my problems. Reacting with water, silanes produce silica, heat, and hydrogen. Reacting with oxygen, they produce silica, heat, and water.

Serendipity! This was exactly what I was looking for. My data supported his hypothesis. His supported mine. This symbiosis ties everything together.

This book investigates archetypal geological effects of silane reactivity. Petrofabrics derived from these beginnings impute an entirely new theory for the energy source and mass contributions to the outer geospheres, in short, a new theory of the Earth.

LORENCE G. COLLINS, Professor of Geology,
California State University - Northridge,
Northridge, California, USA

## CHAPTER I

# EARTH THEORIES

**C. WARREN HUNT**

"Looking back over the way we have come in our discussion of geodynamics we note that, throughout, we were faced with the condition that no coherent theory is known from which all the details follow ... no fundamental equation exists. The only approach to the problem is, therefore, by induction."

**A. E. SCHEIDEGGER 1958**

In the course of writing the book, ENVIRONMENT OF VIOLENCE, it became apparent to me that it was necessary to identify the process by which emanations of gaseous hydrocarbons rise from the Mantle. Worldwide, the alkane hydrocarbons (methane, ethane, etc.) permeate crystalline terranes. Long recognized by many authorities,[5] there has been no explanation for them. Geologists have been steadfast in the belief that mineral hydrocarbons can be generated only from the buried remains of living organisms. As no life is believed possible at the high pressure and temperature conditions of Earth's Mantle, inferences that hydrocarbons are generated there are treated as absurd.

Upon the discovery in about 1960 that emission of successive magmas from seafloor rifts has laid out an orderly series of belts of rock, which preserve successive reversals of Earth's magnetic field, the concept of seafloor spreading was developed. Older ejectites are preserved progressively farther from the rifts of their issuance. This scenario seemed to demonstrate with elegance that mantle rock was upwelling and effusing in a manner that created ocean basins.

[5]  *See EV volume: Hodgson 1981; Gold 1987*

From mantle "plumes" it was now a small step to adduce compensating "downwelling," thus completing the "mantle convection cell." However, the evidence for downwelling is less convincing than that for upwelling. The tangible seafloor lavas with their magnetic stripes are very credible evidence. On the other hand, it stretches the imagination to be asked to regard highly-compressed rock of the Mantle as a churning, viscous fluid driven by the gravity differentials between hot and still-hotter rock. On the other hand, the evidence is incontestable, as already mentioned, that hydrocarbons are dispersed in crystalline terranes, albeit varying greatly in abundance, and in temperature regimes inhospitable to known life forms of the biosphere. No correlation is evident between mantle plumes and the emanating hydrocarbons, although the plume mechanism would seem to be appropriate for the escape of hydrocarbons rising from deeper levels. Neither is there any apparent correlation between the observed dispersion of hydrocarbons in the crystalline Crust and recognized "hotspots."

Starting with only these meager facts, I sought mineral evidence for the requisite hydrogen and carbon in the Mantle or Core. Carbon in the Mantle, in the highly reduced form of carbides, has only recently been recognized.[6] However, carbides frequently occur in meteorites at levels of rock-forming components. They should, thus, be significant in Mantle makeup.[7] Highly-reduced carbides, minerals that carry large energies of formation latent within them, comprise the logical mineral repository for stored carbon in the Mantle. The recent discovery by Leung et al. of the Fuxian kimberlite, northeastern China, demonstrates the repository, and is discussed further in Chapter VII.

Lower-Mantle pressure regimes are generally considered to be 1.3 million atmospheres and rock densities to be near 10.7 $g/cm^3$.

---

[6]  *Leung et al. 1990*

[7]  *See EV volume; Alexander et al. 1990*

Skobelin (Chapter II) questions the high pressure, and Gottfried (later in this Chapter) proposes higher densities. In any case, it is likely that lower-Mantle P/T regimes would be hospitable to carbide preservation, perhaps in association with exotic phases and hitherto unknown molecular states of common minerals.

*Dense metal-carbides in the Mantle could provide the Earth source of carbon for the ongoing emanation of hydrocarbons from the interior of the Earth.*

Hydrogen, the second component, presents a different problem: how could hydrogen, the lightest element, have survived in a core of metal which, itself, according to the current accretionary theory of the origin of the Earth, had settled by gravity to the center of the Planet? The preservation of hydrogen in such a regime must be seen as intrinsically problematical. Volatile elements should have dissipated into space, thus making an hydridic Core impossible.

Notwithstanding this recognized constraint, a number of recent researchers have attributed an hydridic condition to the Earth's Core.[8] Gottfried maintains that inner-Core density exceeds that of iron due to phase change brought on by high pressure, an inherent feature in his interpretation of proto-Earth.

The behavior of hydrogen in metal sponges has received study because of the need to store hydrogen for fuel purposes. Hydrogen enters the metal lattice structure, whence the metal-hydrogen combination is known as an hydride and the metal is said to be hydridic or hydridized. Important new data is emerging rapidly from current research on these materials.

Experimental evidence of the previously-cited authors[8] gives the idea considerable support. Antonov et al. show that hydrogen/iron ratios can be attained that approach 1.0 at

---

[8] *See EV volume; Gottfried 1990; Antonov et al. 1980; Fukai and Akimoto 1983; Suzuki and Akimoto 1989; Badding 1990, 1991*

elevated pressure/temperature conditions, and that such a ratio might characterize the Mantle/Core contact.

The work of Fukai and Akimoto and that of Suzuki and Akimoto lead them to adduce lesser rock density at Mantle/Core contact depth than the density of pure iron [the molten Core]. This suggests to them an environment hospitable to hydride preservation and a deficiency of mass in the outer Core, which they attribute to hydrogen impregnation. The Fukai and Suzuki results are based on laboratory simulations, whereas Gottfried's were inductively reached through cosmology and theoretical physical chemistry.

Badding has done the most recent work under simulated conditions. His results are rather precise determinations of the compression of iron into hydrogen, the rate at which hydrogen impregnation destroys the integrity of crystalline iron, the amount of expansion, and attendant density loss. He concludes,

"Our results show that a large hydrogen component of the core is compatible with current seismological data."

**The metallic planetary Core, thus, has an enormous capacity to store hydrogen, and should easily be able to contribute the hydrogen component of the hydrocarbons that apparently emanate from deep-Mantle levels.**

At low pressures hydrogen atoms entering the crystalline metal lattices add to the weight but not the volume of the metal. With iron this is true up to a pressure of near 3.5 GPa. Above that, the integrity of the metal lattice is destroyed and expansion occurs rapidly. The Badding research finds that outer-Core pressures (near 330 GPa) should lead to densities near 12.5 $g/cm^3$, which compares with recent seismologically determined density of about 12.2 $g/cm^3$.

To now, various handbooks have carried outer-Core densities

of 9.9 to 11.9 $g/cm^3$, an average near 10.7 $g/cm^3$, one-quarter greater than the density of metallic iron. The Fukai and Suzuki school inclines toward the lower figure. Gottfried interprets a 9.4 $g/cm^3$ outer-Core and 30 $g/cm^3$ inner-Core. Badding's figure for the outer Core is, thus, significantly higher, despite the expansion induced by major hydride saturation. Badding interprets that a "structural change" at the atomic level, a "phase change," that is to say, certainly occurs.

Of many metal hydrides known to exist, those of titanium and vanadium are stable at ambient surface temperatures; those of chromium and manganese, metastable; the iron and cobalt hydrides are unstable. At elevated temperatures iron can accept hydrogen almost on a mol-for-mol basis.

Whereas all the cited researchers regard **major hydrides** in the outer Core as likely, **even minor hydrides in that habitat would represent an enormous hydrogen reservoir. Clearly, hydrogen in the Core could be the provenance for hydrocarbons that rise from the Mantle and permeate the lithosphere above it.** Whereas this interpretation must be regarded as speculative, its consistency with the latest research into the subject encourages confidence.

I will now attempt to give an overview of the two divergent concepts of the *origin* of the hydride content of the Core, the Gottfried system and that of Fukai and Suzuki. Proof may ultimately be found in the geological features they engender.

## THE GOTTFRIED THEORY OF THE ORIGIN OF THE EARTH

A German cosmologist and physical chemist, Rudolf Gottfried,[9] writing in the symposium volume entitled "Critical Aspects of the Plate Tectonic Theory," theorizes that because

---

[9] *Gottfried, 1990*

the Earth has an abundance of heavier elements, when compared with a normal planet (a gas giant) or the Sun, it must have lost its complement of lighter elements.[10] Although his physical chemistry is too complex to explain here, the rationale he develops leads to an hydridic primordial Earth. He reasons that the Planet, commencing as a gas giant, would have devolved by the loss of its gas envelope, to the solid Earth we have today.

Gottfried's deductions are carefully quantified from the gas-giant proto-planet of Jupiter-type [both in size and elemental makeup]. Of all atoms comprising the great Planet, Gottfried reasons that 99.7 % would have been hydrogen, the heavier atoms without exception all being concentrated in the planetary core in the form of metal hydrides and other highly reduced complexes.

The astringent density stratification into the "differentiated geospheres" now recognized must have occurred in the gas-giant stage, when chemical segregations were precluded by ubiquitous hydridization. The gas envelope of free hydrogen, being most of the volume and mass of gas giant, is conceived as a sheath above a *solid shell* of "saline hydrides," Gottfried's term for Na-H, K-H, Ca-H, light metal hydrides, Si-H, Al-H, and non-metallic complexes, C-H, Cl-H, F-H, etc. This shell is conceived as the primordial Crust and Mantle. Beneath it a core of heavy metal hydrides would have presaged the present Core.

Gottfried challenges existing theory of the gravity-induced differentiation of the molten Planet into shells of greatly contrasting makeup. He maintains that relative levels of electronegativity and ionization potentials of the differentiated metals are fully respected in his scheme and that resultant crude

[10] *Editor's note: The Gottfried thesis of a primordial hydridic Earth was preceded in 1980 by "Hypothesis of the Primordial Hydride Earth," published in Russian by V.N. Larin of the USSR Academy of Sciences. It is being followed up with a new and more comprehensive book, "The Earth: Its Composition, Structure, and Evolution." Neither book is available yet in English, although consideration is being given to their publication by Polar Publishing. Larin's analysis of the probable hydridic Core preempts Gottfried by ten years.*

stratification by hydridic differentiation of Core, Mantle, and Crust must have occurred within the core of a gas giant before solidification of the silicate geospheres. Differentiation by gravitational settling of heavy minerals from magma is the long-standing interpretation of the origin of the geospheres among geologists. In Gottfried's interpretation this is a flawed and incorrect concept.

> **"The fact that gravitational segregation of individual crystals in a magmatic melt is physically impossible has, however, been appreciated by only a few researchers and the mechanism has received wide acceptance only because there appeared to be no satisfactory alternative. ... We are only just beginning to understand the ways in which igneous rocks are formed ... [and] all previous geochemical interpretations must be re-evaluated.**[11]

Asserting the impossibility of differentiation among highly viscous silicates, Gottfried says:

> **"The only conceivable mechanism able to transport the elements [from Core to Mantle and thence] into the Crust so completely is a[n] hydride one".**

The essence of Gottfried's argument resides in the highly reduced environment, of the hydridic proto-Planet. In that milieu everything would have been mobile; the silicon oxides with their high viscosity would have been absent; the hydrogen envelope would have maintained high pressures on the entire core. Differentiation must, then, in Gottfried's scheme, have preceded loss of the hydrogen envelope. That loss from the proto-Earth, Gottfried attributes to Earth's proximity to the Sun, the ablating effect of atomic particle-flow from the Sun, the so-called "solar wind," dispersing the vulnerable, nebulous gas-sheath into space by direct impacting over a long period.

[11] *Tarling 1981*

As a result of the initial pressures in the gas giant, Earth's Core comprised highly-compressed metallic elements. Pressures were great enough to expand metallic crystalline configurations by forcing hydrogen nuclei into them. Stable, multi-metallic phases of the Core metals in super-dense configurations such as the silicides, $SiFe_x$, $SiMg_x$, et cetera, with densities >30 $g/cm^3$ would then be established in forms that could survive through the ablation of the departing primordial protective hydrogen envelope.

Energy stored in minerals as a result of compression to hydrides and multi-metal complexes became chemically stored energy in the mineral forms of the carbides and silicides ($SiFe_x$, $SiC$, $SiMg_x$, etc.,). The dense core could then remain relatively cool in the manner of an explosive, stable marginally, its metastability being upset only by a major shock. Gottfried pictures the release of this energy during differentiation of the Earth as a process in which:

> **"... vigorous and rapid chemical reactions between the strongly reduced planet body and the hydridized and oxidized ... surface might have been the decisive factors in the activating processes, creating Crust and world ocean as recognized today."**

In the ongoing life of the Planet Gottfried envisions magmatism as being fuelled by release of residual hydridic energy, whereby:

> **"... ultrabasic rocks arose from decomposition of silicides, which only could come from fused metal."**

Gottfried speculates that in the crustal creation process the mass raised to crustal level would have imparted angular-momentum differentials between the expanded outer layers and the Core. This, he suggests, could have resulted in horizontal detachment surfaces, the "Mohorovicic discontinuity" in particular, [the base of Crust/top of Mantle].

# THE EXPANDING EARTH

If the crustal expansion process continues today, Gottfried theorizes, the Core must, at least in part, still be unexpanded with a density of > 30 g/cm³. He goes on to propose a density distribution consistent with an outwardly-decreasing hydridic character and conforming to the known moment of inertia of the whole Earth. His interpretation is consistent with an Earth circumference expansion from a 3 500-km radius [with an all-encompassing "Pangean" Crust] to the present 6 371-km radius and fragmented continents.

Thus, we see that the Core must be host to hydrides and the lower Mantle to silicides, according to Gottfried's theory. Among silicides, a prominent one is the silicide of carbon, known to chemists as silicon carbide (SiC) and to industrial users as carborundum. Chemical interactivity between hydrogen and SiC can yield *hydrocarbons* and *hydrosilicides*, better known as *silanes*. Hydrocarbons of the alkane series (methane, ethane, etc.) are volatile and rich in hydrogen. Their silicon analogues, the silanes (mono-silane, di-silane, etc.) are also hydrogen-rich and volatile.

Ascending buoyantly upward through Earth's Mantle to crustal levels, the silanes and hydrocarbons are capable of entraining solid debris as they go, thus forming the kimberlites and other breccias. As the Core and lower Mantle exhale these chaotic assemblages, their departure has the effect of shrinking the Core and the lower Mantle while expanding the upper Mantle and Crust of the globe.

If the proto-Earth had the thinner and continuous "Pangean" Crust over its entire surface, the expanding globe would leave continental remnants as "shields" preserved between newly surfaced regions. This is, of course, the present Earth aspect. If conservation of mass is assumed, a primordial all-core Earth with a density of 10.7 g/cm³ and a core diameter of 10 240 km could have evolved to Earth's present Core diameter of 6 972 km (68 % Core diameter reduction) by exhalation of the aforesaid volatile components. Concomitant additions to Mantle and Crust would have increased Earth's overall diameter by 25 % to its present size.

This scenario for Core attrition and external expansion can account adequately for present Earth layering. Gottfried's density figure, > 30 g/cm³, if excessive, leaves room for alternative interpretations that lead to the same crustal incrementation.

It is my perception that ongoing intermittent carbide-hydride interactivity is the cause of violent Earth behavior including the creation of sialic crustal rock, earthquakes, volcanism, and fumarolic venting. Silane volatility has a special role in the system by transferring silicon from inner to outer geospheres.

Silicon hydrides, the silanes, commence this transfer at the level of the mid-Mantle discontinuity, 670 to 700 km. Phase changes of the nature of atomic rearrangements, along with the addition of silicon and the liberation of hydrogen, characterize this discontinuity and shallower ones in the upper Mantle. Additional and more profound transformations occur in the higher geospheres, where granites are generated *in situ*. L. G. Collins will develop these subjects further in Chapters IV and V.

At deep crustal levels silanes react readily with water. If they survive to where free oxygen is available, they are quite unstable. (Aston, Ed. 1983) Hydrocarbons, on the other hand, that rise with silanes are relatively much more stable. The reader should note the entirely different behavior of $SiH_4$ and $CH_4$. Sister elements in Group IV of the periodic table, Si and C, nevertheless, bond to hydrogen with opposite polarity. The silicon-hydrogen bond is less stable than the carbon-hydrogen bond; and the silicon imperative for oxygen bonding has strength sufficient to dissociate water. Hydrocarbon, on the other hand, is stable until elevated temperatures trigger its dissociation and, if free oxygen is present, its oxidation. Carbon takes second place to hydrogen for available oxygen; and elemental carbon may be left unoxidized in the form of native carbon, fullerenes, or negatively-charged radicals.

Profound consequences in evolution of crustal geology can be attributed to the different behavior of carbon and silicon. **Mineral transformations, perovskite to spinel, eclogite to gabbro/diorite, gabbro/diorite to granite, express the versatility of silicon. At deeper levels volume incrementation initiates earthquakes and the isostatic rise of diapirs. At shallower levels, where the silane flood makes first contact with water [upper Mantle or lower Crust], oxidation of the silicon releases heat and more free hydrogen. This may initiate volcanism, melt the country rock, form magma, and bring about isostatic rise of mountains.**

**Silica created by shallow-level oxidation may crystallize as quartz sand, remain in hydrous solution, or commingle in melts, creating "intermediate" magma. Alternatively, at these levels unreacted silanes can effect cation replacements of the mafic minerals of gabbros and diorites, converting them into the many sialic rock forms that make up continents.**

**Oxidation of each molecule of methane or silane produces *two* water molecules. These are bulky oxide molecules, which add volume upon being formed in the upper geospheres. Along with the volume incrementation brought about by mineral transformations, the new water and oxidized carbon themselves contribute to Earth expansion. Thus it is that myriad silane and hydrocarbon reactions are the precursors for quiescent and violent endogenic geological phenomena, as summarily described in the EV volume.** These phenomena are the focus of attention in this volume.

Before discussing them, let us summarize the effects of Gottfried's deductions. The first conundrum his theory addresses is (as previously mentioned) the inescapable fact that accretion of cold, primordial, interplanetary debris would have generated so much heat as to have made the preservation of hydrogen either in gaseous or hydride form impossible. His alternative, a mainly-hydrogen gas giant with a small rocky Core, overcomes this objection and leads to the implication that residual hydrogen still must saturate the Core.

The gas-giant origin happily resolves the corollary conundrum of how the concentric geospheres differentiated. As mentioned already, gravitational differentials between mineral particles could not overcome the viscosities of solid [or semi-plastic] rock so as to allow the various components to settle into their present configurations. This conundrum makes an hydridic state during differentiation imperative and implies that residual hydrogen should be expected to be abundant in the Core.

## THE FUKAI-SUZUKI MODEL OF EARTH'S HYDRIDIC CORE

Fukai and Akimoto and Suzuki and Akimoto[12] take a very different view, one that is not inconsistent with accretionary theory. On the basis of experimental evidence, they report that **hydrogen can be forced into hydride form by subjecting iron and water to high pressure.** They theorize that primordial water trapped in the accreting mass would have reacted with metals to yield hydrides and oxides. They think this would leave a ferrous iron oxide by-product; and they claim their experimentation yields it.

Their point, that water could have survived the accretionary process, is emphatically rejected by Gottfried. In fact, many before him such as Fred Hoyle (1983; discussed in the EV volume) have rejected it.

---

[12] *Fukai and Akimoto 1983; Suzuki and Akimoto 1989*

Fukkai and Suzuki sidestep the problem of hydrocarbon emanation, a critical phenomenon that I consider an imperative to be addressed. The well-known imperative of silicon for oxygen seems to militate for it to monopolize oxygen, and then to share any residual hydrogen with carbon before iron would capture any oxygen at all. This is contrary to the Fukkai and Suzuki theories, which, furthermore, fail wholly to address the central issue that Gottfried's hydridic-Earth concept resolves easily: *the impediment represented by rock viscosity to large-scale differentiation within a body of silicates and metal segregations.* Reliance upon gravitational energy to explain differentiation appears to be a gross underestimation of energy requirements.

Furthermore, to energize volcanism, earthquakes, magmatic diapirism, and lithospheric fault movements of all kinds necessitates significant energy release at *selected* sites. Unequal distribution of released energy is quite incompatible with energy from the inexorable dissipation of primordial heat. The Fukai and Suzuki theory never considers this. Neither does it address in any way the general problem of crustal growth, or the corollary, the upward migration throughout time of silicon in its many forms and its concentration in the outer geospheres, especially in continents.

Carbide-hydride systematics, in explaining the source for deep Earth hydrocarbon, fortuitously also provides the long-sought explanation for silicon transpiration, an unexpected discovery of the first order.

## CHAPTER II

# COMPOSITION OF THE UPPER GEOSPHERES
## AND
## THE NATURE OF
## THE "MOHO"

### E. A. SKOBELIN

---

"Many things are incomprehensible to us not because our concepts are weak, but because these things do not fall within the realm of our concepts."

**KOZ'MA PRUTKOV**

"Geologists have given up on the subject, leaving the center of the Earth for mathematicians' amusement."

**R. D. OLDHAM**

---

Geophysical data unambiguously prove the concentric zonal structure of the Earth. This can have only two possible explanations: a difference in the composition or a difference in phase states of material making up the various geospheres. Gravitational segregation, due to one of these, probably created the differentiated geospheres. Gravitational settling of heavy particles is impossible in solid or high-viscosity plastic states of matter and could have occurred only in the early molten stage of the Planet's life.

## Differentiation in a Molten Earth

We must assume that gravitational fractionation into silicate geospheres[13] amounted to differential loss of volatile fractions, especially water, from primary silicate melts. Removal of volatile components must have been most complete for the deepest and densest of the mafic and ultramafic fractionating melts, while being impeded due to higher viscosity in silicate melts.

Experimental data show that, while water is poorly soluble in mafic magma, other gaseous elements are not so constrained and blend into it and form strong chemical bonds over wide ranges of pressure and temperature (PT). These powerful bonds link oxygen with silicon, iron, and magnesium to make the minerals comprising most of the upper geospheres.

## Gaseous Components of the Early Earth

It follows that the existence of residual water and gases and volatile elements in Earth's Mantle is unlikely. Even less likely is that they could account for the enormous volatile exhalations represented by volcanism or those represented by pervasive alteration, mineralization, and ore-formation. All of these processes are known associates of magmatic activity.

Mafic and ultramafic rocks are not host to appreciable water except in fractures, contacts, and secondary hydrous minerals. Otherwise, the water content of these rocks and their late-crystallizing magmatic residues is essentially zero. Liquids and gases found in these habitats must be ascribed to invasion from adjacent terranes.

The conclusion that the Earth's mafic shell is sterile with respect to water and other volatile residues forces us to admit

---

[13]  *All "outer geospheres" are primarily composed of "silicate" minerals. Designated as "sialic," "mafic," or "ultramafic," they fall in the range of 75 % silica for the most sialic down to 35 % $SiO_2$ for extreme ultramafics.*

further that not only gaseous elements and compounds, but also all light and mobile ones as well, must have escaped the Mantle habitat during primary differentiation of the Planet. Included evidently are H, N, C, P, S, F, Cl, Br, I, B, Li, Be, and K, which must have escaped completely (or nearly completely) from Earth's mafic shell, while Na, Ca, and Al only would have been partly retained, and then only in the upper levels. Moreover, large amounts of oxygen and magnesium must have escaped in combinations with other elements. The remaining elements must then have been redistributed and "differentiated" in geospherical configurations.

## Iron in Upper Geospheres

Primary-Earth differentiation is hard to reconstruct because of interactive and often conflicting processes. As a case in point, let us consider the specific evidence of the modern distribution of iron in the Planet. Most of it is in the Core according to R. D. Oldham. But in the upper geospheres we find that the levels of iron *decrease with depth*. That is to say, iron is much less in ultramafic magmas, those characteristic of the deeper geospheres, than it is in the shallower mafic magmas. In part this could be due to iron accumulating in residual liquors or crystallizing out in the tops (the "cupolas") of plutons.[14]

## Other Metals in Upper Geospheres

This evidence fits better to the old thinking of successive geospheres, granite, basalt, and peridotite in that descending order.[15] In addition to the upward increase in iron, it is notable that silicon also rises, whereas aluminum, calcium, and [later] sodium appear. As these elements increase, magnesium

---

[14] *The ultramafic mineral, olivine, is found in a continuous series from pure magnesium silicate [forsterite] to iron silicate [fayalite].*

decreases, and chromium declines nearly to zero.[16]

## Layered Complexes

Another enigma has been the layered complexes, where sialic rock is interleaved amid highly-differentiated mafic rocks.[17] This fact compels us to conclude that Earth's primary differentiation entailed emplacement of an orthogranitic[18] shell over the primordial basalt shell. *Yet another mafic shell may have been extruded over the granite by later segregates from the still-hot planet.*

## Gravity of Inner Earth

Downward from the lithosphere, primary differentiation with depth is attenuated by the progressive decline to zero of the force of gravity at the center of the Earth. Normal gravitational force, downwardly-directed, may be replaced by reversed, upwardly-directed force at depths of 2 700 to 4 980 km.[19,20] Moreover, the "arch effect" created by the planet's solid outer

---

[15] *Dunite [magnesian olivine] with 2-3 % chromite is the deepest in this layering; magnesian pyroxene (representing increasing silicon content [enstatite]) increases until dunite gives way to peridotite. The change in composition continues in the direction of olivines and pyroxenes containing more iron; appearance of plagioclase [anorthite]; disappearance of olivine and chromite; replacement of rhombic pyroxene [enstatite] by monoclinic pyroxene [augite] and calcic plagioclase by relatively sodic varieties [anorthite by labradorite and andesine]. Still higher, free silica appears.*

[16] *In petrographic terms, the changes consist of a gradual upward replacement of dunite by peridotite, [harzburgite, lherzolite], then picrites, norites, and tholeiites, containing higher quartz at higher levels.*

[17] *The red granites within the layered Bushveld complex are an example.*

[18] *A plagioclase-free granite*

[19] *Subhasissen [1983] deduces that maximal conditions of expansion occur at these depths along with an "internal atmosphere." The widely-held concept that 3 500 kb pressures prevail at Earth's center, a figure that is obtained by multiplying mean density (5.51) with radial depth (6 371 km), does not account for the reduction in the force of gravity, much less its reversal, at the center of the Planet.*

[20] *Martjanov (1968) discusses an "arching effect" of the Planet's solid outer geosphere as an additional contributing factor, which may affect further reduction in the pressure at Earth's center.*

shell must also make some contribution to reducing pressure at the Earth's center, possibly by more than an order of magnitude.

Clearly then, 3 500 kb is too high for the pressure at the Earth's center, possibly being excessive by as much as an order of magnitude. Whereas the matter of internal Earth-pressure needs to be re-examined, it may be safely said that, **indisputably, the force of gravity declines at depth in the Earth.**

**Whereas the strength of the gravitational field at inner-Earth levels diminishes, its diminishment does not preclude entirely the potential for primary mineral differentiation. However, it is implicit with the diminishment that the existence of primordial, protoplanetary matter in the Earth's Core, especially volatile substances, deserves serious consideration.**

## Mantle Plasticity

Beneath the solid lithosphere (Belousov and some others call this the "tectonosphere" - the upper hard shell of the Earth), the plastic asthenosphere provides an ideal cover to prevent the escape of fluids from the Earth's Core. This capacity stems from the fact that a plastic, like a fluid, transmits pressure applied to it equally in all directions. As the pressure of gases and fluids rises under the asthenosphere, the plasticity of that layer distributes the force in all directions impeding their penetration above the Earth's Core.

## Lithosphere Rigidity

Earth's primary differentiation would have been halted by cooling and hardening, first at the surface, then progressively inward. The lithosphere-asthenosphere contact would have

shifted to deeper and deeper levels through geological time. Progressive cooling and lithification would, thus, have proceeded more and more slowly in a geometrical progression, varying only by pulses of expansion and contraction.

The occasional reversal of the descent of the lithosphere-asthenosphere contact by a pulse of increased pressure acting on it from below would have raised the pressure-sensitive melting points of contact rocks. Hence, the boundary would have risen and fallen with pressure changes brought on by chemical as well as dynamic events. Pulsating action would bring about fractionation of molten rock components and lead to layered deposition of granites in the Earth's basalt shell and of anorthosites in its peridotite shell. **Granites and anorthosites of this type are the layered deposits that geophysicists recognize as layers of asthenosphere,[21] an interpretation that is preferred over the alternative, changes in the phase state of materials.**

---

### The Nature of the "Moho"

---

The Mohorovicic discontinuity is a worldwide, sharply-defined, seismic-wave-velocity discontinuity. It is generally considered the boundary between Crust and Mantle, and cannot be linked to compositional or phase states of materials. The "Moho," occurring where compositional changes should be gradual, leaves us with only one acceptable interpretation:

### *The Moho is the floor of Earth's primary sedimentary shell.*

Of course, the components of this shell are the contributions of exogeny and endogeny to its lithologies and structures.

---

[21] *The fact that known rocks comprise representatives of all intermediates in the dunite-tholeiite-granite petrologic series is evidence of incomplete differentiation. The existence of incomplete differentiation imputes improbability to the existence of sharp intra-geosphere boundaries, which, in fact, are geophysically observable within the upper geospheres. Complete differentiation would, instead, have produced single-mineral geospheres.*

## The Siberian Platform Model

The Siberian platform provides a unique model of the Crust and Mantle that is important to understand despite being still incompletely studied. Included in its makeup are trapp rock, alkali ultramafics, and granitoid magmatic bodies that are emplaced in various configurations and with varied tectonic intensities.[22] The shallower levels of the platform cover contain the largest concentration of intrusions, the highest grades of metamorphism, and the greatest structural complexity. It is probable that the Mantle beneath the Crust comprising the platform is also permeated by sills and dikes of mafic and ultramafic magmatic rock.

The analogies that are to be drawn from the lessons learned on the Siberian platform are that the upper levels of the Crust are permeated more intensely by magmatic intrusives than lower levels. Metamorphism intensifies upward, that is to say, the primary sedimentary rocks being turned to gneisses and granites, the ultramafics and mafic rocks to amphibolites and greenstones.

## Layering of the Entire Crust

These geological thoughts as to the nature of Earth's Crust are confirmed by geophysics. Seismic studies have everywhere established the presence of numerous reflective horizons, especially at middle and lower crustal levels.[23] These reflections allow the "mobilists" to interpret the lithosphere as stratified tectonically. Seismic analysis finds few steeply inclined reflectors in the upper levels and none at deep levels. The deeper the reflections, the more they are nearly horizontal.[24]

---

[22] *The array of mafic and ultramafic intrusives of the Siberian platform mainly comprise sills in the sedimentary cover. These sills jump in steps to higher stratigraphic levels as they migrate from points of injection. The sources are few in number and mainly situated around the peripheries of the platform, usually where folding interrupts the flat platform structure.*

[23] *Gelfand, 1987; Klemperer, 1987; Mooney, 1987; Thompson, 1987*

[24] *Mooney and Brocher, 1987*

Nearly everywhere in the Crust beneath the sedimentary cover there has been shown to exist increased electrical conductivity, usually accompanying the appearance of stratified structure.[25] The widespread occurrence of *stratified structure* has been abundantly reported within the last three years.[26] The time stratigraphic sections produced by these soundings are so expressive as to leave open virtually no other conclusion.

In a California case seismic prospecting has turned up a packet of rocks, a **"wedge"** in the terminology of seismologists, of clearly sedimentary appearance at depths of 15 to 22 km.[27] The authors ascribe this to **"subduction diving"** in the coastal trench. In other places where finely stratified reflecting horizons are found at deep levels far from any coastal trench, where subduction might be possible, the favored interpretation is **"tectonic stratification"** or **"listric faulting."** [28] These interpretations are taking on the aspect of an **"independent direction for geology,"** a means of diffusing the initial surprise at detection of nearly horizontal structure in the lower and middle Crust where there is no logical explanation for it.

In some instances there are grounds for supposing that sills of various lithologies constitute an acceptable solution to the problem. Thus, the proposal has been made that sill-like intrusives have entered along horizontal faults or "discontinuities."[29] Since these horizontal structures are everywhere present, this interpretation is virtually an admission of the primary sedimentary nature of the Crust, because in the absence of sedimentary structure, horizontal sill emplacement is

[25]  *Banyan; Tuyezov; Rotstein; Green 1987; Mozley 1986*

[26]  *In western Europe: Bois 1988; Galson 1986; Green; Hobbs; Klemperer; Meissner; Louie; Salle; on the Baltic shield: Lund; 1987 on eastern margins of the Atlantic: Durrheim; Pinet; in North America: Clowes; Fuis; Hauser; Valasek; in Australia: Goleby; Wright; Griffin: Wake-Dyster; in Israel: Rotstein 1987; On the Ukrainian shield of the USSR: Tripol'skiy; Sharov 1987.*

[27]  *In California: Trehu 1987*

[28]  *Listric faulting is where a curved fault is steep at the surface and becomes flatter with depth.*

[29]  *Sollogub 1981*

not possible. This is confirmed in many places on the Siberian platform, where all sills have a clearly expressed tabular configuration until they enter unstratified or vaguely stratified rocks, whence they take on an amoeboid form. Thus, **basic sedimentary structure to the deep Crust is implied,** an implication the seismologists may not have intended.[30]

In modern, seismically active areas much study has been devoted to waveguides, the surfaces on which seismic energy tends to travel. In some places the waveguides lead back to the foci of the earthquake energy. Their three-dimensional structures define broad plate surfaces with abrupt transitions from one stratigraphic level to another. The associated evidence of increased electrical conductivity and temperature anomalies supports the interpretation of waveguides as the surfaces of unsolidified sills.

## TOWARD A NEW THEORY OF THE EARTH: SOME GENERALIZATIONS

**Thus, the data lead us to the interpretation of the entire Crust as primary sedimentary in nature. The floor of the Crust, the Moho, is a surface that was once the surface of a solid Earth.** It appears that this surface is covered in most places by ancient terrigenous rocks, but in others, where it is seismically less responsive, the Moho may be overlaid with ancient lavas and flood basalts of the early Earth.

My conclusions conflict with those of other researchers because they seek to address the broad spectrum of geological information. I take into account all phases of geology, including deep structure and lithology, magmatism, volcanism, breccia pipes and ore formation, tectonism, and so forth. All of these must be linked and account fully for the evidence that others have attempted to explain through theories of transmagmatic

---

[30] *The sedimentary habit of sills is well established on the Siberian platform, where sills clearly mimic their sedimentary host configurations. Where they leave the sedimentary rocks and enter vaguely-stratified volcanic rocks, they immediately take on the irregular, often amoeboid, structural form characterized by their new host.*

fluid transfer, heterogeneous [and diamond-bearing] Mantle, hypothetical subduction of oceanic plates, and so forth.

These deliberations are substantially limiting, especially in the search for **sources of endogenous fluids, energy for the tectonic-magmatic process, and mechanisms of mantle magma formation. They must exclude an abiogenic origin for hydrocarbons and oil accumulations that emerge from basement platforms, and as contributors to volcanic effusions.**

The question of phase states of materials in the upper geospheres is very important for determining the depths of the boundaries between the lithosphere and the mobile, upper Mantle beneath it, the asthenosphere, and the depths of the basalt and peridotite shells. The extreme complexity of these problems can be judged by the outstanding research that has yielded interpretations that still have not been submerged by the anarchy of the "new global tectonics."[31]

The ambiguity of interpretation of geophysical data forces the use of indirect methods and leads to large divergences among estimates of depths to the base of the lithosphere on continental platforms.[32] On the one hand, some researchers think the asthenosphere extends everywhere and is even composed of several stratified members.[33] On the other hand, the asthenosphere is interpreted by some workers to be developed only in certain places, and absent or poorly developed elsewhere, notably on platforms. In the latter case in the territories of the former USSR it is limited to the regions of the Carpathians, the Caucusus (and southern Caspian, Baykal, Kuriles, and Sikhote-Alin', whereas over the main platform area these strata are absent or poorly developed.[34]

[31] *Bott 1974; Gutenberg 1963*

[32] *From 100-150 km, Artjushkov 1979; to 250-300 km, Dobretsov 1981; to 300-500 km, Alekseyev 1978; to 1 000 km, Belousov 1967, on continental platforms. A 1.5- to 2-fold reduction in lithosphere thickness is usually believed to prevail beneath ocean basins.*

[33] *Alekseyev 1977, This is the situation only in the Carpathians, Caucusus (and southern Caspian), Baikal, Kuriles, and Sikhote-Alin' of the USSR.*

[34] *Guterman 1977; Yegorokin et al. 1984*

Perhaps more reliable as a source of information on the depth of the floor of the lithosphere (Belousov's "tectonosphere") are data gained by analysis of **deep-focus earthquakes, which cannot originate in plastic material but are, nevertheless, linked to some kind of stress in solid rock.** The depths of foci of such earthquakes are 700 to 720 km, figures that give an indication of the local thickness of the lithosphere.

And thus, the question of lithosphere thickness is quite open. Our interpretations should be recognized to be more intuitive than scientific and to be constrained only so as not to contradict our limited knowledge.

**Under the ocean basins lithosphere thickness, contrary to traditional thinking, should be greater than under the continents, most likely close to 700 km.** The evidence for this comes from foci of deep-focus earthquakes, which are clearly linked to island arcs and abyssal trenches that bound ocean basins and continents. **Minimum lithosphere thickness, which is, perhaps, only 100 km, should be found within young folded regions on continents and beneath mid-ocean ridges.**

# THE ENERGY OF INNER-EARTH:
## SILANE AND HYDROCARBON SYSTEMATICS

### C. WARREN HUNT

In the opening Chapter I raised the question of the presence of hydrides in the Core and their potential for interacting with silicides, especially the silicide of carbon (silicon carbide) in the lower Mantle. Chemical reactions had been proposed independently by Gottfried and myself (as reported in the EV volume) to overcome the deficiencies of contrary deductions, such as those advanced by Fukai and Suzuki.

A conundrum posed by Gottfried's theory that might have been aired is the whereabouts of Earth's oxygen during Gottfried's "highly reducing" gas-giant stage of proto-Earth. It was pointed out that water preservation is precluded [notwithstanding the contrary views of Fukai and Suzuki], because its presence would be anathematic in "highly reducing" conditions. The heavier metals, including silicon, are well-accommodated in Gottfried's system because they form direct links with hydrogen, as "hydrides." Exclusive hydride-metal compounds, the planetary norm in Gottfried's scheme, would have left no place for the large amounts of oxygen that must have been present. It is imperative for the survival of Gottfried's theory that this conundrum be resolved.

To accord oxygen any place at all in the system, silicon in large quantities must have remained linked to it in silicate form. Happily for resolution of this constraint, silicates or slightly less oxygen-rich silicon minerals, can themselves bond to hydrogen. Under great pressure these should be as stable as metal hydrides.

**I suggest that slightly-oxidic compounds of rock-forming metals comprised major components of Gottfried's hydridic proto-Earth and accommodated its abundant oxygen.**

The addition of these silicon-oxygen radicals to Gottfried's hydrides in a process of hydridic differentiation makes plausible the evolution of the mafic silicate Mantle from an hydridic proto-Earth. Copious degassing concomitant with differentiation could then have provided the hydrosphere.

The present situation in which hydrogen and hydrocarbons [but not gaseous oxygen] rise from the Mantle is a demonstration of two important points. First, there must be pressure-deficient zones, ["PDZs"], transient lesser-pressured zones permeable to ambient fluids under sufficient pressure. Second, the PDZs must be pervaded with fluids.

Migrating gases receive much attention hereinafter. Here, it will suffice to express briefly the fundamental reactions involved in carbide-hydride systematics.

The opening reactions are those between hydrides of the Core and neighboring dense carbides of the Mantle. The products are volatile; and their sudden evolution triggers diapirism. This must be quiescent in the lower Mantle, where earthquakes do not occur. At mid-Mantle levels, 710- to 670- km depths, the interaction turns explosive; and earthquake energy is released. Boundary conditions that bring this about are taken up in Chapter V.

Progressive cation replacement by silanes results in the creation of less dense minerals, Evolving from this interaction, bubbles of hydrocarbons and their silicon analogues, the silanes commingle with rock fragments and rise buoyantly and cataclysmically. The gas permeates available porosity, mobilizes and lubricates the permeated breccia, and provides the driving energy for upward advances in bursts. Such explosive diapirs describe kimberlites. They entrain pulverized wall rock in slurries within their gasified liquid medium.

Skobelin in Chapter II, "Outer Geospheres," expresses doubt as to whether a volatile bubble could penetrate the plastic Mantle. He suggests that the buoyant upward force would be dissipated in all directions. Gold (1987) already answered that question (see EV volume) in terms of the following mechanism: **The internal pressure in a gas-filled chamber is nearly the same top to bottom. Lithostatic pressure adjacent to the chamber is far greater adjacent to the bottom than it is adjacent to the top. This causes lithostatic pressure to pinch off the bottom of the chamber, while chamber pressure holds open the top. An elongate bubble originates in this way, which only differs in shape and magnitude from the bubbles that rise in water coming to a boil on the stove.**

**The diapiric rise of an "Earth bubble" of hydrocarbons and silanes should be a quiescent rise so long as viscous flow allows gradual [but not necessarily slow] plastic-rock response. However, with any reaction between wall rock and silanes in the diapir, heat will be released and lead to inspissation of the fluid phases of the slurry, an internal viscosity increase. Incrementation of the $Si^{4+}$ or $SiO_2$ content of wall or diapir rock would embrittle the rock, reduce its ability for plastic response, and impede future diapir progress. A bursting-out against this impedance best explains earthquakes in the domain of plasticity of the Mantle, where shear stress buildup is impossible.**

**Silane migration into shallower Earth levels may be impeded by its own reactivity at three stages. Collins and I describe this reactivity in Chapters IV and V as a first stage of perovskite to spinel reaction, next an eclogite to gabbro or diorite reaction, and finally the crustal gabbro or diorite to granite reactions. Silane reacting with water produces silica, hydrogen, and heat. When silanes transform minerals, they**

yield heat and water as a by-product. Extreme heat may be produced in this way to create magmas and the energy for volcanism.

In these later stages of silane activity, in which mafic cations are replaced by silica and silicates, diorites and gabbros become felsic and silica-rich, and heat and hydrogen are released, the presence of potassium with water yields granitic magma. If water is limited, myrmekoid mineral transformations result (a subject that is treated thoroughly in the Appendix, p370).

The high viscosity of silicic magmas leads to volcanic eruptions in which rhyolites, pumice, obsidian, and steam are prominent. Once a conduit is open, "bi-modal volcanism," basaltic lava flow following rhyolitic eruption from a single vent, follows. Bi-modal pluton emplacement, side-by-side mafic and granitic intrusive bodies may also occur, where magma does not have enough energy to erupt.

Where temperatures allow precipitation of silica as, for example, at the periphery of a melt, an envelope of silica may block further upward progress of the melt and lead to explosive rupture of the chamber. Phreatic explosion in this manner underground, in which violent enlargement of a Mantle chamber occurs, may characterize some earthquakes. At shallower levels the collapse of gas- and steam-filled chambers also may result in earthquakes [see Skobelin, Chapter VI].

Volcanic repetition of the surge-melt-solidify-explode cycle continues so long as silanes, generated at deep mantle levels, rise to provide the necessary silicon. When earthquakes occur without fault offset or volcanic emission at the surface, they may be thought of as effects of underground volcanism.

### Fig. III-1
### BASALTIC LAVA FLOWS OVER SILICEOUS
### ASH DEPOSITS.
Top: conformable contact.  Middle: eroded (unconformable) contact.
Bottom: water-laid ash beds with overlying lava in upper left
background.

Deep drilling is now turning up enormous surprises, the super-deep drillhole on the Kola Peninsula, Russia, the deepest man has been able to drill [almost 13 km], being the first in this respect.

Free water and porosity were found in deep, horizontally oriented fracture zones. The porosity averaged 2.7 % and reached as high as 4.3 %. The operators interpreted the presence of fracture porosity that supplemented interstitial porosity from the fact that actual water influx was greater than they anticipated on the basis of measured interstitial porosity.

Gas was also found in the Kola deep well, dissolved for the most part in the water, at many levels. Hydrogen, carbon dioxide, nitrogen, methane, and helium are reported in that order of decreasing abundance. Reports on the well history say that "in some instances drilling fluid [was] flowing out of the well 'boiling' with hydrogen," a subject that is taken up in more detail in Chapter VII.

The above facts are published. Another fact of profound importance appears not to have been made public by the Russians. It is reported to me by Mr. John Frey, a geologist who inspected the core. The unreported anomaly is that **the Kola drillhole had drilled out of the mafic shield rocks in which it started and, below six or seven km, had penetrated light-colored quartz-rich granite to the 13-km depth.** The operators had vaguely reported "significant mineralization," presumably in this lithofacies.

**This revelation has profound significance for the concept of silane systematics as advanced here. The growth of continents *from below* by addition of *silicon to the basal rocks of the lithosphere*, a process never before recognized as a real possibility, is what silane systematics should be doing. The petrological evidence, as presented in Chapters IV, V, and VIII, explains the process as one of mineral transformation wherein granite is produced from various mafic proto-lithofacies.**

Other surprises have been turning up in the geophysics and drilling of the San Andreas fault zone, California. Geophysical work in the Parkfield sector shows without doubt that porosity is high, 12 % being interpreted as appropriate for the 3-km width of the fault zone. This is interpreted to penetrate through the lithosphere, and, thus, to much greater depth than the Kola drillhole. This subject is taken up in more detail in Chapter VI.

It is suggested by these facts that water permeates the Crust and, perhaps, penetrates the upper Mantle. This water may be either surface water filling void space below the water table or water created by combustion of hydrogen released in silane interactivity with the rocks. Its distribution should be uneven and oriented to natural, deep, fracture systems, including but not limited to the "Benioff zones" (Chapter VI) and the "lithospheric rifts" (Chapter II). Ambient gas and vapors would be expected to enter such rifts preferentially; and it is there that the near-surface, felsic rocks should be generated.

Within the Mantle, where silane- and hydrocarbon-saturated diapirs rise, the dissolved silica that enters melts gives them high viscosity. Where rock transformation occurs, permeability is likely to be reduced by quartz (or coesite or stishovite) deposition as well as by mineral volume incrementation. Continuing inflow of rising vapors, whether in Benioff zones or lithosphere rifts, may split an impeding barrier hydraulically. A rapidly propagating fracture of this sort may be thought of as an artifact of underground volcanism; and it would allow entrained breccia to be propelled upward or outward at great velocity. Rapidity is important for an hydraulic fracture to propagate through plastic mantle rock. With rapid pressure increase, the high inertia of the impacted rock precludes a response by plastic deformation.

One unexplained impediment remains. Crustal rock is horizontally stratified [as Skobelin points out], not only in sedimentary basins, but also in crystalline terranes. The fracture porosity in the Kola super-deep well and the KTB pilot drillhole (Germany) show this feature in cores very well (Chapter VII). The explanation for horizontal fracturing is provided by Bailey (1990), as follows: Upon transcending the ductile rock of the upper Mantle and lower Crust, an advancing gas-saturated breccia encounters brittle upper Crust. This impedes its upward progress, and the upward-fracturing that has taken the breccia through ductile rock is deflected into lateral hydraulic fracturing ("fracking," in oil industry terms).

Silane in the advancing breccia may react with water, producing silica sand, which then is "fracked" outward into the newly opened cracks. Sand-filled sills and crypts may be created in this manner, entirely accommodating sand produced by reacting, inflowing gas.

In cases where diapiric breccias are hydraulically forced into lateral crypts by the "fracking" action, the continued buildup of volatiles and accompanying magmas takes a mushroom form. This raises the surface in a welt, a "cymatogen" in the terminology of Lester C. King (1983). If inflow exceeds lithostaic pressure, the mushroom may burst, driving entrained sand diapirically through the brittle Crust. Skobelin describes this mechanism in Chapter VI, as the means for emplacing magmatic sills. Hunt describes deposits of surficial quartz sand of this provenance in Chapter VIII.

Volcanic release through the top of a cymatogen allows subsidence of the surface. This may create a graben on the former crest. If this feature has much length, it is called a rift or trench. Tension fractures associated with the rift work downward as the cymatogen subsides, releasing the pent-up products of silane oxidation with characteristic volcanic violence from their crypts.

Two energy sources operate in this process, the first being decompression of compressed and dissolved gas, the second, gravity inversion, the buoyant rise of relatively light material. In the latter case, lower-density silane oxidation products (quartz and water), are for a time held in hydraulically fractured crypts. With breaching of the roof of the welt, pressure is released, and slurried sand rises, reversing the gravitational inversion and allowing the roof to subside. This may appropriately be termed **"crustal convection."**

Let us consider the options for the produced lithologies of crustal convection. As all must be high-silica products, there are five fundamentally different products that may enter the lithosphere:

**Magmatic melting and mixing: plutons**
**Mineral replacement at an atomic level: transformed rock**
**Explosive volcanic rock: entrained in gas or liquid**
**Precipitation from water: chemical silica**
**Quartz sand: slurries**

**Magmatic melting and mixing** occur when the exothermic silane combustion is too rapid for heat dissipation into the country rock, which then melts.

**Mineral replacement at an atomic level** ensues from the pervasion of porous terranes by silanes or by dissolved silica resulting from the reaction of silanes with water.

**Explosive volcanic entrainment in gas or liquid** results from the sealing off of a chamber of accumulating gas. Dissolved silica deposits impervious layers, which then lead to explosive volcanism. If aluminum and some other metals are available, clays and open-lattice, low-temperature zeolites may be emitted.

**Precipitation from water** develops from dissolved silica in a hydrous carrying-fluid. Deposits may comprise quartz veining, chert deposition, or interstitial cementation within porous rock.

**Quartz sand slurries** result from silane reaction with water at depth. Quartz precipitates and collects as water-saturated sand in a forcibly expanding crustal chamber. When a disturbance occurs that ruptures the caprock, the slurry debouches in response to prevailing hydrodynamic pressures. The quartz sand may be extruded on the surface as a body of pure quartz sand, or it may stall and lithify as quartzite in dikes.

Each of these four topics is treated in successive chapters. In the broad scope it is appropriate to point out at this point that **all** of these are the processes that have formed continents and mountain ranges. Mountains, of course, may rise on continents or on the ocean floors.

## THE REACTIVITY OF HYDROGEN WITH CARBIDES, SILICIDES, AND GERMANIDES

In dealing with the Group IV elements and their hydridic effluents, the alkane hydrocarbons, silanes, and germanes that occur worldwide in all terranes, including crystalline ones, we should review the basic chemical equations. Only the first member of the hydrocarbon, silane, and germane series needs to be discussed. These are methane ($CH_4$), monosilane ($SiH_4$), and monogermane $GeH_4$). The second, third, and higher members of each series, ethane, propane, etc., disilane, trisilane, etc., digermane, trigermane, etc., have similar behavior and molecular configurations, which may be expressed as $nMH_{4+n}$

## Hydridization Reactions

$$4H_2 + SiC \quad \rightarrow \quad SiH_4 + CH_4 \qquad (1)$$

$$4H_2 + SiSi \quad \rightarrow \quad 2SiH_4 \qquad (2)$$

$$4H_2 + SiGe \quad \rightarrow \quad SiH_4 + GeH_4 \qquad (3)$$

Silanes and germanes are volatile, like the alkane hydrocarbons. Following are their reactions with water and oxygen as well as the iron-water reaction (after Fukai).

## Oxidation Reactions

### Reactivity between Water and Monosilane, Germane, and Iron

$$SiH_4 + 2H_2O \quad \rightarrow \quad SiO_2 + 4H_2 \qquad (4)$$

$$GeH_4 + 2H_2O \quad \rightarrow \quad GeO_2 + 4H_2 \qquad (5)$$

$$(1+2/z)Fe + H_2O \quad \rightarrow \quad FeO + 2/zFe_zH^* \qquad (6)$$

$^*z$ is the unknown quantity of hydrogen injected into Fe.

Reaction (4) is vigorous and spontaneous; reaction (5) less so.

### Reactivity between Oxygen and Methane, Monosilane, Germane, and Iron Hydride
#### - under conditions of limited oxygen

$$CH_4 \quad + 2O_2 \quad \rightarrow \quad 2H_2O + C \qquad (7)$$

$$SiH_4 \quad + \ O_2 \quad \rightarrow \quad 2H_2 + SiO_2 \qquad (8)$$

$$2GeH_4 + 4O_2 \quad \rightarrow \quad 2H_2GeO_3 + 2H_2O \qquad (9)$$

$$4FeH_z \quad + 3O_2 \quad \rightarrow \quad 2Fe_2O_3 + 2zH_2 \qquad (10)$$

- under conditions of abundant oxygen
(reactions go to completion)

$$CH_4 \quad + 2O_2 \quad \rightarrow \; 2H_2O + CO_2 \qquad (11)$$
$$SiH_4 \quad + 2O_2 \quad \rightarrow \; 2H_2O + SiO_2 \qquad (12)$$
$$2GeH_4 \; + 4O_2 \quad \rightarrow \; 2H_2GeO_3 \; + \; 2H_2O \quad (9)$$
$$4FeH_z + (3+z)O_2 \; \rightarrow \; 2Fe_2O_3 + 2zH_2O \qquad (13)$$

Reactions (7) and (11) are everyday occurrences in our environment. Reactions (8) and (12) are only seen under laboratory conditions. Reactions (10) and (13) follow in my judgment from (6). Monosilane spontaneously oxidizes at room temperatures, explosively so at 500°C. Germane oxidation reaction (9) is less vigorous under room temperature and atmospheric conditions but explosive above 300°C.(Glockling 1969)

## Electronegativities

$$Si = 1.8$$
$$H \; = 2.1$$
$$C \; = 2.5$$

These figures reflect the different behavior of the silicon and carbon atoms in hydridization: In SiC the silicon acts as a metal, giving up electrons to carbon. When hydrogen intervenes, it must accept electrons from silicon but gives electrons to carbon. Despite Si and C occurring in the same column of the periodic table, their behavior is quite different, even opposite under some conditions.

In a Si / H bond  Si is positive; H is negative.

In a Si / C bond  Si is positive; C is negative.

In an H / C bond  H  is positive; C is negative.

## Bond Energies

| | | | | |
|---|---|---|---|---|
| Si | - | O | : 460 | kJ/mol |
| C | - | H | : 414 | " " |
| C | - | O | : 355 | " " |
| C | - | C | : 334 | " " |
| Si | - | H | : 314 | " " |
| Si | - | Si | : 196 | " " |

## Heat of Formation

| | | | |
|---|---|---|---|
| $SiH_4$ | = | 11.9 | kCal/g |
| $SiO_2$ | = | 201.34 | " " |
| $H_2O$ | = | 68.4 | " " |
| $CO_2$ | = | 94.4 | " " |
| $CH_4$ | = | 19.1 | " " |
| SiC | = | 1.43 | " " |

From these data may be deduced the heat released upon oxidation of silane and methane, as follows:

$$SiH_4 + 2O_2 > SiO_2 + 2H_2O$$

Heat:     $-11.9 + 201.3 + 2 \times 68.4 = 326.2$ kCal/g/96 mol-eq-wt

$$CH_4 + 2O_2 > CO_2 + 2H_2O$$

Heat:     $-19.1 + 94.4 + 2 \times 68.4 = 212.1$ kCal/g/80 mol-eq-wt

Total combustion of silane yields 54 % more heat than total combustion of methane on a weight basis and 28 % more on a mol equivalent basis. Partial combustion of silane (silicon oxidized; hydrogen released) yields 62 % of the total latent heat. By comparison, partial combustion of methane (hydrogen oxidized; carbon liberated) is exothermic to the extent of 44 % of total available heat of combustion.

These figures are presented to allow the reader to apprehend that silane and hydrocarbon emanation may result in several levels of intensity and depth within the Earth at which heat release may occur. The sites of release depend on the complexities of oxygen and water availability, pressure and temperature conditions, and local mineral susceptibilities to transformation.

## CARBON AND SILICON: AMBIVALENCE

The Group IV elements of the periodic table, carbon, silicon, and germanium, having atomic numbers 6, 14, and 32, straddle the boundary between metallic and non-metallic behavior. Their diverse and extraordinary performance during volatile migration from inner to outer Earth is the subject of this section.

Among the three, carbon is best known to man because of its activity in life processes. The very existence of life is linked to the bridges carbon builds between otherwise incompatible elements, symbiosis on the atomic level. Symbiosis is the hallmark of higher levels of living forms. Larger creatures may be appropriately thought of as biotic colonies as to their internal organization. Externally they depend on the providence of their world on a day-to-day basis; internally they depend on microbes. Providence referred originally to convenient access to daily bread. It has come to include paper, petroleum, cement, glass, and silicon wafers.

In any case, these five commodities used in daily life illustrate the range of ambivalent dependence of human life on the surrounding inorganic and organic world. Carbon is an atomic component in the first three, silicon of the last three. Carbon is the more visible element in our surroundings. But silicon is the foundation for our continents, and now is known as the key to information technology. Thus, our awareness of the fundamental importance of silicon has grown recently.

Increased involvment of silicon in our lives is to be expected as our understanding of it increases. One could say that besides its role as the foundation of continents, wherein its electrons are welcomed by oxygen, silicon has a reverse "humor." When linked with hydrogen in silane form, silicon forces the hydrogen to take on its negative charges. In this way, enormous latent energy is invested in a silane molecule.

The most profound property of silicon from the geological point of view is the fact that pervasive silanes can transform minerals and bring about continental growth. Silicon substitution for heavier metals causes convective overturning in the Crust and Mantle by mass inversion. It sets up conditions for earthquakes, crustal dilation and collapse, and for orogenesis. Silanes take on the role of transformers of the planet, silently, relentlessly, atom by atom, crystal by crystal, expurging heavier metals, implanting light silicon radicals.

Thus, carbon is a good actor, ambivalent but benign in imparting the chemicals for life, energy, workable building materials, carbon fibers, diamond, and so forth. Silicon, by contrast, is an element of oblique behavior. Its ambivalence leads to obdurate materials, the already-mentioned siliceous impediments to volcanic explosion, massive ash falls that create deserts and smother life, granitic massifs that emerge, buoyantly, like arctic pingos above growing keels of transformed rock.

In SiC the silicon behaves as a metal, giving up electrons to carbon, which, in the presence of silicon, behaves as a non-metal. For the SiC bond to be broken by hydrogen the latter must accept the electrons from silicon but gives them up to carbon. The two basic gases generated by hydridization of silicon carbide perpetuate the opposite charge situation that prevailed in SiC.

Dissociation of hydrocarbon yields negatively-charged carbon radicals. These persist until oxidizing conditions destroy them. Dissociation of silane only happens when silicon can maintain its metallic identity with positive charge. The integrity of silicon as metal is preserved throughout, while carbon flip-flops. This ambivalence is key to the distinct behavior differences between hydrocarbon and silane.

These differences arise from the contrasting behavior of oxidized and reduced forms of carbon (CO or $CO_2$ vs. $CH_4$) and that of the silicon counterparts ($SiO_2$ vs. $SiH_4$). All the carbon forms are gases; the reduced silicon is a gas while the oxidized compound is a solid that only vaporizes at 2 230°C. This results in markedly different geological behavior for the sister elements.

Ease by which an element can switch from carrying extra electrons (reduced state) to electron deficiency (oxidized state) represents a low threshold of chemical stability. If stability can be upset by transfer of static charge, the inductive driving of ions by electromagnetic field-fluctuations (electrophoresis) could affect their chemical states. Earth magnetic field-fluctuations, which occur at low frequencies (0 - 1 000 Hz) may be effective under some circumstances to upset the equilibrium of silane- and hydrocarbon-generated systems.

The connection between earthquakes and very-low-frequency (VLF) and extra-low-frequency (ELF) electromagnetic pulses is known, if not fully understood. Earthquakes are known to occur in times of enhanced VLF and ELF activity.(Adams 1990; Parrot 1990; Yoshino 1991) It is germane to this subject to mention the little-known fact that living forms have been shown to maintain about themselves ELF fields, which can be affected by geomagnetic fluctuations. The well-known fact of animal behavior becoming erratic in times of earthquakes attests to this linkage. Clinically, positive low voltage electromotive fields have been measured in the vicinity of the human cerebral cortex; and negative potential is maintained in the extremities. These are, apparently, for sensory and body regulation purposes.(Becker 1990) From this it is evident that the interference by a natural field might bring out bizarre behavior in animals.

Such life processes are beyond the scope of this book. We leave the matter with the thought that carbon and silicon hydrides, oxides, and native elements are fundamentally unstable. Calling to mind the meteorological metaphor so often quoted in discussions of chaos theory, a butterfly in the next county is imagined to trigger a chain reaction that creates a thunderstorm. A new order is then established out of meteorological chaos. For our solid Earth environment, a benign and mundane occurrence like a magnetic storm must be regarded as a possible trigger for endogeny. Ambivalent silicon is the butterfly.

The remarkable chemistry of silicon, while only beginning to be understood, is unfolding through information technology as a litany of versatility. The astonishing behavior observed in our present tiny sampling of the potentials of the element implies that a vast field lies ahead for explorative research.

Silicon combines chemically with 64 elements, and it forms alloys with another 18 of 96 stable elements.(McGraw Hill 1982) Its key characteristic for our purposes is affinity for oxygen, an attraction so great that the hydride "burns" in water spontaneously, extracting the oxygen, releasing the hydrogen, and giving off heat comparable to that released by the combustion of carbon in oxygen. The release of silane energy produces many profound effects in the mineral and biotic world.

Where carbon chemistry still holds major surprises [as demonstrated by the recent discovery of carbon 60, a material of extraordinary properties (Chapter VII)], silicon may have even more. **Silicon is the cryptic element from which continents** are built, the wonder metal of electrochemical performance. In future years it should prove itself the element best suited to serve as a chemical arbiter, an intendant for yet undiscovered facts of nature.

Why has the hydride form of silicon not been recognized heretofore? Its spontaneous combustibility on contact with water or oxygen is the simple answer. At crustal levels beyond the reach of our drills, all emanating silicon in silane is converted to oxides before we can detect it. We have, thus, been blinded to the very existence in nature of silanes.

Our discussion has reached into silicon's role in diapir advance, its essential contribution in chamber creation and a source of volcanic ash, and its critical role in the arcane process of earthquake energy release. In coming pages I and other contributors will describe these characteristics in "hot spots," fumaroles, and other cryptic places.. In Chapter VII I will analyze the "mother" gases that yielded the mixes we find as ambient or trapped in crustal rocks. All of these involve geology that is unsatisfactorily explained by existing theories but relatively easily understood as products of silane systematics.

## REGIMES OF OXIDATION AND HEAT RELEASE

The idea of plastically flowing mantle rock upwelling in a plume and downwelling around the central upwell is the popular model invoked at Hawaii and other "hot spots." The mechanism is physically awkward and, perhaps, impossible. Gaseous emanation of silanes, which oxidize upon reaching first water in the "vadose zone," can produce all the features of "hot spots" without invoking Mantle convection.

## Regimes of Product Output

The heat released in various oxidation reactions is comparable for silanes and hydrocarbons. Only the products vary considerably. **Three basically different output regimes depend on available quantities of water and oxygen.** We will call them Regimes #1, #2, and #3 and describe each.

The products of the three regimes are:

> **#1:** SILICA + HYDROGEN + METHANE + INERT GASES
> **#2:** SILICA + ELEMENTAL CARBON (or CARBON RADICALS) + HYDROGEN + INERT GASES
>
> **#3:** SILICA + CARBON DIOXIDE + WATER

*Regime #1:* A mixture of silane, methane, and inert gases, on its first encounter with limited water will yield these first regime products. It is not known what form the nitrogen would take, as it readily combines as a silicide and as a hydride (ammonia). The scenario occurs under anoxic conditions, thus preserving the methane, newly evolved hydrogen, and inert gas. Hydrogen has been added; silica produced. Under high confining-pressure even with high temperature, hydrocarbons should survive. Impedance to further migration may result from silicification of rock by mineral transformation or silica precipitation; and earthquakes may mark the breaching of such barriers by the ongoing high-pressured gas flow.

If these residual gases gain access to the atmosphere, a fiery eruption should be expected along with ejection of siliceous ash and lapilli. After a flow of gas of this type subsides, magma that was melted during subsurface combustion of silanes may flood through the conduit to the surface. This is the explanation for the frequent occurrence of white ash beds capped by basalt flows in volcanic terrains. **(Fig. III - 1)**

*Regime #2*: The second regime also occurs under anoxic conditions. It occurs under lower pressures than regime #1, but at temperatures in excess of 1 250°C. Silane behaves as in regime #1. But hydrocarbon is thermally dissociated [cracked as in industrial carbon black production]. Hydrogen is liberated from both silanes and hydrocarbons. This creates another highly-combustible residual gas which is apt to cause a fiery volcanic eruption as with regime #1. In this case the fiery ejecta would be accompanied by soot.

It was pointed out in the EV volume that soot is a prominent component in the deposits of the Cretaceous/Tertiary boundary. Whether it results from volcanism or extraterrestrial impact, is a question that has lead investigators into many remote geological field scenes over the last twelve years without producing consensus.

*Regime #3*: This regime of products will occur in an oxygenated subsurface environment. In the extreme case everything is oxidized that can be. Emission will not be fiery. In many mining districts with volcanic associations, rocks and minerals are oxidized types, and carbon is prominent. This is the situation in the Carlin district, Nevada, where soot and petroleum are found with disseminated gold in the very-near-surface mineralization; and metal sulphides are found only below hundreds of metres of oxidized surface ore.

In general there is a continuous gradation of partial conditions between regime **#3** and the others. Variations are due to quantity of oxygen available, original mix of gases, and opportunity for chemical activity between various volatile components and wall rock. Ejection pressures great enough to force ash, lapilli, and hydrous fluids to the surface also permeate conduit wall rock with the hydrous fluids, thus forcing dispersal into the country rock of silica, hydrocarbons, elemental carbon, and metal sulphides.

## IN SUMMARY
## MAJOR IMPLICATIONS OF SILANE BEHAVIOR

Profound geological implications have been imputed to the migration from depth of the volatile hydrides, especially the silanes and hydrocarbons. Their existence is contradictory of the theory of Earth accretion. For this reason, the Gottfried proposal of an hydridic proto-Planet in which differentiation of Mantle and a Pangean Crust could occur is favored and, in fact, appears to be a necessary precedent for explaining the observed geology. The systematics for carbide-hydride interactivity amount to an entirely new theory of the Earth. Let us summarize their aspects:

### Heating

1a. Emplacement of huge amounts of silicon by replacement processes and the release of enormous amounts of heat in the upper Mantle explain diapirism, magmatism, and melting of the asthenosphere, which may then

1b. extrude on the surface as "flood basalts," or

1c. effect mineral transformations in secondary stages, release yet more heat in the Crust, and cause melting (anatexis) and magma generation.

### Volcanism

2a. Bursting upward, the volatile bubbles directly cause earthquakes,

2b. volcanism, and

2c. "hotspots" in the crustal surface.

## Hydrothermal Silica

3a. *In situ* alteration of mafic plutonic rock to silicic lithologies (granites) on batholithic scales (Collins 1988) amounts to the generation of a granitic batholith or continental shield rock.

3b. Pervasive hydrothermal silicification may result from dissolved endogenic silica.

## Isostatic Orogeny: Crustal Dilation/Collapse

4a Uplift by dilation of entire regions of seafloor, continent, or of more localized surface welts results from gaseous intrusion and consequent mineral transformation. Subsidence results as pressure decreases due to gas escape.

4b. Mountain ranges rise with the removal of heavy metals from primordial mafic rocks by *in-situ* replacement.

## Residual Gas

5a. Residual gases in rock pores in the upper Crust can be interpreted as products of reactions between mantle and lower crustal gases and mineral cations.

5b. Carbonaceous deposits derived from deep-Earth gases include:
   (i) Soot, fullerenes, and particulate carbon,
   (ii) diamonds, and
   (iii) graphite,

All of these can be explained as products of various levels of oxidation and heat application to the volatile intruders, silanes and hydrocarbons.

## Particulate Silica

6a. Silica from silane oxidation provides the silica resource for erupted tuff and lapilli.

6b. Quartz comprising high-silica sand and quartzite reflects endogenic provenance.

## Carbon Radicals

7. Carbon in native or radical forms is derived by partial destruction of hydrocarbons and likely contributes carbon to the coalification of biomass as well as to clastic sedimentation on the surface.

8. Germanium, which is found erratically distributed in coals, and which must have originated at mantle levels, gives support to the idea that some of the carbon in the coal also must come from such depths.

There are many minor additional behavioral results, some of which are discussed in future pages. Banded-iron ores and aluminous deposits may qualify. But the foregoing are the continent builders. The entire panoply of major endogenic processes that have puzzled geologists over generations becomes understandable in terms of carbide-hydride systematics. In the EV volume, crustal dilation and collapse were discussed using the example of the Canadian Rocky Mountain orogeny. Mantle chamber explosion was exemplified by Gros Brukkaros. Fault mechanisms were taken up in connection with the much-studied [and much-misunderstood] San Andreas fault.

# WHY MOUNTAIN RANGES RISE

The perception of Rudjer Josip Boscovich in 1850 that mass deficiency at depth below the roots of mountains compensates for their surface mass was the first recognition that mountains rise above surrounding terrain due to buoyancy. Their density is less than that of surrounding terrain. This process is known as "isostasy," and it characterizes the mode of origin of all serrate mountain ranges. These terrains are often still rising and subjected to vigorous erosion. In geologists' parlance the topography is "youthful."

Lighter plutonic rocks, the granites, are the highest in silica content, and "intermediate rocks" [e.g. "mafic" andesites and diorites] have silica contents intermediate between granites and "ultramafic," rocks, the heaviest of plutonic rocks. Ocean bottoms are mainly underlain by mafic and ultramafic rocks, continents by granites and intermediates. Superpositioning of the lighter rock ["continental crust"] along the margins of continents above heavier rock ["oceanic crust"] implies that the continental crustal rock is the younger deposition unless tectonic emplacement is invoked.

The origins of continents and their rocks have been addressed by five generations of geologists since Boscovich's time. Despite a plethora of erudite literature, the results have been much less than satisfactory. Earth "differentiation," the settling of crystals in magma, is inhibited by magma viscosity and rendered ineffectual on the scale of continent growth. The sliding around, or "drifting," of continents, even if true, could not build mountains. The attempt to measure stress buildup in one imagined collision between "plates" [see Cajon Pass drillhole, Chapter VI], only succeeded in proving beyond doubt that no such stress is present and, thus, that no collision is occurring.

The provenance of silane emanation and combustion affords a complete resolution to the enigma of the origin of continental Crust. Introduction and concentration of silicon at crustal levels by silane migration, release of heat upon silane combustion, mineral transformations by cation exchanges, melting of country rock adjacent to silane permeation, and creation of "intermediate" magmas with components of immigrant silica and former country rock, all contribute to mountain building. This is so whether the mountains are batholithic, volcanic, or sedimentary. Siliceous magmatic rock on the surface characterizes batholithic mountains; siliceous extrusive rock on the surface and magmatic in the subsurface characterize volcanic mountains; and magmatic rock only in the subsurface characterizes mountains made up of sedimentary rock.

Where magmatism is induced by excessive silane immigration, the edges of a melt include late-assimilated blocks, the xenoliths. These are blocks surrounded by melt, but which remain unmelted as cooling progresses. They may flow with the melt as it responds to tectonic pressures. They may respond to hydrodynamic injection into lesser-pressured nearby sites in sills or dikes. As silane infusion continues, a magma mass can grow to the size of an entire mountain range, a "batholithic" size.

When mafic and granitic plutons have been found adjacent to one another, it has led to the interpretation that the hotter mafic melt has caused silicic country rock to melt, thus creating a granite magma. However, if silane pervasion resulted in mineral replacements within a pluton so as to generate granite, the same result would have been achieved without any contribution from country rock.

Mafic magma is the more fluidic, silicic magma the more viscous melt. Extrusion of mafic magma on the surface is a common prelude to the diapiric rise of an intermediate or granitic batholith. This is why lavas and small intrusives of mafic lithologies are found mantling and flanking granitic batholiths. The Sierra Nevada batholithic mountains are a good example of this order of igneous emplacement.

Tectonic pressures on magmatic masses before and after congealment are created by the buoyancies intrinsic to their lesser densities. Sharp contacts may be created where magma is unable to melt the cool peripheral wall rocks. Gradational contacts result when wallrock is fractured and, then intruded by melt rock created by heat release during mineral transformation. Such magmatic permeation typically tapers off with distance from the affected area. The Temecula terrane [described by L. G. Collins in Chapter IV] is exemplary of these systematics.

Mass deficiency created by silane transformation is the control and regulator of the level to which a mountain range rises isostatically above a crustal welt. The welt, or "cymatogen," itself represents deep-seated diapirism. Active emergence indicates continuing infusion of silanes. These functional facts of carbide-hydride endogeny are simple and basic to the interpretation of mountain building and to the distortion of the planetary geoid. (Frontispieces both illustrate this)

Orogenesis is a consequence of three basic processes whereby new crustal rock is created: mineral transformation, magmatism, and volcanism. After mineral transformation commences, if magmas develop, their advancing peripheries meet cooler country rock. They may penetrate it, congeal, and create a seal that impedes continuing flow of gas from below. Ongoing reaction of silanes may brecciate the seal. Such a melt breccia along with country rock at the top of an overpressured chamber creates andesite, the rock type "intermediate" between the melted mafic precursor and the silica-rich end-products of silane pervasion.

In combination with the foregoing scenario or, alternatively, in one where steam, hydrogen, and hydrocarbons accumulate behind the unmelted seal, a fiery, explosive, volcanic event is apt to follow. Silica and intermediate melt are exhaled in the familiar form of volcanic glass, ash (a misnomer), tuff, lapilli, tripoli, perlite, and other pyroclastics.

Timing of an explosive event depends on the ongoing rate at which silanes, hydrocarbons, and hydrogen migrate from depth to supply the silane-water reaction. The complex rock mechanics of any given pervasion and chambered site determine how well gas is entrapped, and hence, the level of pressure buildup and timing of explosions.

Lava flow should ordinarily follow explosive gas emission. A cycle of explosion, lava emission, chamber sealing, and dormancy should be the normal sequence. The photos in **Fig. III-1** show several typical occurrences of this type.

## OXYGEN, IRON, AND MAGNESIUM

**L. G. COLLINS & C. WARREN HUNT**

The banded iron ores contain iron-rich chert layers alternating with iron oxides (hematite and magnetite), iron carbonate (siderite), calcium-iron carbonate (ankerite), calcium-magnesium carbonate (dolomite), and dominantly ferrous-iron, hydrous silicate (greenalite). Transportation of these minerals and their deposition in repetitive beds is itself enigmatic. Modern environments comparable to those of the preCambrian deposits are non-existent.(Stanton 1972)

Two facts of the banded-iron ore genre are pertinent to the question of origin and need to be borne in mind. **(1)** The mineral suite contains only vanishingly small amounts of alumina amid abundant iron oxide and interbedded silica

[mainly chert]. **(2)** The deposits have dimensions up to 160 km in length, their deposition implicitly occurring in long trough-like depressions. The trough deposits are most easily visualized as volcanic exhalations from vents aligned along a fracture system in which silane-bearing gases emerged with hydrous fluids containing carbon dioxide, iron, and magnesium.

These facts lead to an interpretation of volcanic exhalation of metal and silane products; and they wholly negate the concept of these sediments as derivatives of fluvial erosion from a continental source. An erosion-sedimentation sequence from weathered rocks should also bring in clays rich in alumina. (James 1966) The common inclusion of volcanic fragments in banded ores and the frequent areal and temporal association between them and metavolcanic rocks also lend support to an exhalative origin. (Cloud 1965)

Ratios of Mg/Ca are greater than unity in carbonate beds associated with the preCambrian banded-iron ores, whereas, by comparison these ratios are less than one in sedimentary iron ores of younger age. Thus, the preCambrian banded-iron ores have a special origin that cannot be linked to an erosion-sedimentation cycle that may explain the younger deposits. (Rubey 1951)

In the EV volume the probability was established that these strata were deposited in an early atmosphere having little or no oxygen but relatively high carbon dioxide content (0.03 atm. instead of the present 0.0003 atm.). These conditions would enable transport of ferrous iron in solution. Once transported, if iron were precipitated locally in a basin by oxygen released from photosynthetic bacteria or algae, then sedimentary banded-iron ores could result. In that case, however, rocks associated with banded-iron ores should have carbon residue from bacteria or algae. As no such traces of carbon are observed, an exhalative, silane-generated endogenic source is more plausible.

Two conflicting processes operated in Proterozoic times. On the one hand the silane and hydrocarbon releases from inner Earth tended to lock formerly-nascent oxygen into mineral form while producing carbon dioxide and copious amounts of water. On the other hand, photosynthesis tended to convert atmospheric and hydrospheric carbon dioxide to mineral form, returning oxygen to the atmosphere.

Where photosynthetic oxygenation predominates today as the natural means for maintaining our oxygen-rich atmosphere, conditions in Proterozoic time were different. Periodic reducing conditions prevailed, as shown by the precipitation of ferric-iron oxides and ribbon cherts. The ion, $Fe^{3+}$, is too soluble to have precipitated without such reducing atmospheric conditions. The repetitious bands of chert and iron indicate pulses of exhalative volcanism.

Reversals of atmospheric oxygen caused by this sort of volcanism may also help to explain faunal proliferation and demise. The Ediacaran fauna of Australia, 800 my ago, is regarded as an effect of atmospheric oxygenation. Knoll (1991) explains it as a result of $^{13}C$ enrichment in the Proterozoic carbonate sedimentation. He interprets sedimentary enrichment to show that the biota has sequestered $^{13}C$, which was then deposited preferentially, leaving the atmosphere depleted. An equally plausible explanation would be emission of hydrocarbon gas from Mantle levels. Where Knoll defines $^{13}C$ deficiencies of -5.5 $^o/_{oo}$ as "normal" for the Mantle, in fact much greater deficiencies occur. For example, Leung (1990) reports values of - 25 $^o/_{oo}$ in the Fuxian kimberlite (Chapter VII); and hydrocarbons with levels of $^{13}C$ of -48 $^o/_{oo}$ are found in gneisses of the Carswell gastrobleme (Chapter IX).

A later flowering of life is demonstrated by the Burgess fauna of 550-my age. The soft-body imprints of nineteen entire phyla are represented at the type locality in the Canadian Rockies. Fifteen of these phyla went extinct before our Cenozoic Period.

From the other four has evolved all present life of the planet.(Gould, 1989)

The Burgess fauna occurs in a talus at the foot of the Cathedral carbonate bank, a reefal growth on a submarine fault-block in Cambrian times.  The same carbonate bank sixty or so km to the south is the site of the Mt. Brussilof magnesite deposits, where endogenic magnesium, which had been stripped from Mantle minerals by silane transformations was conducted upward in a lithosphere fracture and redeposited.(Chapter IX)

Thus it is that solutions to many of geology's most perplexing problems emerge with the concept of active volcanic effusion of endogenic silane and hydrocarbon gas.  Carbon of this provenance offers a simpler explanation for biotic proliferation as well as for atmospheric $^{13}C$ depletion in Ediacaran and Burgess time.  The banded-iron ore enigma and magnesite deposition are understandable as products of enhanced emanation of silanes and hydrous fluids enriched silica, magnesium, and iron.

## PLATE TECTONICS vs EARTH EXPANSION
### C. WARREN HUNT

*"INITIATION OF SUBDUCTION ON EARTH BY LARGE-SCALE MANTLE OVERTURN:  The difficulty of identifying forces large enough to create new subduction zones on the Earth has inhibited the development of a complete dynamic theory of plate tectonics.  Ridge-push or gravity-sliding forces cannot overcome the fault plane friction that results from the weight of the overriding plate on the subducting plate, [and] ... a sinking thermal diapir of the size generated by an unstable thermal boundary layer cannot generate sufficient viscous drag on a lithospheric plate to initiate subduction."*

**... FROM D. L. HERRICK &
E. M. PARMENTIER (1991)**

"... the hypothesis of seafloor spreading was greeted with enthusiasm, and thereafter [the oceanographers] explained most of their results in conveyor-belt terms. But despite much later information the concept remains assumptive and its mode of operation speculative."

**LESTER C. KING (1983)**

"Recent studies indicate three problems with the concept of continental drift as an incidental corollary of plate movement: (1) Slab pull cannot drive plates with continental leading edges, (2) There is no low velocity zone under shields, and (3) Continents have "roots" 400-700 km deep. These problems imply that if continental drift occurs, it must use mechanisms not now understood, or that it may not occur at all, plate movement being confined to ocean basins."

**PAUL D. LOWMAN, JR. (1985)**

S. Warren Carey (1976) in many books and technical papers for over thirty years has vigorously advocated Earth expansion and continent growth as inescapable facts of geology. In our book the workings of crustal expansion mechanisms are considered, and the energy and material contributions of carbide-hydride interactivity are advanced to explain what Carey described. The effectiveness of the concept for explaining observed geology appears far superior to the offerings of plate tectonics ("PT").

This comes about because the basis of PT is a montage of deductions mainly taken from geophysical contexts. It is often skillfully patched together with fragmental knowledge of ocean basin geology, it seems to appeal more for its elaborate vocabulary of tailor-made buzz words.

Whereas the original idea of plate tectonics came from the [correct] recognition that rock is being added to the Earth's surface at the mid-Atlantic ridge, it was then reasoned [incorrectly] that rock coming up in one place meant there had to be rock descending elsewhere. It seemed logical that the site where rock descended, which was named a "subduction zone," should coincide with the loci of deep-focus earthquakes, the "Benioff" zones. These features in some places slant under the edges of continents and island arcs. That made it an easy step to draw the conclusion that oceanic Crust was being forced under continents and/or sliding under due to having greater density than the continental rock.

Rapidity for the process could be estimated from the age of associated rock and continental fault movements. And from there it was another easy assumption that continental "plates" slithering about on a global scale could account for all manner of enigmas of geology so long as specific identification of products and effects were viewed from a distance.

Thus it was that a concept enunciated by Alfred Wegener in the 1920s and quietly retired in the 1930s, was exhumed in the 1960s to live again. "Continental drift" meant separation of **floating** continental masses originally. Its subdititious reincarnation, "plate tectonics," has the newly-contrived aspect of **driven** continental "plate" movement, the drive being provided by frictional shear of convecting mantle rock beneath a driven plate.

The opening quotation by Herrick and Parmentier explains that inner-planetary heat is wholly inadequate as an energy source for the task of moving continents. The idea persists, however, in the absence of any other means for plate driving. If rising mantle rock is deflected at the Moho, its deformation within the solid planet obviously must require investiture of enormous amounts of energy, far more than the small gravitational differentials that might cause the mantle rock to rise in the first place. This differential is no more than the density contrast between very hot rock and less hot rock.

Where a mantle plume is perceived not to be deflected at the Moho but to reach the surface at a spreading center, the concept is compounded by the imaginary cycle of rock cooling on the surface, contracting to higher density again, and then sinking back into the Mantle beside the plume, which continues to rise. Plume drag on a continent seems forgotten in this scheme, as the subducting mass dives under the continent directly in contact with it at a Benioff zone.

The plume drag scenario appears less than adequate for its task from an energy point of view, especially in consideration of the fact that continents have deep tooth-like roots that clearly suggest immutable permanence. At the continental margins the concept of oceanic rock underthrusting the continent is also unconvincing by reason of energy deficiency. The density increase achieved by cooling is very small, certainly far too small either to start subduction or to force underriding action. Likewise, the density differential of the cooled slab compared to the continental margin, appears deficient for its task of sinking the slab into high-density mantle rock! In any case, the sinking slab and subduction intuitions are unsupported by observable fact, as no emerged, new, crustal rock can actually be shown to be sinking back through the Crust. Rather, all such rock appears to be resting comfortably not far from where it was extruded. The outer geosphere has simply expanded.

Within the lower Crust and upper Mantle greater-density rock at lower temperatures is transformed to lesser-density rock with higher temperatures. The less-dense hot rock then rises isostatically. The origin of the heat in such rising masses is inexplicable unless a source for new heat generation can be identified. Mantle inhomogeneity coincides with the topology of "hot spots," but these sites are not coincident with the positions where mantle convection is required to occur to defend the concept of continental drift.

Physical principles governing heat flow in homogeneous media cannot accommodate the creation of the observed surface heat differentials. Rapid Mantle convection was conceived to fill this deficiency. Requiring high rates of deformation of viscous rock, the concept is an excursion in science fiction that evokes rock behavior that is certainly not known to be possible with present knowledge of the Mantle. Attempts to defend the concept on the basis of mantle inhomogeneity have not produced convincing insights.

Thus, unequivocally, **lack of an explanation for the known uneven heat distribution in the lithosphere is not encouraging to the idea of mantle convection as the motor for plate tectonics.** On this point alone plate tectonics fails to live up to its billings. There are others, however.

First, consider the inconvenient but well-known fact of rock plasticity at mantle and lower-crust depths. Deep earthquakes give a seismic signal similar to that obtained in brittle fracture. This has led to the earthquake sources being interpreted as frictional effects of stress release on faults between blocks of *brittle* rock. With no more license than this, the conceptions of subduction and plate movements are taken to be explained. The inconvenience occurs when one recognizes that the milieu of these earthquakes is plastic rock, which cannot maintain shear stress leading up to brittle fracture.

**The remaining fracture mechanism by which an earthquake can be explained in these circumstances is hydraulic fracturing ("fracking"). This process is one of expansive rupturing by excessive hydraulic pressure. After pressure buildup, sudden rupture produces a fracture that propagates rapidly, *simulating brittle fracturing*. No energy release that can be interpreted as subduction slippage or plate migration need occur.**

Another difficulty with relating fault action to wandering,

crustal plates emerged from research in the drilling of the deep Cajon Pass drillhole in the San Andreas fault zone (the "SAF"), southern California. The drilling was to measure the stress buildup, which proponents of PT confidently expected to find in the SAF. Such stress buildup is essential if the "Pacific plate" is, in fact, impacting against, and diving under, the "North American plate."

**The Cajon Pass drillhole completely failed to detect stress. Neither was any temperature or pressure buildup detected in the fault zone. These unexpected results prove beyond doubt that subduction of the "Pacific plate" under the continent cannot be occurring at that site. Since there is no evidence of a collision in progress, nor any pressure-temperature regimes consistent with stress-release faulting, the entire preconceived notion that fault movement causes earthquakes fails.**

When nothing was found to support the preconceptions of plate tectonics, the facts of the investigative drilling seem to have been quietly buried. There has been no recognition that plate tectonics is a discredited myth, nor any reconsideration in the literature of the theories of earthquake energy, fault mechanics, or plate tectonics. Quite to the contrary, the order of the day has been "business as usual" with no wavering on the important matter of faith in PT, stress buildup, or brittle stress release as the source of earthquakes! Geology in disarray!

Another favorite "proof of PT" is the clear evidence that the Hawaiian Islands and some other volcanic island chains have erupted in a linear order, northwest first, southeast later. This is taken as virtual proof that the *Crust is sliding northwestward* over a stationary "hotspot," the top of a convecting "Mantle plume." The Emperor seamount chain at the northwest end of the Hawaiian chain represents a still earlier series of volcanoes in the same basic chain but on a different azimuth. The interpretation is that the *Crust slid north to produce them in Cretaceous time.*

The age order of volcanoes is impressive but not necessarily proof of crustal slippage. Alternatively *the effusing volcanic source may have moved south and then southeastward under the Crust.* A variation of this would work on the Hawaiian Islands themselves: successive emissions exhaling through a lithosphere fracture and progressing along the fracture as previous vents became plugged with earlier injectites. Magma congealing during any pause in the emissions would divert subsequent emissions to new vents along the fracture.

**The simple idea of a series of volcanic vents being active in succession hardly warrants the grand interpretation that the whole Crust is slipping over a "hot spot." A static Crust with a simple lithosphere fracture would be able to provide the observed effects. The underlying gas plume might either be moving or stationary.**

As shown by the test conducted by drilling in the San Andreas fault zone, geologic studies on the continents provide little to support and much to discredit PT. Starting out in the mid-1960s with enthusiastic and hopeful speculation, the advocates of plate tectonics, after twenty-five years of intensively vetting their idea, have been unable to achieve any proof or clear purview of the PT mechanism in operation. It is still at the starting gate.

In the coming pages I and contributor Collins address subjects where carbide-hydride systematics provide satisfactory explanations to major geologic phenomena over which the geological community has been wearied by the jaded idea of plate tectonics for the last thirty years.

## DEPRESSION OF OCEANIC CRUST
## DUE TO LOADING

In carbide-hydride systematics large masses of siliceous rock, as well as some mafic magmatic rock, are introduced into the Crust and debouched upon its surface. Being less dense than the mafic ocean Crust, the effusive continental rock-masses ride higher due to isostasy. At continental edges, deltaic load-transfer, thrust-slippage of nappes out upon the oceanic Crust, and glacier-like rock flowage all contribute to build a seaward-thinning wedge of continental rock. The continuing piling-on of new mass upon the surface causes depression of the oceanic Crust and the Moho beneath it.

Through geological time, this enormous loading depresses proximal oceanic Crust, more at the very edge of the continent, and progressively less, farther out on the continental shelf. Since loading is uneven, the dips within the wedge of depressed Crust and on the Moho vary; and this is one characteristic by which depressed oceanic Crust can be identified. **Differential loading and encroachment of continent-derived rock are the causes of landward slopes of oceanic Crust and Moho beneath the lips of continents.**

Dynamic underthrusting of a continent by oceanic Crust is nowhere demonstrable, whereas in many places the oceanic Crust proximal to a continent is depressed and slopes beneath the continent because of the loading of continental rock upon it. Underthrusting is absent.

Seismic reflections obtained in petroleum exploration off the coast of California provide an example of such loading. Meltzer & Levander (1991) attribute this seismically quiescent subscape to subduction.**(Fig. III-2)**

Commencing from a point near the foot of the continental slope (90 km or so offshore), the profile progresses eastward almost to the coastline. A very prominent reflector beneath the continental Crust is traceable from a starting depth at the edge of the continental shelf of 6 km almost to the coastline, where it has plunged to an 18-km depth. This is interpreted as the

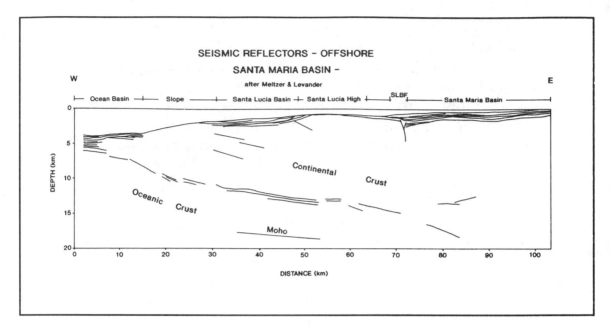

### III-2

Prominent reflections define near-surface basin sedimentation and an undulating horizon interpreted as the inter-crustal contact between continental and oceanic "plates." The undulating contact is too irregular to be an active subduction surface, and brings to mind a surface warped by uneven loading, an effect of isostatic subsidence under a depositional load of continental debris. The Santa Lucia high, flanked by basins of sediment on either side, suggests igneous plutonism, an eruption that bulges the inter-crustal contact as a magmatic and/or diapiric conduit might do. The SLBF (Santa Lucia Bank Fault) is a normal growth fault (grows upward with depth, and time), a natural structure at the edge of a basin in which sediment is accumulating.

contact between the base of continental Crust and top of subducting oceanic Crust. Structural attitudes in the wedge of continental rocks imply folding, presumably a consequence of drag by the under-slipping oceanic plate. The Moho is some six to seven km deeper than the alleged subduction surface.

This interpretation is unnecessary and inappropriate. First, the shallow seafloor known as the Santa Lucia Bank in the middle of the profile is a block faulted horst with an igneous core. The tilted block is underpinned by an upward conical bulge of the inter-crustal discontinuity. This suggests igneous intrusion that pierces both the oceanic crust and the overlying continental wedge. Block faulting and intrusive injection imply a structural horst in *east-west tension*, features wholly incompatible with the east-west compression of an actively-impacting and subducting oceanic "plate."

Some would call the contact between oceanic and continental rocks a "Benioff zone," apparently forgetting that Benioff zones are defined by seismicity and a steep enough dip to take the zone into the Mantle. This contact is emphatically not a Benioff zone.

The wavy contact between the continental and oceanic plates on this profile would create powerful resistance to subductive slippage at each change of dip, and fragmentation of the opposing plates would ensue. *Such fragmentation is not apparent on the seismic interpretation; and subductive movement is highly unlikely.*

Thus, I demur vigorously against the imputation of subduction on the basis of these data. Neither the structures shown by the Meltzer and Levander seismic profile nor geological common sense support the bizarre attribution of this subscape to subduction. Much more likely, the fold and fault structures in the continental rocks above the reflector show that **depression of the oceanic Crust at the edge of the continent has occurred, by loading of continental rock upon the edge of the oceanic Crust. Erosional debris from the adjacent landmass and its re-sedimentation on the continental shelves with concomitant intrusion and extrusion of plutonic rock into and upon the surface of the accumulating sedimentary pile are fully capable of creating this depositional wedge on the continental shelf and depressing the underlying oceanic Crust and Moho. Low-angle thrust-faulting, gravitational in origin, is to be expected with the emplacement of such a peri-**

*continental mass as this; but not as the primary contributor to its structure.*

*Lateral encroachment is the mechanism of growth by which continents expand through geological time. Their onlapping of the passive oceanic margin isostatically depresses the oceanic Crust, as the profile, Fig. III-1, illustrates with such elegance with respect to the shelf terrane off central California.*

D. I. Gough (1986) has studied the matter with respect to British Columbia and concluded that no subducting oceanic plate appears geophysically to be underriding the intermontane region, the 400-km belt between the Coast Range on the west and the Rocky Mountain trench on the east. He says that this belt shows a high heat-flux, high density, low electrical-conductivity, and a thick, low-velocity, asthenospheric layer at the top of the Mantle under a thin crust. Asthenosphere is attributed to partial melting, which Gough deduces as consequential upon "mantle heat upflow," a heat plume, that is to say. Emphatically, Gough says that neither a Benioff zone nor a subducted plate is geophysically visible beneath continental British Columbia.

Gough's evidence renders absurd the much-touted idea that a subducting oceanic "plate" is underriding the adjacent terrain of the western United States. There is no substantive evidence in geology or geophysics in this region that demonstrates the existence of a "diving slab" beneath the continent. This terrane is considered by Lester C. King in his brilliant critique, "WANDERING CONTINENTS AND SPREADING SEA FLOORS ON AN EXPANDING EARTH." He says (p63) that "...between the Juan de Fuca and the mainland...stratified turbidites now dip east, [thus] showing that the present Juan de Fuca Ridge rose later, [after they were deposited],... about 3 my ago." This makes the ridge too young to fit into the subduction pattern that is promulgated as consequential upon the sea floor magnetic geochronology for the region.

So much for the "field evidence." What is the possible behavior of a "plate"? Can it be pushed, dragged, or pulled?

"Pro-motive propulsion" of rock plates driven by active compression from behind is not possible because internal cohesiveness of a "plate" that is subjected to such compression is insufficient for the force to be transmitted through the plate to where it can oppose any sort of obstacle without internal disruption. Certainly, the resistance to driving a plate against an inclined plane representing the basal surface of a continent would deform, divert, and fragmentize the plate. Pro-motive piston-type compression from behind to propel a slab must be regarded as a mistaken concept.

Shear friction from mantle turnover, a dragging action of underflowing asthenosphere, is now the favored concept for the driving of a subducting plate. To perceive this as working requires imagining that tangential shear stress transmitted to the bottom of a plate by undersliding Mantle is transferred to the plate's upper surface, where the opposing drag of the continent resists plate motion. The plate must behave with structural integrity to transfer the force, no slippage within the plate being allowed, whereas at the top contact with the continent, perfect slippage must occur. No plate would have such structural integrity. The idea is implausible.

The remaining option is tension applied as a field force to draw a slab. Whereas this is not possible from endogenic force fields, it conceivably could result from exogeny. Gravitational attraction from a close-passing asteroid could attract plates differentially and draw a continental mass over an oceanic one. However, this mechanism is not susceptible to general application; and it would yield overriding, the reverse effect from subduction.

Implausible as subduction is shown to be, underriding motion on a *nearly horizontal* plane is not inconsistent with the field evidence of gravity-driven thrust faults. If a continental margin were raised slightly, a detached plate could glide out upon oceanic Crust. Then the overthrust source terrane could subside, and give the [false] appearance of subduction. However, as this scenario could only hold for a nearly horizontal subduction surface, and near-horizontal angles are not what subduction, or "diving," is conceived to explain, the comparison with low-angle faulting is not appropriate.

At increasing angles of dip, subduction rapidly becomes mechanically implausible. At a 30° dive angle almost 20 % of the driving force is directed, as compression perpendicular to the underridden plate and across the discontinuity. At a 60° dive angle, over 80 % of available force is so directed. So much for "pro-motive" force in plate movement.

Passive force is still to be dealt with. Gravity differentials that motivate isostasy provide some excess mass that could facilitate sinking. Subduction would then be a combination of pro-motive shear drive transmitted through a "plate" and isostatic submergence. However, the differential density does not appear to comprise a field force sufficient to sink surface rock downward past rock of increasing density far into the relatively-high-density Mantle. **The notion of gravity-driven subduction is both impossible and [fortunately] unnecessary.**

## THE POTENTIAL FOR ENDOGENIC VIOLENCE IN OUR ENVIRONMENT

With a fresh memory of wild oil wells in Kuwait and dire warnings of pollution from the partial combustion of their emissions, it may be a useful exercise to compare the man-made "event" with a release of inner Earth gas.

The great crater of Gros Brukkaros, which was described in the EV volume as a volcano without volcanic rocks, must have flowed gas and liquid mist exclusively. Its orifice is approximately 14,000,000 times the combined orifices of the 600+ Kuwait oil wells. Since each of the well casings choked back flow significantly, by contrast with the Gros Brukkaros orifice in which impedance to "open flow" would have been only the load of entrained mass in the escaping gas, the actual flow that opened the crater in the first place and held it open must have been a multiple of the fourteen million figure. In addition, where the Kuwait wells are producing from two-km depth, Gros Brukkaros may have been flowing from 10, 20, 30, or more km depth, where initial reservoir pressure would have been much greater.

Thus, the natural endogeny of exhaling silanes and hydrocarbons and their derivatives could easily give us a blast that would toxify, if not wholly deplete the Earth's atmospheric oxygen supply in a matter of days or months. In the EV volume I pondered the occurrence of repeated oxygen depletions represented by deposition of the enormous preCambrian iron formations. Release of silane and hydrocarbon gas from the Mantle could explain these quite well.

The recent suffocation of 1,500 Cameroonian lakeshore dwellers after a belch of $CO_2$ gas from Lake Nyos is a lesson in the effectiveness of this kind of toxification. **Archeologists and paleontologists looking for explanations for sudden disappearances of human or animal populations should be wary of underestimating the powerful forces of endogeny. Past extinctions may have been *largely* the result of endogenic gas release.**

## Chapter IV

# ORIGINS OF GRANITE:
## THE PROBLEM AND POSSIBLE SOLUTIONS

### INTRODUCTION

C. WARREN HUNT

"...until the origin of granite is truly understood, a large part of geological science has no reliable foundation."

• • •

"Geologists .. at times lean too heavily on .. collective wisdom .. and the judgment of others rather than on the evidence."

**J. A. RODDICK, 1982**

• • •

"For the large-volume silicic batholiths so commonly associated with subduction zones, such as found in the Sierra Nevada or the Andes of South America, it is very probable that *other sources of volatiles exist that are related to the large volumes of andesites and basalts being generated from below the Crust.*"

• • •

"The origin[ation] of granitic melts in the Crust requires several weight percent water. .It is interesting to note that *Earth appears to be the only terrestrial planet with a thick continental crust ..., and ... the only one covered with water.*"

**WHITNEY, J.A. 1989**

Systematic silicon emanation from mantle depths as silanes and the immense energy release that occurs with their oxidation are processes that together achieve resolution of Roddick's conundrum. The simple elegance of carbide-hydride systematics is extended by the demonstration that it readily explains *in situ* mineral transformation. This work, on which L. G. Collins has labored for 25 years, provides glue for the carbide-hydride theory. The power of the concept in this way is driven home effortlessly.

After two hundred years, from the days of Hutton and Cuvier, the enigmatic source of Earth's pro-generative energy, the problem that has eluded geologists from earliest days may be resolved by this concept. If so, future attempts to unravel planetary history will be much facilitated.

## ROCK   TRANSFORMATION IN SITU:
## MAFIC   TO   FELSIC

### L. G. COLLINS

## New Jersey: Amphibolites

In the vicinity of some abandoned magnetite iron mines at Dover, New Jersey, on the Hibernia anticline and Splitrock Pond syncline, magnetite is concentrated in amphibolite units, which occur in an interlayered preCambrian series of metasedimentary and meta-igneous rocks (**Fig.IV-1**).[35] The amphibolites have the appearance of altered basalt lavas of surficial provenance, but exhibit metamorphism that is generally considered only possible at great depths in the Crust.

---

[35]  *Sims (1958).  My project was inspired by this regional geology, in particular the association of the amphibolite units with two large folds, the Hibernia anticline and the Split Rock Pond syncline.  Paul Sims of the U.S. Geological Survey had mapped the amphibolites as lenticular bodies, thickest on the noses of folds, thinning toward the limbs, and even disappearing. Lenticularity was the aspect that attracted my attention.  I wondered whether amphibolite compositions might differ between the noses and limbs.*

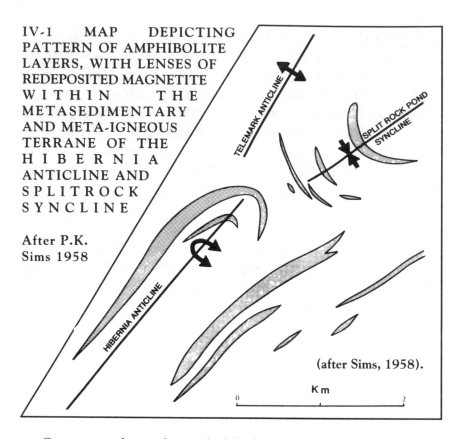

IV-1 MAP DEPICTING PATTERN OF AMPHIBOLITE LAYERS, WITH LENSES OF REDEPOSITED MAGNETITE WITHIN THE METASEDIMENTARY AND META-IGNEOUS TERRANE OF THE HIBERNIA ANTICLINE AND SPLITROCK SYNCLINE

After P.K. Sims 1958

TELEMARK ANTICLINE

SPLIT ROCK POND SYNCLINE

HIBERNIA ANTICLINE

(after Sims, 1958).

Km

0          2

Current geologic theory holds that in a tight fold the more malleable rock units tend to move plastically to lower-pressured sites on noses. Thus, they are predicted to thicken on noses and thin on limbs, thus gaining lens-shaped cross sections. My field investigation, however, did not confirm this scenario on the Hibernia anticline. Tracing eight individual amphibolites, I found their thicknesses remained constant as I approached the axes and did not pinch out or disappear on the limbs. This inversion of expectations and failure of my working hypothesis compelled me to reconsider the bases of Sims' conclusions.

As his work was directed toward understanding of the magnetite ore structures, he had only traversed the tectonic structures at infrequent intervals. He had not recognized that gradual change of the amphibolite layers with tectonic position, as forecast by prevalent theory, does not, in fact, occur.

My working hypothesis anticipated recrystallization accompanied by plastic flow during folding. This should have allowed some elements to migrate to noses while others were left behind in the limbs. Whereas these amphibolites lacked evidence of deformation or mineral substitution, there was still a chance that differential chemical migration might have prevailed. I proceeded, therefore, with collecting more than 900 samples along the eight amphibolite layers over more than 40 km of strike length around the nose and limbs of the anticline and syncline.

I found no differential migration of elements to structural axes in the amphibolites. Instead, there was clearly an association between tectonic shearing and recrystallization with mineral and chemical change, an association that intensifies as magnetite ore bodies are approached. Tectonic deformation, while minimal on noses, has created open avenues on the limbs. Through these passages hydrous fluids have passed, leaching cations and recrystallizing plagioclase and ferromagnesian silicates.[36] Iron in the magnetite ores is accounted for as metal leached from formerly iron-rich augite, hornblende, biotite, and now-destroyed hypersthene.(Collins 1969a)

In contrast to my explanation, which was based on studies in the host terrane, Sims' work had been primarily in the mines. His theory for the magnetite concentration entailed migration of iron in hydrothermal fluids carried from an unknown, deep source [perhaps an iron-rich magma]. He regarded the wall rocks in the mines as passive containers for the introduced magnetite. By contrast, and attendant on iron relocation, my work established the exodus on a large scale of calcium, aluminum, magnesium, and iron from tectonically sheared amphibolites.

---

[36] *Where the amphibolite units graded into magnetite ore, plagioclase and the ferromagnesian silicates (hypersthene, augite, hornblende, and biotite) were found sheared and recrystallized, and hypersthene destroyed. Augite, hornblende, and biotite had less iron in them; and coexisting plagioclase, formerly "An$_{55}$" (55 % calcic; 45 % sodic composition), was recrystallized so as to become "An$_4$" (96% sodic; 4% calcic composition). The term "An" followed by a number, such as An$_{40}$, is shorthand for the percentage of calcium in sodic + calcic cations in plagioclase, the anorthite percentage of albite + anorthite. An$_{0-10}$ = albite; An$_{10-30}$ = oligoclase; An$_{30-50}$ = andesine; An$_{50-70}$ = labradorite; An$_{70-90}$ = bytownite; An$_{90-100}$ = anorthite.*

## New York: a Granite

In a similar study of magnetite concentrations in pyroxene granite[37] near Ausable Forks, New York, I had again recognized the subtraction of large volumes of iron across distances as great as 600 feet on either side of magnetite concentrations.(Collins 1959, 1969b) In that process dark, iron-rich, pyroxene granite was converted to white, iron- and calcium-poor biotite granite with auxiliary garnet near magnetite concentrations. Pervasive replacement and recrystallization had occurred in this situation, not only in amphibolites but in granite bodies as well.

## Temecula, California, a Mafic Dike

The New Jersey and New York information reminded me of a problem I had encountered years earlier on a California terrane. My initial exploratory trip to rocky Santa Margarita Canyon in 1967, revealed a monotonous granite and diorite scene that was dissuasive of further interest. Closer inspection, however, had turned up unexpected complexity.

A black, fine-grained dike 10 to 70 cm in width transecting coarse-grained, dark-gray diorite and white granite[38] appears to demonstrate injection of mafic magma *after* diorite and granite were in place. But fragments of the dike, wholly enveloped by diorite, require the presence of the dike *before* diorite flowed in around them.

If one took the second assumption, that the diorite and granite were injected after the dike, still more puzzles were created. For example, the dike abutted granite with sharp contacts in some places and with gradational contacts in others. This can be taken to suggest that the granite flooded in as a magmatic melt, freezing with sharp contacts in fractures within

---

[37] *Clinopyroxene, hedenbergite.*

[38] *The dike is now recognized as fine-grained amphibolite. Its composition and petrographically visible "remnant zoning" in plagioclase imply precursor andesite.*

older rocks. Alternatively, the granite can be interpreted to have formed by pervasive replacement of the mineral fabric of older mafic rock, leaving the enveloped fragments bypassed for obscure reasons. Thusly, gradational contacts could result without melting. I favored this last interpretation.

These contradictory relationships had been sufficiently puzzling for me to have sent a student to the canyon for a senior-thesis study of the outcrop. The student's micropetrography revealed an unexpected abundance in the Temecula granite of a mineral association named myrmekite[39], at levels up to three percent (**Fig. IV-2**). The theory most favored to explain myrmekite holds it to be an "exsolution product" after magmatism, a separation of components of country rock during cooling.[40] I accepted this established wisdom for the Temecula granite uncritically at first. However, I had not considered the conflict that was created by the discounting of my original deduction, that Temecula granite originated by *in-situ* mineral replacement instead of by crystallization from a magma. Pre-existent diorite as well as the black dike had been penetrated and replaced. Subsequent work was to establish Temecula Canyon as the very clearest example of *in-situ* granite formation. The exsolution origin of myrmekite was discredited in the process.

---

[39] *Myrmekite is spoken of as a mineral but is, in fact, an association of two minerals, a doublet of plagioclase and quartz. Myrmekite studies spanning 117 years have resulted in at least six major hypotheses on its origin. All of them tended to trivialize its contribution to rock-forming processes. My work refutes all the earlier theories and shows myrmekite to be an important indicator [in some places, in fact, it is the only clue] to transformation in situ of mafic rock to granite.*

*A comprehensive explanation of the labyrinthine trail of evidence for this extremely important process, which takes place at microscopic and sub-microscopic levels, is to be found in the Appendix, page 370.*

[40] *A segregation of plagioclase and quartz from high-temperature potassium feldspar (K-feldspar) crystals during the waning stages of solidification of a granite magma.*

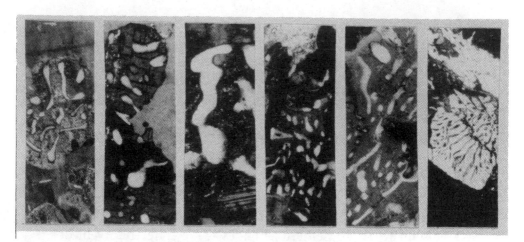

**IV-2: PHOTOMICROGRAPHS OF QUARTZ VERMICULES IN MYRMEKITE FROM TRANSFORMED PLUTONS.**
The varied habits of crystallization are well illustrated by these photomicrographs. The tendency for clustering of the vermicules is particularly impressive.

## New Jersey: a Transition in Gneiss

In yet another study on the Hibernia anticline and the Splitrock Pond syncline, I examined a sheared, 30-metre wide, biotite-hypersthene gneiss, which I had previously traced and sampled.[41]

It was this study which first raised my skepticism of the magmatic origin of myrmekite by exsolution from K-feldspar. The gneiss was thought to demonstrate sedimentary origins, because its quartz-rich zones contain graphite. This high-temperature, metamorphic form of carbon is generally interpreted as the chemical remnant of organic matter. While graphite is not normally found in igneous rocks, it often occurs in metasediments such as quartzites and marble, both of which occur in the subject study area as gneissic interlayers and lenses.

---

[41] *Micropetrography showed that this gneiss had a variable composition, ranging from a biotite-hypersthene-plagioclase composition to gneiss that, variously, comprised hypersthene-plagioclase, biotite-plagioclase, biotite-garnet-plagioclase, biotite-garnet-plagioclase-quartz, or biotite-garnet-sillimanite-K-feldspar-plagioclase-quartz. The K-feldspar-bearing gneiss also contained myrmekite.*

My examinations showed that myrmekite occurred only where the gneiss was strongly sheared and recrystallized in fold limbs. The myrmekite coexisted with garnet, sillimanite, K-feldspar, and graphite. The gneiss in the nose of the folds is unsheared and is associated with biotite, hypersthene, and plagioclase.[42]

It seemed reasonable to correlate recrystallization of original biotite-hypersthene-plagioclase gneiss in sheared zones with release of iron, magnesium, and aluminum, which were transferred to garnet while potassium and aluminum would go into K-feldspar and excess aluminum into sillimanite. Given these conditions, **the K-feldspar in this terrane is not a primary mineral, crystallized from a magma, but a secondary mineral formed in a sheared [but solid] metasedimentary rock at temperatures below the melting interval for granite. The associated myrmekite must also have formed during the time in which the K-feldspar was crystallized.[43]**

Unfortunately soil and vegetal cover obscure the 50-m interval in which occur the later-stages of replacement and recrystallization, that is to say, the transition from undisturbed biotite- and hypersthene-bearing rock near the noses of folds to garnet- and sillimanite-bearing rocks in the limbs of folds.

---

[42]  *The elemental compositions of these minerals have a bearing on their history of mineral regeneration in the sheared zones. In addition to silica in each of the minerals, biotite contains potassium, magnesium, iron, and aluminum; hypersthene contains iron and magnesium; and plagioclase contains calcium, sodium, and aluminum. The plagioclase $An_{80}$ in the unsheared gneiss proved to be unusually rich in aluminum (nearly 30 percent aluminum oxide), whereas in the sheared rocks, the recrystallized plagioclase ($An_{40}$) contained only about 15 percent aluminum oxide.*

  *The elements in the silicate minerals in the unsheared rocks are the same as those found in the silicate minerals in the sheared rocks. For example, sillimanite is an aluminum-rich mineral containing nearly 63 percent aluminum oxide; garnet is an iron-magnesium-aluminum silicate; and K-feldspar contains potassium and aluminum.*

[43]  *This myrmekite occurs where garnet has replaced hypersthene, the earliest stage of gneiss recrystallization. This myrmekite clearly originates in a replacement process, thus supporting my hypothesis.*

## In Summary: Evidence for Transition

In summary, the fold limbs exhibit mineral transitions from iron-rich hypersthene in unsheared massive rock to iron-rich garnet and coexisting sillimanite in sheared and recrystallized rock. Therefore, the hypersthene and plagioclase of the primary gneiss would be logical sources of the iron and aluminum in the garnet and excess aluminum represented by the sillimanite. That is to say, depletion of aluminum in plagioclase provided aluminum to the sillimanite.

In my investigation it was apparent that the original biotite-hypersthene-plagioclase gneiss is mafic and black, whereas the recrystallized rock is sialic and nearly white. Comparing chemical analyses of sheared and unsheared rocks, it is apparent that much silica and potassium came into the sheared rocks while much calcium, aluminum, iron, and magnesium were subtracted. The silica formed quartz and contributed silica to recrystallized silicate minerals.

The principle of uniformity has impelled much geological thinking in recent decades toward "closed systems," processes where elements are recycled but stay within the systems and merely change their forms. External input is absent. My description of a direct correlation between the composition of unsheared biotite-hypersthene-plagioclase gneiss and its sheared and recrystallized equivalent, biotite-garnet-sillimanite-K-feldspar-plagioclase-quartz gneiss, requires an external source of silica [although not iron (as Sims thought)]. Thus, the system must have been open, not closed.

Where, then, did the released elements go? Some of the iron went into garnet and magnetite concentrations (as in the recrystallized, sheared, amphibolite units), but much of it has disappeared from sight. Magnesium was transferred to garnet or remained in biotite. Some calcium was transferred to recrystallized plagioclase, but large amounts can not be

accounted for. The disappearance of these elements, presumably carried toward the surface in hydrous fluids, is strong evidence that the system must have been open.

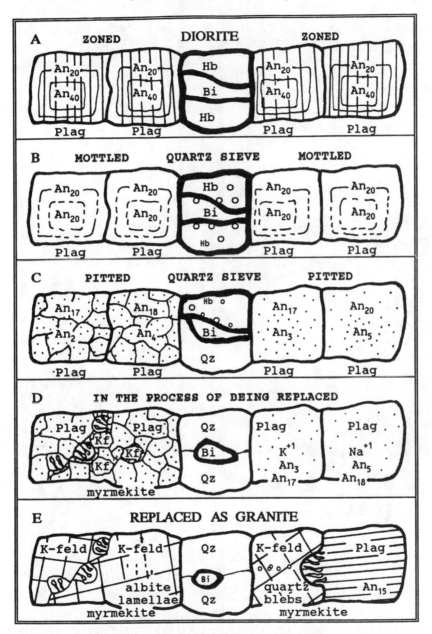

**IV-3: SCHEMATIC DIAGRAM TO ILLUSTRATE THE FORMATION OF MYRMEKITE**

Showing progressive change from dark gray diorite to white granite with myrmekite.

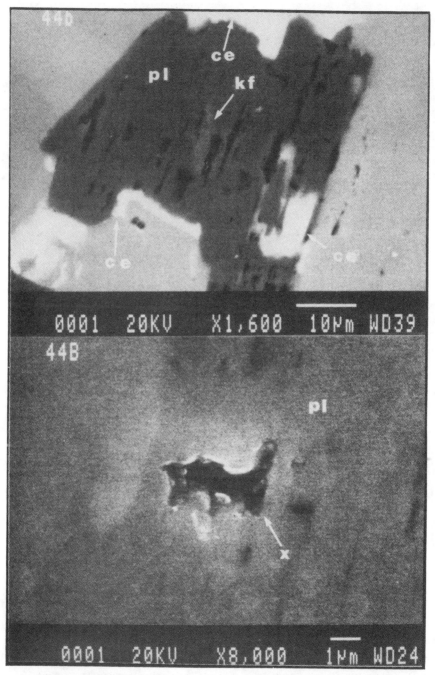

**IV-4: PHOTOMICROGRAPHS SHOWING PLAGIOCLASE (pl) INTERNALLY PITTED AND RIMMED WITH K-FELDSPAR (kf) ADJACENT TO PITS.**

Bright rims are celsian; x marks a pit in plagioclase enlarged to x8,000 magnification.

## Return to Temecula

Recalling now my student's studies of the diorite and myrmekite-bearing granite in the Temecula area, the tabletop-smooth outcrops, the stream-polished bedrock on the canyon floor, the absence of weathered rock, and the clear exposures of all geological relationships, I theorized that I should be able to trace in detail the transition from myrmekite-free to myrmekite-rich rock.

In fact, this hope materialized perfectly. With a diamond saw for sampling, I found all the transition stages, early, middle, and late, that I predicted were present.(**Figs. IV-2,IV-3,IV-4**) It was now certain that the myrmekite did not form by exsolution from K-feldspar but by replacement of plagioclase by K-feldspar.

## Myrmekite by Replacement of K-feldspar

The realization that myrmekite was formed where K-feldspar replaced plagioclase had far-reaching implications. It meant that the granite of the Santa Margarita Canyon was derived by replacement of the diorite.(Collins 1988a,b) Then, **because myrmekite is common in granites (whether granite pods of small size, measurable in cm, or masses of batholith proportions, 100 to 1 000 km in diameter) a replacement origin for all granite bodies must be a possibility.** Without denying that some granites form by magmatic processes, my hypothesis expressed the potential for large granite masses to be developed by replacement *in situ*.

The failure of the magmatists' reasoning is attributable precisely to its reliance on experimental work done on closed systems, in "bombs" in laboratories. The precise numbers to the third decimal point that are obtained from these experiments encourage a false sense of security by the experimentalists in their results. Imagined proofs of important points are, in fact,

illusions that contribute little that is relevant to natural processes.

My criticisms are not that the laboratory results are inaccurate, but that **the natural environment remains an "open system" a vast "unenclosure" that cannot be matched by the isolated conditions in a "bomb."**

It is significant that no one has ever produced myrmekite in a "bomb," although other intergrowths[44] typical of magmatic rocks have been produced in this way. This alone suggests that the laboratory experiments in closed systems do not represent what occurs in nature where myrmekite is formed.

**My data show that many [but not necessarily all] granites, including major batholithic ones, contain myrmekite and have evolved from mafic rock by replacement processes. These replacement processes are the result of hydrous fluids that move through fractured rocks in open systems. They do not involve solid-state diffusion through great distances.[45]**

An hypothesis, if it has merit, must be able to make predictions. From the characteristics of the sheared and recrystallized amphibolites, gneiss, and granite in New Jersey, New York, and California, I have made the successful predictions as described in later sections. The formation of myrmekite and of granite, *in situ*, is consistent with Skobelin's concept of evolving geospheres and follows directly from the Hunt concept of the migration of silicon hydrides from the Mantle. These hypotheses dovetail surprisingly well.

---

[44] *Feldspar-quartz intergrowths produced in a "bomb" include micrographic and granophyric textures. In these magmatic textures the quartz blebs are runic or triangular in shape, and the host feldspar (plagioclase or K-feldspar) is uniform in composition. In contrast, the quartz in myrmekite occurs as tapered vermicules intergrown in plagioclase, which, itself gradually changes composition as the myrmekite is approached.*

[45] *The maximum amount of solid-state diffusion is one-half the diameter of the average fractured crystal. That distance may be no more than one millimetre, extending from the center of a broken grain to its rim. I agree with the magmatists that large-scale, solid-state diffusion is impossible and suggest that most of the movement of elements from replaced rocks is effected by diffusion into boundary fluids.*

## The Cordierite-Myrmekite Association

Having established conclusively to my own satisfaction that the transformation of sodic feldspar [plagioclase] to potassic feldspar [microcline or orthoclase, "K-feldspar"] by ion exchange, attended by deposition of quartz vermicules, results in the creation of the myrmekite association, I sought an understanding of the fundamentals of the process. This entailed considering the controlling interrelationships between the densities of minerals, the ionic radii of the replacing and replaced metals, and the space adequacy for the transformation to proceed.

**Densities:** Plagioclase has a density ranging from 2.61[46] to 2.72[47] in contrast with potassic K-feldspar, 2.56.[48] The purport of these density differences is that more alumina and silica are present in any given volume of plagioclase than in K-feldspar.

**Ionic radii:** The variation among ionic radii, wherein a large potassium radius (1.33 Å) dwarfs that of sodium (0.95 Å) and calcium (0.99 Å), establishes the constraint that space must be made within the plagioclase crystal before potassium can enter it. Manifestly, potassium cannot be added to the surface of plagioclase, because it is solidly surrounded by other rock-forming crystals. The plagioclase lattice cannot expand to accommodate the potassium.

**Space adequacy:** Creation of space for potassium within plagioclase lattices requires then that other metal components such as aluminum, calcium, and sodium must first be expelled. As it turns out, this loss creates the space for the internal growth of K-feldspar and myrmekite.(**Fig. IV-4**).

**Cordierite:** Recognition of this apparently simple but fundamental fact led logically to the deduction that volume-for-

[46] *Pure sodic: albite plagioclase, $NaAlSi_3O_8$.*

[47] *Pure calcic: anorthite plagioclase, $CaAl_2Si_2O_8$.*

[48] *Potassic: microcline or orthoclase, $KAlSi_3O_8$.*

volume replacement of plagioclase must be capable of producing other silicate minerals having a density similar to that of K-feldspar (2.56). If another such mineral could be found associated with myrmekite, its presence would validate my theory.

*Earth's mineral inventory contains only one mineral that fits the requirements: cordierite.*[49] On that basis, I searched the literature for magnesium-rich, cordierite-bearing terranes in which the author also listed coexistent myrmekite.

Cordierite is a high-temperature, magnesium-, iron-, and aluminum-rich mineral that commonly occurs adjacent to igneous intrusions [for example, where sedimentary shales have been "cooked" by magmatic heat to coarse-grained, cordierite-bearing hornfels]. This assemblage forms in an essentially closed system by component elemental exchange. Coexistent and co-produced minerals may include other high-temperature, magnesium-, iron-, and aluminum-rich minerals [garnet, sillimanite, various spinels, etc.].[50] These minerals also may be created by regional metamorphism.

I suggest a third environment of cordierite formation, one in which cordierite is formed by recrystallization and replacement of plagioclase at temperatures that are high but still below melting temperatures.

For example, if a magnesium-rich gabbro is sheared to create an open system, hydrous fluids can enter and interact with pyroxenes, hornblende, and biotite, partly or completely replacing them with quartz. Deletion of calcium and sodium from the plagioclase creates voids in its lattice. That allows magnesium and iron released from the ferromagnesian silicates to enter the altered plagioclase and convert it to cordierite.

---

[49] *Cordierite, $(Mg,Fe)_2Al_4Si_5O_{18} \cdot nH_2O$, has a density of 2.53 for magnesium-rich varieties and up to 2.78 for iron-rich varieties (Deer et al. 1962).*

[50] *Garnet, $(Mg,Fe)_3Al_2(SiO_4)_3$; sillimanite, $Al_2SiO_5$; spinel, $(Mg,Fe)Al_2O_4$.*

Plagioclase crystals in this regime may variously be replaced by cordierite, by relatively sodic plagioclase (the situation locally at Temecula), or by K-feldspar and myrmekite.

## A Finnish Terrane

One such terrane was reported in gneisses from Finland by Anna Hietanen (1947). Sketching from her photomicrography, she shows relict minerals preserved as inclusions within cordierite. Many of her cordierite crystals have cores filled with sillimanite needles. In the zone between core and rim in each of these crystals cordierite prevails; in the rims, the cordierite crystals are full of quartz blebs. This pattern is exactly what would be predicted if cordierite replaced a former, "zoned plagioclase" crystal.[51] A transition zone from Hietanen's gneiss to nearby gabbros should exhibit gabbro in the process of being converted to cordierite-garnet-sillimanite gneiss.

## A Colorado Terrane: the Idaho Springs Gneiss

A terrane of cordierite-garnet-sillimanite gneisses occurs in the Idaho Springs Formation in Colorado. A fault separates the Elk Creek pluton, a gabbro, from a cordierite-garnet-sillimanite gneiss. According to my theory, the cordierite gneiss could be a replacement product of the gabbro.

The USGS authors, Moench and others (1962), Sims and Gable (1967), Gable and Sims (1969), Gable (1980), in a series of papers published over eighteen years, present a completely

---

[51] *Zoned plagioclase crystals in igneous rocks commonly have relatively calcic, aluminum-rich cores and relatively sodic rims. Sillimanite inclusions in cordierite cores occur where too much aluminum is contained in the calcic plagioclase than can fit into the cordierite lattice. The excess aluminum effectively creates the sillimanite inclusions.*

*The intermediate zone has just the right ratio of aluminum and silica to make only cordierite. The rims, on the other hand, have relatively sodic plagioclase and too much silica to fit into the cordierite structure. The excess silica in the cordierite crystallizes as quartz blebs.*

different and much more complicated interpretation of the history of the Idaho Springs Formation. It includes:

(a) original deposition of sedimentary strata,
(b) interlayering of the sediments with mafic volcanic rocks, both sills and lavas,
(c) deep burial,
(d) regional metamorphism,
(e) folding into anticlines and synclines,
(f) intrusion by igneous plutons causing contact metamorphism adjacent to the plutons,
(g) uplift and retrograde metamorphism, and
(h) erosion and exposure at the Earth's surface.

The cordierite gneisses were considered to be metamorphic derivatives of magnesium-, iron-, and aluminum-rich sedimentary members of a varied sedimentary pile now known as the Idaho Springs Formation, derivations from former arkoses, graywackes, pelites, and impure sandstones, particularly.[52]

However, one such layer, rich in cordierite-gedrite[53] contained too much magnesium, and too little calcium, sodium, and potassium for any known sedimentary source. The authors admitted the uncertainty of its origin.

Three episodes of metamorphism in this "officially approved" version of geological history occurred in three markedly different environmental conditions. During the earliest episode (the one involving regional metamorphism), garnet was formed and made stable by the high temperatures and pressures of deep burial.

---

[52] *A sandstone with abundant feldspar grains is an arkose; a graywacke is a sedimentary rock containing about equal portions of clay, quartz, and feldspar; a pelite is a rock that is composed mostly of clay; the impure sandstone is dominantly quartz, but contains some magnetite and feldspars.*

[53] *Gedrite, an amphibole,*
$$Na_{0.5}(Mg,Fe)_2(Mg,Fe)_{3.5}(Al,Fe^{3+})_{1.5}Si_6Al_2O_{22}(OH)_2$$

The second episode was intrusion by magmas. Magmatic heat at the contact with metasedimentary rocks supposedly caused cordierite to replace some of the garnet.

The third episode (the retrograde metamorphic event)[54] occurred during uplift. Deformation permitted hydrous fluids pervasively to generate low-temperature alteration minerals within the crystal lattices of pre-existing high-temperature minerals.[55]

Igneous rocks of the second episode of Idaho Springs evolution intruded the supposed metasedimentary rocks of the first episode. The Elk Creek gabbro and the Pisgah granodiorite were among these intrusions. The intrusions were considered to have welled into the noses of folds after folding.

On this basis one would conclude that the intrusions would be younger than the Idaho Springs Formation, and that they could have caused adjacent sedimentary rocks to recrystallize as cordierite-bearing gneisses.

My theory, however, interprets the Elk Creek gabbro and the Pisgah granodiorite as the primary source rocks from which the gneisses were derived, and hence, **older** than the gneisses. The gabbro and granodiorite would have been sheared, replaced, and recrystallized in that order. The fact that many of the cordierite-bearing rocks are too far from igneous intrusions to have been thermally altered by them suggests that these rocks cannot be products of contact metamorphism.

From these arguments and relationships, I predicted that gabbro remnants, which would have been by-passed in the granitizing process should be preserved in the cordierite gneiss across the fault from the main Elk Creek gabbro mass.

---

[54] *Retrograde metamorphism converts high-temperature minerals to hydrated low-temperature minerals.*

[55] *Garnet alters to chlorite; cordierite alters to pinnite; feldspars alter to fine-grained muscovite.*

This prediction proved correct: gabbro lenses are engulfed in cordierite gneiss. To follow it up, I sampled the cordierite gneisses, not only near the Elk Creek gabbro, but also near the Pisgah granodiorite and elsewhere in the region. Sims and Gable had reported that the central, massive, coarse-grained, Pisgah granodiorite was not uniform in composition but also contained mafic-rich layers that alternated with felsic layers. I found the mafic layers and felsic layers and traced them to the contact with the Idaho Springs Formation, where mafic and felsic gneisses occur in the same positions on the opposite side of the supposed igneous contact, all the rock types having parallel strikes and dips.

On the nose of the syncline south of the southern tip of the massive Pisgah granodiorite, more of these relationships are present, but with no mapped igneous contact at all on the Sims and Gable map between massive granodiorite and foliated gneisses.[56]

**Rock types across the supposed contacts are variations of the same original units, their differences merely textural and mineralogical. The sharp contacts depicted by Sims and Gable are fictitious; only "gradational contacts" [an oxymoron] are to be found.**

*As Goethe said– "We see what we know."*

Sims and Gable had noted that some gneisses contained cordierite while others did not. Those with cordierite also commonly contained K-feldspar whereas those containing sodic plagioclase but lacking K-feldspar did not contain cordierite. Where they attributed this to primary differences in the **sedimentary environments**, I attribute it to chemical and mineralogical difference in the primary **igneous rocks**.

---

[56] *In the nose, mafic gneissic layers contained hornblende, pyroxene, biotite, and calcic plagioclase led northeastward along the limb into cordierite-garnet-sillimanite gneisses with myrmekite. Tracing felsic, biotite-plagioclase gneisses in the nose northeastward to the limbs, I observed biotite-sillimanite gneisses that lacked cordierite.*

If original igneous rock contains abundant biotite (a source of potassium), other ferromagnesian silicates, and calcic plagioclase, then the recrystallized rocks have K-feldspar and cordierite. If the original igneous rock contains sodic plagioclase and only small amounts of biotite, then K-feldspar and cordierite are sparse or absent from the recrystallized gneisses.

In the case of the problematical cordierite-gedrite gneisses (mentioned earlier) their chemical compositions compare with those of the hornblende-pyroxene gabbros, suggesting their derivation from that source.

An apophysis at the southeast side of the Pisgah granodiorite is drawn by Sims and Gable as a dike extending from the main granodiorite mass and transecting supposed metasedimentary rocks. If the granodiorite were an intrusive magma (younger than surrounding metasedimentary units), the term, "apophysis" would be proper. However, with felsic and mafic layers in the Pisgah granodiorite situated in felsic and mafic layers of the adjacent gneisses, this "dike," with undeniably massive appearance intact, is in reality, a part of the pluton that missed being shattered and sheared in the folding process.

Then, in an area not mapped by Sims and Gable, immediately to the east of the apophysis,[57] *the complete series of gradational stages from diorite and gabbro to massive Pisgah granodiorite is displayed in outcrop.*

In the nose of the anticline, where the Elk Creek diorite-gabbro is least disturbed, and its original massive texture preserved, quartz, the ensign of transformation, is absent. Similarly in the massive, largely unreplaced Pisgah granodiorite (on the nose of the syncline) only small percentages of quartz occur. By contrast, toward the gneisses where the igneous rocks are deformed, the ferromagnesian silicates in the igneous rocks begin to have quartz sieve textures.

**Progressively closer to the gneisses, magnetite granules begin to appear, and total quartz content increases.** Therefore, completion of the replacement and recrystallization process would mean that these igneous rocks are conceptually the metamorphic end-products of the supposed impure sandstones, graywackes, and arkoses formerly interlayered with cordierite gneisses! Some of these rocks contain up to 65 % quartz and 10 % magnetite, and their textures resemble closely the igneous rocks in their early stages of replacement and recrystallization. Clearly these field and mineralogical relationships between the igneous rocks and the gneisses corroborated my predictions.

I addressed the question of whether the igneous rocks are really younger than the gneisses (Sims' and Gable's interpretation) or older (my interpretation) by Rb-Sr age-dating. I chose samples of the gneisses and the least disturbed areas of gabbro and granodiorite. The Pisgah granodiorite proved to be 1.82 (±0.13) by, the gneisses 1.67 (±0.05) by, and the gabbro too disturbed (by introduced fluids) to give an accurate isochron.

Scatter of the gneiss data gives plus-or-minus errors of 0.13 by, presumably the effects of hydrous fluid movements through the open system. The range of possible error is too large to rule out the possibility that the Pisgah granodiorite could be the same age as the gneisses. However, the age data lean toward support of my hypothesis, and warrant the statement that all relationships point to the igneous rocks being older than the recrystallized gneisses in the Idaho Springs terrane of Colorado.(Collins & Davis 1982)

[57] *Several diorite and gabbro units had been mapped by Sims and Gable, but these units were always shown as isolated pods in the cordierite and sillimanite gneisses. The pods were considered igneous bodies (metamorphosed sills or lava flows) distinct from the Pisgah granodiorite. The small outcrop area that I found east of the apophysis shows that here, the massive biotite granodiorite of the Pisgah pluton grades into biotite-hornblende granodiorite and then into the same kinds of rocks (hornblende and pyroxene diorite and gabbro) that occur in the mafic pods. The mafic rock, where dominantly pyroxene diorite and gabbro, projects northeast from this outcrop along the foliation-strike to the remnant island-masses, the mafic, pyroxene-bearing pods in the gneisses. Similarly, the more felsic, biotite-hornblende and biotite parts of the Pisgah body, which are also rich in calcic plagioclase, become cordierite and sillimanite gneisses toward the northeast from this small outcrop area east of the apophysis.*

Interesting relationships become apparent from this study:

## First, metasedimentary vs endogenic silica:

The Idaho Springs Formation is mapped over a huge Rocky Mountain region. It has exclusively [until this study] been considered metasedimentary. Discrediting of the interpretation of a sedimentary origin raises the question of whether not only this formation but other preCambrian metasedimentary terranes are, in fact, sheared, recrystallized, and replaced igneous rock.

## Second, the conversion of mafic igneous rocks in large volumes to quartz-rich gneisses:

The widespread occurrence of this phenomenon is supportive of the Hunt hypothesis that silanes emerge from the Mantle and react to produce heat and silica that can effect replacement of sheared mafic igneous rocks. Replacements of these mafic igneous rocks to form quartzose gneisses of the Idaho Springs Formation on a volume-for-volume basis would have required large quantities of silica. There is no other recognized provenance of silica adequate to provide so much material.

## Third, interpreted telltale remnants of biotite gabbro in gneiss:

At the boundaries of the mafic intrusions with the cordierite-garnet-sillimanite gneisses, remnants of biotite gabbro occasionally are found in which plagioclase is 90 to 100 % anorthite (nearly pure calcic plagioclase). In some of these gabbro-remnants the biotite comprises nearly 30 % of the rock. Biotite, being relatively soft and cleavable, renders biotite gabbro that can be sheared, replaced, and recrystallized more easily than a hornblende or pyroxene

gabbro. Moreover, the high-aluminum content (up to 36.7 % $Al_2O_3$) characteristic of the plagioclase found in biotite gabbro results in recrystallized rock rich in aluminous minerals, such as garnet, cordierite, and sillimanite.

The aforesaid biotite gabbro remnants are, thus, anomalous. Notably, they occur distant from shear zones (only traces of biotite are found in gabbro adjacent to shear zones, the supposed "contacts" between igneous rocks and the gneisses). This suggests that highly-biotitic gabbro has been destroyed and replaced by biotite-lean granitic rocks near shear zones.

The generalization follows, that **the scarcity of biotite gabbro in preCambrian terranes is a consequence of the creation of granites from precursor biotite gabbros.** Early-crust gabbros with potassium-rich biotite and high-calcium, high-aluminum plagioclase are imagined to have been transformed by later preCambrian times into granites. The volume of granite in the Earth's Crust nearly doubled in preCambrian time.(J. L. Anderson 1983)

The question of how new granite made room for itself in the solid lithosphere is central. Magmatic intrusion with assimilation of country rock does not satisfy field conditions because of the gradational contact conditions previously described. *In situ* replacement is essential; and its progress must have been permitted by volume preservation. No volume problem exists where a granite was originally biotite-rich gabbro that was sheared, recrystallized, and replaced volume-for-volume.

The common occurrence of relict, sheared gabbro enclaves in granites supports the interpretation of replacement of former gabbro by an equal volume of granite. Coarse quartz vermicules in myrmekite in the granite also support

this interpretation. Where granite bodies were emplaced above melting temperatures, magmatic granite is still the appropriate interpretation. Melting would, of course, have destroyed all evidence of former myrmekite and other features of replacement.

### Fourth, open and closed systems:

Conventional wisdom for the evolution of cordierite-garnet-sillimanite gneisses [based on experimental laboratory studies and the resultant phase-equilibria diagrams] suggests that the following balanced, mass-for-mass equation represents behavior in closed systems:

biotite + sillimanite + quartz = cordierite + K-feldspar + water

This reaction is imagined to occur in a temperature range of 640°C to 665°C and pressures of about 2.5 to 3.0 kb. (Hoffer 1976)

I do not promulgate such phase-equilibria diagrams or the above equation because my studies of the cordierite-bearing rocks in Colorado show that the experimental work and phase-equilibria diagrams do not apply in open systems such as that one.

**Emphatically, cordierite in the Colorado rocks was not formed from reactions between biotite, sillimanite, and quartz in a closed system, but by replacement of aluminum-rich plagioclase in an open system. Geology will only allow a qualitative approach to theory, that is to say, the approach of observation, until ways are found to duplicate in the laboratory the crustal environment in which cordierite is formed by replacement.**

## Fifth, strontium and rubidium chemical balance changes:

Igneous rocks and gneisses give evidence of destruction of primary biotite to form K-feldspar. Recrystallized biotite-cordierite gneisses generally exhibit a slight enrichment in potassium and a two-fold enrichment in rubidium.[58] Simultaneous transformation of primary calcic plagioclase to cordierite, sodic plagioclase, or K-feldspar with myrmekite results in a 9-fold subtraction of calcium[59] and a 5-fold subtraction of strontium.[60] Changes in isotopic ratios result, which are not predictable from magmatic processes. The strontium isotopes, $^{87}Sr$ and $^{86}Sr$, behave similarly and would normally be expected to be carried out of the system by hydrous fluids at similar rates.

My hypothesis, however, predicts that they will move at rates that are unequal because the minerals that contain them are affected unequally in the replacement process.(Collins & Davis 1982) Rubidium is concentrated (with potassium) in biotite. When its isotope $^{87}Rb$ decays to daughter isotope $^{87}Sr$, that elevates the $^{87}Sr/^{86}Sr$ ratio in the biotite as well as the host rock. Conversely, absence of rubidium in plagioclase allows little $^{87}Sr$ to be produced, and results in a depressed $^{87}Sr/^{86}Sr$ ratio in the plagioclase.

In hydrous replacement, biotite and other ferromagnesian silicates tend to break down and be replaced by quartz before the plagioclase is affected. This early destruction of the biotite lattice releases strontium with high $^{87}Sr/^{86}Sr$ ratio and allows removal of more $^{87}Sr$ from the system than is removed later when the more abundant common strontium with low $^{87}Sr/^{86}Sr$ ratio is released.

---

[58] *Average $K_2O$ in igneous rocks increases from 2.3 % to 2.8 % in average gneisses; Average Rb in igneous rocks increases from 63 ppm to 120 ppm in average gneisses.*

[59] *Average CaO in igneous rocks decreases from 6.3 % to 0.7 % in average gneisses.*

[60] *Average strontium decreases in igneous rocks from 530 ppm to 111 ppm in average gneisses.*

Retention of rubidium and differential movements of $^{87}$Sr and $^{86}$Sr isotopes disrupt Rb-Sr systematics in various members of the Idaho Springs Formation in this way.

### Sixth, samarium and neodymium chemical balance changes:

In similar fashion, rare earth elements are affected differentially by the early destruction and replacement of the ferromagnesian silicates relative to the later destruction and replacement of plagioclase. Among the affected rare earths, isotopes of neodymium and samarium, $^{143}$Nd and $^{147}$Sm, display differential movements similar to those of strontium. Samarium tends to follow calcium out of the system more than does neodymium. Therefore, the Nd-Sm systematics for age-dating analyses are also disrupted.

### Seventh, evolution of two-mica ("peraluminous") granites:

Destruction, replacement, and recrystallization of a biotite-bearing gabbro produces a cordierite-garnet-sillimanite gneiss that is also rich in aluminum. If this gneiss should subsequently become hot enough to melt, the magma would, on cooling, crystallize to form a granite known as a "two-mica, peraluminous" granite.[61](**Fig. IV-5**)

Normally granites rich in aluminum are interpreted to have been derived from the melting of aluminum-rich sediments. (White & Chappell 1977) **My Colorado studies show that two-mica granites need not evolve from the melting of aluminum-rich sedimentary rocks. Instead, peraluminous granites may evolve by transformation from gabbros in situ.**

---

[61] *Peraluminous means that more aluminum is in the rock than can be totally contained in the feldspars. Therefore, some aluminum is extra and goes into other aluminum-rich minerals, such as biotite, muscovite, and garnet.*

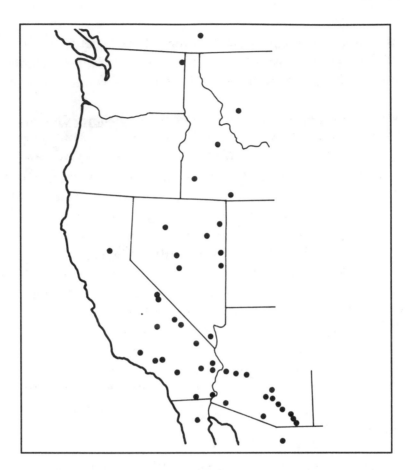

**IV-5**
**DISTRIBUTION OF PERALUMINOUS GRANITES IN THE**
**WESTERN USA AND ADJOINING MEXICO AND CANADA**
**(in part after Miller & Bradfish 1980)**

## MIGMATITES AND APLITE-PEGMATITE DIKES

Generally, geologists interpret migmatites and aplite-pegmatite dikes as derivatives of melts. Granitic chemistry makes these the first rock types to melt when temperatures rise and the last to crystallize when temperatures decline.(Tuttle & Bowen 1958) Where dikes and pods with felsic compositions and textures branch dendritically into fractures in rocks

collateral to granitic bodies, it has seemed logical, therefore, to conclude that fluidic emplacement must have occurred at temperatures in excess of melting temperatures.

The common presence of collateral migmatites in metasedimentary gneisses adjacent to granitic plutons encourages the idea that the heat of plutonic intrusion has caused local, partial melting in the adjacent metasedimentary terranes so as to create the pods and lenses. This deduction, encouraged by the usual color contrast, cream-colored granitic pods against dark wall rock, appears to reflect marked chemical dissimilarity. Divergent textures and sharp, cross-cutting contacts reinforce the illusion. These contrasts are so strong that migmatites or aplite dikes are often treated as examples of (1) the injection of magma from an outside source or of (2) localized partial melting and diffusion of melted components into fractures and other pressure-deficient sites.

These assumed relationships can be duplicated in the laboratory. Relying on such synthetic analysis and the illusory field relationships that seem to support a magmatic origin for aplite and pegmatite, many geologists extend the magmatic hypothesis to all aplite-pegmatite dikes, granitic masses, or migmatite zones. They even treat the magmatic concepts as established and no longer needing to be questioned.

Comments made by reviewers of two of my own papers are typical: "This view would seem to ignore the transitional migmatites that seem to preserve the initial melting stages of the gneisses." "Petrologists specializing in igneous rock have their reasons for believing that the K-feldspar in ordinary granites crystallized from a 'granite minimum' type of melt, rather than having been introduced by wholesale solid-state metasomatism." I have consistently disclaimed the position that replacement is by "wholesale solid-state metasomatism." Nevertheless, the reviewer is not to be dissuaded, and reflexively imputes solid-state diffusion to my interpretation.

Other examples of an abundant literature in which magmatism is treated as a foregone conclusion are cited by Roddick (1982):

"I think today we can accept as fact that the vast volume of materials of the granite clan which are intruded into moderately high crustal levels, attain their position in the form of silicate melts."(Fyfe 1973, p. 273)

"For most of these [plutonic] series, magmatic differentiation is generally accepted as the main factor responsible for the formation of the different rock types."(Albuquerque 1971, p. 2792)

"The variation of rock types from gabbro to granite may be best attributed to differentiation of a granodiorite magma to account for the main sequence of rocks."(Clemons & Long 1971, p. 2739)

"Field relations indicate that the major granitic rocks - were emplaced in at least two magmatic phases."(Fullagar 1971, p. 2856)

"Intrusive relations of the granitic bodies indicate that they were liquid."(Wiebe 1970, p. 114)

Siloam granite in Georgia "is undoubtedly of igneous origin, clearly cross-cutting the original stratigraphy."(Jones & Walker 1973, p.3653)

Roddick, thus, calls attention to the faulty logic whereby all aplite-pegmatite dikes are attributed to magmatism. In subsequent paragraphs I describe some that have been generally accepted as magmatic in origin, but which are far better explained as developments of silane replacement.

Let the reader not assume that I regard all aplite-pegmatite as the product of replacement, however. There **are** many cases where melting temperatures have been exceeded, and where the aplite-pegmatite dikes **do** have a magmatic origin. Let the reader recognize that silane infusion and transformation do not conflict with magmatism, but rather, that magmatism occurs as a consequence of the heat released with copious silane reaction, and that the end products of transformation and magmatism are often confusingly similar.

Similarity of end products is the problem; and the misperceptions consequential upon superficial geological work entrench the mis- and over-application of magmatic interpretations. The silane-replacement hypothesis is not in conflict with magmatic hypotheses. It *is* the agency for rock transformations at less than melting temperatures.

## Faulty Logic

Before looking at particular examples, let us consider some of the invalid criteria that have been used for determining that migmatite and aplite-pegmatite dikes are magmatic in origin. For example, *sharp and cross-cutting contacts* are often evoked as evidence for intrusion. The logic, however, does not necessarily follow. Sharp and cross-cutting contacts can be attained by faulting or by plastic flow of solids. Moreover, replacement can parallel a contact, rather than proceed at right angles to it. This happens when a "crushed" rock is converted to granitic composition up to its sharp contact with uncrushed mafic rock.

*Rotated foreign blocks, the so-called "exotic" blocks,* are also described as proof that the enclosing rocks were liquid. The falsity of this tenet lies in the common occurrence of such rotated blocks in fault zones. Uncrushed, disoriented blocks "float" in a sea of crushed fragments. Therefore, the crushed rock that surrounds the rotated blocks is permeable and

vulnerable to replacement on an atom exchange level at temperatures below melting.

*Fine-grained boundaries against coarse crystals* are sometimes cited as evidence for magmatic intrusions of aplite-pegmatite dikes. The fine-grained boundaries are said to be "chill zones," an interpretation of textural change that may hold true for basaltic dikes, but may fail when applied to granitic aplite-pegmatite dikes. This is because a fine-grained boundary texture can also result from replacement of crushed rock; or it can be created by deformation, which grinds coarse crystals to fine grains. Subsequent annealing and recrystallization may then destroy the evidence of crushing.

*Water content is another common influence on crystal size* in aplite-pegmatite dikes. It has nothing to do with heat transfer in a melt.(Swanson 1977) Rapid loss of hydrous vapor during cooling can cause rapid nucleation and result in fine crystal sizes, whereas trapped hydrous vapor can promote crystal growth to large sizes at temperatures below melting. And tectonism that renews movement along shear planes can impede crystal growth.

Thus, the problem of magmatism versus *in-situ* transformation is complex and not susceptible to hasty or prejudicial interpretation. **Figures IV-6 and IV-7** illustrate the formation of granitic veins and dikes in typical tectonically-stressed terranes.

**IV-6
PHOTO SCENE
SHOWING A
TYPICAL HABIT OF
APLITE-GRANITE
VEINING**
In sheared tonalite,
Josephine Mountain
pluton, San Gabriel
Mountains, California.

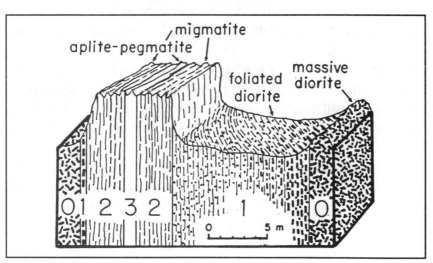

**IV-7
SCHEMATIC DIAGRAM FOR MODE OF EMPLACEMENT OF
APLITE-PEGMATITE DIKES**
In the Kernville pluton, southern Sierra Nevada, California.

## Aplite-pegmatite, the Pala Gem-bearing Dikes

Let us now examine some appropriate localities of aplite, pegmatite, and migmatite development. One area of granite aplite-pegmatite dikes which has been extensively studied is the lithium gem-bearing Pala pegmatites northeast of San Diego in southern California. These aplite-pegmatite dikes are zoned and have been described as having been formed from a granitic magma.(Jahns, 1954; Jahns & Tuttle 1963)

Some meager evidence for magmatic provenance is found in original temperatures of formation of these rocks. Fractionation temperatures of the isotope, $^{18}O$, are established between coexisting mineral pairs. These temperatures range from 700° to 730°C in the aplite walls to 520° to 565°C in the coarse-grained, gem-bearing cores.(B. E. Taylor et al. 1979)

Without taking issue with these data, nor doubting that magmatic crystallization of these crystals occurred in liquid above the melting interval for granite [There is little room for disagreement that crystallization of *certain minerals* occurred in liquid above the granite melting interval], **I *would* still draw attention to the fact that *there is no evidence to support the proposition that <u>whole aplite-pegmatite dikes</u> crystallized from liquids introduced to the granite magma from external sources.***

The Pala pegmatite dikes cut massive gabbro and tonalite. Their pinkish-white color stands out in marked contrast to the black mafic host rocks. It seems almost bizarre to suggest that the two rock types are related or that the pegmatites formed *in situ.* But that, in fact, is what my detailed study shows.

By tracing the pegmatites along their strikes, I have observed that they grade progressively to granitic and aplitic textures and then to gneissic textures, in which the rock contains remnant mafic minerals of the former gabbro or tonalite along with

abundant quartz. Where these rocks are granitic or gneissic, abundant myrmekite with coarse quartz vermicules is scattered throughout the rock.

Where gabbro and tonalite are dominant, I find only the early stages of replacement. The mafic silicates are partially replaced by quartz. The K-feldspar has not intensively replaced the plagioclase. This rock is still black and grades along strike into quartz-rich gneiss [the Bedford Canyon formation(?)].

**Coarse pegmatites are localized where hydrous fluids came through the primary gabbro and tonalite, leaching out their mafic components. This resulted in volume loss and vug creation into which large crystals grew. The existence of vugs also meant that free oxygen could be present to oxidize introduced silanes. This oxidation created heat and raised temperatures above the melting interval for granite.**

In this hot hydrous environment the crystals that grew in the pegmatite were arranged in zones controlled by the gravitational field. Outside these melt zones, however, the temperatures were not hot enough to produce a silicate liquid. There, the replacement textures preserve the textures of the original altered gabbro and tonalite.

## Pegmatites in the Vetter Pluton

In the San Gabriel Mountains north of Pasadena, California, this same scenario occurs in pegmatites associated with the Vetter pluton and the San Gabriel anorthosite.(Collins 1988a) Tracking from areas of quartz-feldspar intergrowth in graphic granite, I have progressed into aplite-pegmatite dikes associated with the Vetter pluton. From the granite, where the crystals of feldspar are coarse-grained, and the quartz forms runic and triangular quartz blebs, the texture along strike changes to myrmekite-bearing aplite-pegmatite dikes and then into gneissic

bands in which remnant mafic silicate minerals of the original igneous rock still remain. The graphic textures and uniform composition of the feldspar imply crystallization from a melt, whereas the lateral gradation to myrmekite and coexistent feldspars with non-uniform compositions suggest formation by replacement processes at temperatures below melting.

## Pegmatite in the San Gabriel Anorthosite

Also in the San Gabriel Mountains there is an anorthosite body with thick granite-pegmatite dikes (3 to 5 m in width and more than one km in length). These pegmatites contain muscovite, biotite, and trace amounts of almandine garnet. Some of the feldspar crystals are 10-cm long and exhibit graphic textures in which runic quartz is optically parallel and extends through both K-feldspar and albite plagioclase crystals. Because no granite occurs anywhere near the anorthosite, the pegmatites have always been assumed to have originated from an unknown granite magma at depth. I suggest that they have formed *in situ*.

Where the aplite-pegmatites occur, the host anorthosite contains localized biotite clots up to three cm in diameter. These places would be zones of weakness and easily sheared. The introduction of hot hydrous fluids could release potassium from the biotite to produce the K-feldspar and to change the dark mafic anorthosite to the light, felsic, granite dikes. I have walked these dikes from their coarse zones laterally to where the crystals become fine grained. Across this distance the felsic rock becomes gneissic and contains myrmekite and remnant biotite. This supports the hypothesis that the pegmatites have been derived *in situ* from the original biotite-bearing anorthosite.

## Aplite-pegmatite Dikes in the Cargo Muchacho Mountains

In the Cargo Muchacho Mountains of southeastern California, west of Stud Mountain, a dark, massive, mafic quartz monzonite contains both hornblende and biotite.(Collins 1988a; Henshaw 1942) Progressing from south to north, one passes from slightly deformed quartz monzonite with elongate local shear zones into areal exposures of light-gray rock in which biotite and hornblende provide the gray color, and plagioclase has been replaced by K-feldspar and myrmekite.

Still farther to the north, similar elongate zones of the same dimensions become cream to white instead of gray. All ferromagnesian silicates have disappeared, and only quartz and feldspar remain. The former gray zones have been converted to white and cream-colored aplite and pegmatite with remnant myrmekite.

The transition from gray rock enveloped in the massive quartz monzonite can be traced continuously and gradationally northward into gneissic rock and then into white aplitic granite. The granite occurs in elongate patches up to 30-m wide and more than two-km long in which angular, blocky remnants of the dark monzonite or gray monzonite occur in the midst of aplitic and granitic dike swarms.

**Clearly, the fine-grained gray rocks surrounded by massive, dark, quartz monzonite are in the first stages of transformation, proto-pegmatites, so to speak. Remnant fragments of mafic rock enveloped in the dikes show foliation that is never rotated and always aligned with respect to the foliation in the surrounding wall rocks. Thus, the dark fragments have not been carried in by a granitic magma from some outside source; and the granitic dikes must have formed *in situ* by replacement around the remnant mafic fragments.**

# ALUMINUM ENRICHMENT

## THE CASE OF THE CARGO MUCHACHO MOUNTAINS

Aluminum enrichment occurs in the Earth's crust in several ways. One common process is weathering of granitic rocks to produce laterite soils.[62] Another way aluminum enrichment occurs is through the conversion of mafic rocks to granites *in situ* by hydrous fluids. I had observed this latter process of enrichment for the first time near Dover, New Jersey, where a biotite-hypersthene-plagioclase gneiss is changed into a garnet-sillimanite gneiss in the limbs of the Hibernia anticline and the Splitrock Pond syncline.[63] Calcium, magnesium, and sodium were carried away during recrystallization and replacement. Aluminum, being less soluble and mobile, tended to be segregated as sillimanite and garnet, which remained behind as the rock was converted to granitic compositions.

I reasoned that if myrmekite is associated with the alumino-silicate mineral, sillimanite, perhaps it might also be associated with like minerals, polymorphs of $Al_2SiO_5$, such as kyanite and andalusite. In the Cargo Muchacho Mountains of southeastern California my students had collected kyanite and andalusite crystals from a muscovite-rich phyllite, which had been mapped by Henshaw in 1942 as a member of the Vitrefax Formation. It occurred in metasedimentary layers that were interpreted as former sandstones, graywackes,[64] and aluminum-rich shales, which had been metamorphosed by the magmatic heat of proximal intrusive granites, granodiorites, quartz monzonites, and diorites.

---

[62] *In the tropics the abundant rain dissolves the soluble elements, including silica, out of the rocks, leaving a residue, the "laterite soils" rich in aluminum oxides and hydroxides and minor iron oxides and hydroxides. When these soils harden, they become bauxite rock, the primary aluminum ore for industry.*

[63] *In-situ rock transformation, mafic to felsic, as discussed earlier.*

[64] *A graywacke is a sedimentary rock containing about equal proportions of clay, sand, and feldspar grains.*

Henshaw's report did not mention myrmekite, although it contained the unusual interpretation that leucogranite[65] had formed out of dark granite. My theory implied that such replaced rocks could contain myrmekite.[66]

Henshaw's suggestion that the leucogranite was formed from the dark granite stemmed from his finding gradational stages of the conversion from primary biotite to magnetite. My prediction of myrmekite stemmed from my former observation that biotite, which has been replaced by quartz and magnetite, is often associated with myrmekite in replaced rocks. I wrote Henshaw, asking him about myrmekite. He did not remember myrmekite being present, and his samples and thin sections had been discarded or lost.

Curiosity impelled me to reconnoiter several canyons that transected the Range. In random samples, 44 specimens in all, and covering every igneous lithology I could find, myrmekite was abundant, especially in the granite and leucogranite!

This raised the question of an origin for the phyllitic rocks of the area. Were the kyanite- and andalusite-bearing phyllites really metasedimentary rocks or were they sheared, replaced, and recrystallized granitic rocks? Could I find a transition point between sedimentary and meta-sedimentary lithologies where myrmekite first appeared?

The desert terrain afforded excellent exposures, although desert varnish concealed mineralogy. I could trace the rocks

[65] *Leucogranite is a white granite containing less than five percent ferromagnesian silicate minerals.*

[66] *Henshaw described a gneissic leucogranite at the north end of the mountains, which contained less than two percent biotite and trace amounts of magnetite. He mapped a dark granite south of the leucogranite, which contained up to ten percent biotite and trace amounts of hornblende. Between these two granites is a zone of leucogranite dikes that cut the dark granite in closely-spaced, parallel sheets. These sheet-dikes increased in abundance and became closer together northward toward the main, massive, leucogranite mass. Henshaw placed the boundary between the two granites at that point where he estimated that more than fifty percent of the rock was leucogranite.*

continuously for kilometres without changes being hidden by overburden. My findings, from almost 350 thin sections, turned out to be a startling reversal of prior perception, a far different origin for the Cargo Muchacho granites than Henshaw had interpreted.

The first unusual discovery I made was a transition along strike from what appeared to be a magmatic quartz monzonite to a schistose metasedimentary-appearing lithology.[67] Detailed sampling of the massive biotite granite [to the west of the quartz monzonite] revealed unexpected replacement features. This granite contains K-feldspar and myrmekite, indicating that it replaces a still-earlier rock. The identity of the earlier rock has been destroyed by replacement processes.[68]

Progressing toward the kyanite- and andalusite-bearing rocks from the former granite terrane, biotite is increasingly replaced by quartz and muscovite[69] until it has been obliterated. Next the K-feldspar and plagioclase are replaced by muscovite until they are all converted to muscovite [K-feldspar being the last to

---

[67] *This feature comprises a quartz monzonite pluton in the eastern part of the Cargo Muchacho Mountains. Adjacent, to the west of the pluton, is an older diorite mass. Its greater age is shown by scattered, angular, blocky fragments of the dark, locally black diorite being contained in the adjacent, massive, gray, quartz monzonite. Random orientation of interlocking crystals in the quartz monzonite, normal zoning in plagioclase from calcic cores to sodic rims, sharp and cross-cutting boundaries of quartz monzonite against diorite, and occluded, angular, diorite xenoliths support interpretation of the quartz monzonite as crystallized from magma.*

*Northward toward the supposed-metasedimentary Tumco Formation, the quartz monzonite gradually becomes foliated and recrystallized. Intensity of deformation increases northward to the point that former, angular blocks of included diorite are flattened or elongated pencil-like (up to 0.5 cm thick and 80 cm long). The rock becomes so layered and schistose or gneissose as to simulate a metasedimentary petrofabric. In the transition zone, K-feldspar crystals gradually appear in the now-gneissic rock to become coarse metacrysts (augen) bordered by myrmekite. "Augen" is a German term for relatively large crystals that look like "eyes" in the rock, in this case 1 to 3 cm in length.*

[68] *Some of this granite has been affected by later changes in which its texture and mineralogy have been altered so as to resemble the "metasediments" mapped by Henshaw as the Vitrefax Formation.*

[69] *Muscovite: $KAl_2(AlSi_3O_{10})(OH)_2$, white mica.*

---

be totally replaced]. Fine-grained, muscovite-quartz phyllite[70] remains.

Then, even the phyllite changes. Progressively, toward the center of the terrane phyllite is transmuted into massive quartzite with vugs lined with kyanite crystals. The final rock product at the center of the replacement and recrystallization zone is nearly-pure massive quartzite! Locally the quartzite is speckled with black tourmaline, magnetite, and apatite. All these progressive changes, from the myrmekite-bearing granite to rock fabrics traditionally considered diagnostic of metasedimentary origins, have developed by replacement. Aluminous phyllites (the kyanite- and andalusite-bearing rocks) and the quartzite have evolved *in situ* by replacement from granite, which, itself evolved *in situ* by replacement from mafic igneous rock.

In an outer envelope, against the granite and surrounding the kyanite-bearing zones, the granite locally contains up to 40 percent disseminated epidote.[71] This outer envelope represents a repository of calcium and aluminum extracted from the original granitic rocks during replacement. In the central quartzite zone the stripping of calcium was almost complete, but some aluminum remained in kyanite, andalusite, muscovite, and tourmaline.

Elsewhere, distant from the kyanite-bearing rocks, epidote veins lace the granite with intricate detail (veins up to 5 cm wide, the total volume of epidote content up to 30-40 % of the rock). The calcium and most of the aluminum of this epidote is not indigenous to host rock lithologies in the vicinity of the

---

[70] *Phyllite is a generic term for highly aluminous, micaceous silicate rock, which in the Cargo Muchacho Mountains locally contains kyanite, ± black tourmaline $[Na,Ca)(Li,Mg,Al)-(Al,Fe,Mn)_6(BO_3)_3-(Si_6O_{18})(OH)_4]$, ± magnetite $[Fe_3O_4]$, ± apatite $[Ca_5(PO_4)_3(F,Cl,OH)]$, and, rarely, a metacryst of andalusite $[Al_2SiO_5]$.*

[71] *Epidote, $Ca_2(Al,Fe)Al_2O(SiO_4)-(Si_2O_7)(OH)$.*

epidote veins, and generally must have come from an outside source. The outside source is logically the plagioclase that was replaced by K-feldspar and myrmekite and the hornblende that was replaced by quartz in the nearby, recrystallized, former quartz monzonite that became granite. The abundant epidote veins are a strong argument for the mobility of aluminum under certain conditions, in contrast with the usual view of aluminum as geochemically immobile.

At no place in the Cargo Muchacho Mountains did I find the successive stages of formation of myrmekite across a short distance so exquisitely expressed on a small scale as at Temecula, California. The zones of replacement are broad, and the stages of conversion from one rock type to another are spread out across hundreds of metres. Only by the examination of several hundred samples along the strike of the transition in the foliated zone was convincing evidence produced that showed how these rocks were derived from originally-massive igneous lithologies.

Cursory examination would inevitably lead one to interpret the scene as an igneous magmatic contact with sedimentary-appearing rocks metamorphosed along the contact. Even Henshaw thought that he could see remnant cross-bedding in the sandstones ["We see what we know."]. Close examination, however, shows that these structures are not cross beds but fractures, and the thin sections show that the quartz grains are not residual sand grains, but quartz crystals grown during replacement of ferromagnesian silicates and feldspars.

My observations in the Cargo Muchacho Mountains do not preclude kyanite and andalusite from forming in other terranes in which aluminum-rich metasedimentary rocks have been deeply buried at high temperatures and pressures. But under those conditions, vugs could not be preserved. The kyanite and andalusite in the rocks in the Cargo Muchacho Mountains must have formed at relatively shallow depths in which hydrous fluids moved through sheared rocks to extract elements and create vugs.

The foregoing descriptions cover the replacement environments that produce sillimanite, kyanite, or andalusite, the minerals that best represent extreme aluminum enrichment. Another environment of aluminum enrichment is that of the "peraluminous," two-mica- and garnet-bearing granites. Peraluminous granites occur in many parts of the world. But they are particularly abundant in a wide swath in western North America, extending from southern California, northward through Nevada, Utah, and Arizona, through Oregon, Washington, Idaho, and Montana into Canada.**(Fig. IV-5)**

Many of these granites contain myrmekite, and in some, the quartz vermicules are very coarse. The common interpretation for their derivation is that they result from the melting of preCambrian or Paleozoic, continental, aluminum-rich, metasedimentary rocks. The coarse quartz vermicules, however, support my hypothesis that they are derived in many cases, at least, from biotite-rich gabbros and diorites, and not from metasedimentary rocks. **The peraluminous granites contain enclosures of diorite and gabbro, but they lack enclosures of metasedimentary rocks. These relationships clearly support an origin of the enclosures as remnants of the source rocks from which they were derived.**(Collins 1989)

# POLONIUM HALOS AND MYRMEKITE IN PEGMATITE AND GRANITE

## Introduction

During development of theory on the origin of myrmekite, I became aware of an interesting application of these studies to the seemingly-unrelated enigma of polonium ("Po") halos.

Whereas most geologists have long accepted a "magmatic" origin for large granite plutons, "batholiths," in which crystallization is imagined to have been attenuated over millions of years (Leake 1990), their magmatic view was technically challenged by Robert Gentry, a physicist with a yen to confirm the biblical tale of Genesis.(Gentry 1965, 1970, 1974, 1983, 1988) He suggested that the Po halos, which are found in the minerals biotite and fluorite in granites and associated pegmatites, prove that the host rocks were created, not from a magma, but almost instantaneously during Day 1 of the Genesis Week.

The radioactive elements emit particles of varying energies. Po halos are zones of atomic disarrangement caused by alpha particle emission. It is self-evident that the emitted particles must be younger than the granite minerals, biotite and fluorite, in which they occur. The radiating particles take paths that pass through neighboring minerals. Having varying penetrative powers, various particles produce characteristic radial zones of destruction that are diagnostic of particular decaying elements.

Gentry advanced six premises having to do with the radioactive behavior of Po, on the basis of which he reasoned that, if the time for the granite and late-stage pegmatites to crystallize were millions of years, then, regardless of the original quantity of polonium isotopes in the magma, their lives would be too short for any to have remained for the final crystallization of biotite and fluorite. Gentry is thus arguing that Po could not occur at all in the minerals[72] where the Po halos are found.

| 238U DECAY SERIES | | | 235U DECAY SERIES | | | 232TH DECAY SERIES | | |
|---|---|---|---|---|---|---|---|---|
| Symbol | Particle Emitted | Half-life Period | Symbol | Particle Emitted | Half-life Period | Symbol | Particle Emitted | Half-life Period |
| $^{238}U$ | a | $4,51 \times 10^9$ yrs. | $^{235}U$ | a | $7.13 \times 10^8$ yrs | $^{232}Th$ | a | $1.39 \times 10^{10}$ yrs. |
| $^{234}Th$ | ß | 24.1 days | $^{231}Th$ | ß | 25.6 yrs. | $^{228}Ra$ | ß | 6.7 yrs. |
| $^{234}Pa$ | ß | 1.14 minutes | $^{231}Pa$ | a | $3.43 \times 10^3$ yrs. | $^{228}Ac$ | ß | 6.13 hours |
| $^{234}U$ | a | $2.35 \times 10^5$ yrs. | $^{227}Ac$ | ß | $2.2 \times 10^1$ yrs. | $^{228}Th$ | a | 1.90 yrs. |
| $^{230}Th$ | a | $8 \times 10^4$ yrs. | $^{227}Th$ | a | 18.6 days | $^{224}Ra$ | a | 3.64days |
| $^{226}Ra$ | a | $1.62 \times 10^3$ yrs. | $^{223}Fr$ | ß | 22 minutes | $^{220}Rn$ | a | 54.5 seconds |
| $^{222}Rn$ | a | 3.82 days | $^{223}Ra$ | a | 11.2 days | $^{216}Po$ | a | 0.158 seconds |
| $^{218}Po$ | a | 3.05 minutes | $^{219}At$ | ß | 0.9 minutes | $^{212}Pb$ | ß | 10.6 hours |
| $^{214}Pb$ | ß | 26.8 minutes | $^{219}Rn$ | a | 3.92 seconds | $^{212}Bi$ | ß | 60.5 minutes |
| $^{214}Bi$ | ß | 19.7 minutes | $^{215}Bi$ | ß | 8 minutes | $^{212}Po$ | a | $.3 \times 10^{-7}$ sec |
| $^{214}Po$ | a | $1.5 \times 10^{-4}$ sec | $^{215}Po$ | ß | $1.83 \times 10^{-3}$ sec | $^{208}Pb$ | - | stable |
| $^{210}Po$ | ß | 22years | $^{215}At$ | a | $10^{-4}$ seconds | | | |
| $^{210}Bi$ | ß | 5 days | $^{211}Pb$ | ß | 36.1 minutes | | | |
| $^{210}Pb$ | a | 140 days | $^{211}Bi$ | ß | 2.16 minutes | | | |
| $^{206}Pb$ | - | stable | $^{211}Po$ | a | 0.52 seconds | | | |
| | | | $^{207}Pb$ | - | stable | | | |

IV-8

**TABLE SHOWING THE DECAY CONSTANTS FOR THE RADIOACTIVE ELEMENTS**

72  *Gentry's six premises are:*

(A) *Polonium isotopes ($^{218}Po$, $^{214}Po$, and $^{210}Po$) are the last three of eight decay products that are formed when $^{238}U$ radioactively decays to $^{206}Pb$.*

(B) *The half lives of the three polonium isotopes are relatively short: $^{218}Po$ = 3.05 minutes; $^{214}Po$ = <200 microseconds; and $^{210}Po$ = 140 days (Table IV-8).*

(C) *The decay of each isotope produces alpha particles, the radial ejection of which damages the lattice structures of biotite and fluorite creating visible spherical shells (rings in two dimensions) of damage.*

(D) *The radii of these rings are proportional to the energies of release of the alpha particles, are different among themselves, and hence, distinguishable for each isotope.*

(E) *A concentration of $10^8$ to $10^9$ decayed atoms is needed to create visible damage rings.*

(F) *Absence of rings indicative of Po precursors ($^{238}U$, $^{234}U$, $^{230}Th$, $^{226}Ra$, and $^{222}Rn$) and the presence of high $^{206}Pb/^{207}Pb$ ratios are interpreted to mean that isolated Po halos could only have been created from concentrations of Po atoms that were isolated from precursors and encapsulated within biotite and fluorite crystals (presumably by the Creator). A separate progenerative uranium source would, thus, be precluded and unavailable for emplacement of the Po ... or so Gentry thinks.*

From this base, Gentry drew the conclusion that Po halos in biotite and fluorite must indicate nearly instantaneous, "fiat" creation of granite during Genesis Week.(Gardner 1989)

Numerous attempts have been made to counteract Gentry's claim and to show that Po halos are formed by less dramatic processes. None of these has been fully satisfactory.[73]

Gentry has met the counterclaimants with additional arguments, pointing out that:

(A) There is no evidence for hydrothermal fluid injection, which might bring radioactive precursors into position to create the isolated Po halos, since the mineral samples containing Po halos are from fresh, unweathered rock.

(B) Distribution of the beta-particle-emitting lead isotopes is inadequate to explain the presence of short-lived $^{218}$Po and $^{214}$Po nuclei.

(C) No remnants of uranium or other precursors occur in the biotite and fluorite crystal nuclei to support the contention that the Po halos are variants of uranium halos.

**Gentry, therefore, continues to challenge the geologic community to prove that the Earth is older than 6,000 years.**(Feather, 1978; Gentry, 1983, 1988)

---

[73] *Attempts to explain Gentry's conundrum include the following:*

*(A) The halos were created by hydrothermal fluids that carried or injected the polonium into the crystals (Joly 1917; Kerr-Lawson 1927; Henderson 1939; Henderson & Sparks 1939; York 1979; Chaudhuri & Iyer 1980).*

*(B) The halos were formed from Po ions in fluids that were released during the weathering of uranium-bearing minerals (Meier & Hecker 1976).*

*(C) The large numbers of $^{210}$Po halos can be explained by diffusion of beta-emitting Pb isotopes (Hashemi-Nezhad et al. 1979).*

*(D) The halos were not really formed by disintegration of polonium but are merely variants of uranium halos (Moazed et al. 1973).*

## New Evidence Against Gentry's Hypothesis

Richard Wakefield (1987-88, 1988) demonstrated that some rocks near Bancroft, Ontario, Canada, which contain Po halos in biotite, are uranium-rich, calcite-biotite veins associated with granite pegmatites. Careful documentation of the geological relationships show that these veins and pegmatites must have been introduced in preCambrian time, long after the primordial origin of the regional terrane. Such an age spread is, of course, incompatible with Gentry's interpreted timescale.

Wakefield's study, however, did not satisfy Gentry, who [illogically] disregards the time relationships implicit in the crosscutting of veins and dikes and the sequences of events in their metamorphism.(Gentry 1983, 1988, p325-327) Moreover, in Gentry's hypothesis creation of metasediments, metavolcanic rocks, and metamorphosed intrusive gneiss complexes are all permitted from Day 1 to Day 3 of the Genesis Week. His model makes any Po-halo-bearing granite or pegmatite primordial, regardless of any complex history of the terranes in which they are found.

Therefore, although the ancient age of the Po-bearing rocks may be resolved for geologists, Gentry remains unconvinced and persists with his argument: *if the minerals that give the Po halos crystallized in granites and pegmatites from melts after millions of years of cooling history, the lives of the various isotopes are insufficient for them to have survived. Gentry's hypothesis, by this reasoning, must be correct.*

He wants the geologic community to produce evidence that shows how Po halos in granite can form by natural instead of supernatural processes. No one has done this up to now.

## Odd Circumstantial Facts

When I looked closely into Gentry's Po halo studies, I noted three odd circumstantial facts contrary to his hypothesis. The first oddity is that his *Po halos all occurred in granites and granite pegmatites, never in any other rock types, excepting locally near Bancroft.*

In that area, the uranium-bearing calcite veins cross-cut the granitic rocks, and Po halos are absent from mafic rocks whether the mafic rocks are older or younger in age than the granites. This is true even when biotite is relatively abundant among the mafic minerals.[74]

It is odd that Po halos are found only in certain, supposedly primordial, biotite-bearing granites and not in all primordial, biotite-bearing rocks. The Creator was evidently very selective about where he tucked the short-lived Po!

A second oddity I discovered was that *all of the granites in which Gentry found Po halos also contain myrmekite.* Myrmekite is a replacement mineral intergrowth, a fact that suggests that Po halos may not be present in all granites but only in granites formed by replacement processes. Conversely, perhaps only granites containing myrmekite should exhibit Po halos. In reality, even these granites do not always contain Po halos. Again, the Creator was very selective.[75]

The third oddity I noted is that *Po-halo-bearing rocks are always associated with uranium concentrations.* Gentry describes several uranium-rich localities that contain Po halos, including some in Finland, Sweden, Germany, Canada, and New England.(Gentry 1988, page 36; Wiman 1930)

---

[74] *These Po-halo-free rocks include biotite-bearing gabbros, diorites, and tonalites, as well as their volcanic equivalents. On the basis of field, chemical, and microscopic textural relationships, all of these mafic rocks are crystallized from magmas at high temperatures.*

[75] *Some granites are derived from melted sedimentary or rhyolitic volcanic rocks, some by fractionation within magma (either partial melting or fractional crystallization), and some by partial melting and rising of magma that leaves behind a mafic residue (restite). Granites derived from melts do not contain Po halos.*

Why, one must wonder, would an all-powerful Creator choose to put Po halos only in rocks that contain abundant uranium? If Gentry is correct, and Po halos have no association with uranium, why does the Creator not put the Po halos also in granites free of uranium concentrations?

Polonium, being one of the daughter products of the radioactive decay of uranium, is expected to be found near uranium concentrations. To suppose a supernatural origin for the Po-halo-bearing granites is irrational.

A fourth oddity became apparent to me later as I delved into the detail of Gentry's hypothesis. *Only selected Po halos, those for the isotopes $^{218}Po$, $^{214}Po$, and $^{210}Po$ (products derived in $^{238}U$ disintegration) have been found by Gentry.(Dutch 1983; Gentry 1983) Potential Po isotopes of other mass numbers ($^{215}Po$, $^{211}Po$, $^{216}Po$, and $^{212}Po$) are produced in the $^{235}U$ and $^{232}Th$ disintegration series (Fig. IV-8), but Po halos of these types are not found.*

Why might the Creator not make Po halos of these other types, since all are available in nature? Does this unusual omission imply some simple natural cause for the selective presence of certain Po halos?

## The True Origin of Polonium Halos

Biotite and coexisting myrmekite are both formed during replacement processes in a granite. It follows that Po halos in biotite that coexists with myrmekite must also be attributed to replacement processes.

The properties of radon are germane to this understanding. Radon is the radioactive decay product of $^{226}Ra$ that evolves into $^{218}Po$. As an inert gas, it moves freely through cracks in rocks unimpeded by reactions with minerals lining the cracks.

Evidence for this ease of radon travel is noticeable in water wells prior to earthquakes. The creeping rock-movements associated with seismically-active terranes open avenues for radon-bearing water to move into lower-pressured pore space and to the surface. Therefore, on the basis of this mobility, we would expect radon to move into a shattered and sheared habitat of diorite or gabbro that was in the process of being converted to myrmekite-bearing granite.

As $^{222}$Rn is the precursor for $^{218}$Po, this polonium isotope is the first one to be formed in the decay process. Although the half life of $^{222}$Rn is relatively short (3.82 days), enormous numbers of $^{222}$Rn atoms are released because of the abundance of $^{238}$U in the Crust. For this reason $^{222}$Rn concentrates as a dissolved element along with silica in hydrous fluids, which then migrate in response to tectonic pressures into porous sites in the mafic crustal rocks.

Two factors favor diorite or gabbro sites for the formation of Po halos in conjunction with myrmekite-bearing granite and pegmatite development:

(1) Biotite (a common mineral in diorite or gabbro) is cleavable easily on the planar "leaves" of biotite "books." These cleavage surfaces constitute porosity into which hydrous fluids carrying radon gas can move.

(2) In both biotite and fluorite the crystal lattices contain sites where negatively charged fluoride ions ($F^{-1}$) or hydroxyl ions ($OH^{-1}$) can be accommodated. These lattice sites are relatively large, and provide space where similar-sized ions can enter and take up lattice positions.

When atoms of $^{222}$Rn decay to form in succession $^{218}$Po, $^{214}$Po and $^{210}$Po, the three polonium isotopes exist as negatively charged ions, $Po^{-2}$, whose sizes are similar to the fluoride and hydroxyl ions. In this way polonium isotopes are naturally

accommodated and concentrated into fluorite ($CaF_2$) and biotite in granitic rock that is subjected to shear stress.

Thus, polonium was deposited in new crystals that grew from voluminous hydrothermal flushing of sheared and fractured, formerly-solid, mafic rock. Quartz and exotic minerals replaced mafic minerals, creating granite and pegmatite in their place. Two receptive mineral structures, **(1)** biotite cleavability (giving permeability to radon-carrying fluids) and **(2)** open-lattices in both biotite and fluorite crystals, explain why those minerals become repositories for polonium, and why the tracery of Po halos prove its ephemeral presence. The large volumes of hydrothermal fluids involved in this process are compatible with rapid growth of large pegmatite crystals of fluorite, biotite, and other minerals.

In the wall rocks near such shattered zones, where small, primary biotite crystals may be disseminated in the original diorite or gabbro, the stresses can shear them, and thus allow introduced fluids to aid recrystallization, by the annealing of microfractures, and secondary growth that enlarges the crystals. Since the large "books" of biotite in pegmatite and the small crystals of biotite in adjacent granite both develop through replacement processes at temperatures below the mineral melting intervals, these biotites and fluorites, whether growing or recrystallizing, provide ready-made lattice sites for rapid precipitation of polonium ions. Simultaneous growth of this kind of biotite and fluorite along with movement of dissolved $^{222}Rn$ atoms into the crystals enables the rapid accumulation of Po isotopes. These concentrations then decay to produce the Po halos.

The volumes of radon that emerge from deep in the Earth's Crust, dissolved in hydrous emanations, can be tremendous where uranium is abundant. Concentrations of this ambient radon can provide the enormous numbers of atoms needed to produce the Po decay halos. Radon emanating from a uranium

source is a continuous chain of disintegration episodes that can provide a constant supply of new gas to a diorite or gabbro body as it is transformed into granite or pegmatite.

From these insights it follows that Po halos in a granite need not have been produced in a short time. Some halos may have formed early, some later. Rapid entry of radon and precipitation of polonium could occur if a gabbro or diorite site were made porous and depressurized by tectonism.

The frequent coexistence of Po halos in biotite with myrmekite in plagioclase and microcline of the same rock fabric gives a clear indication that a progressive replacement process in solid, unmelted rock has taken place. No magma is involved in this process.

Finally, the presence of $^{218}$Po, $^{214}$Po, and $^{210}$Po halos only in granites and pegmatites and the absence of $^{216}$Po, $^{215}$Po, $^{212}$Po, and $^{211}$Po halos in these same rocks become understandable. Three rationales attest to this logic:

First, the half lives of radon isotopes for the different Po isotope precursors are diagnostically different. For $^{222}$Rn in the $^{238}$U series [the source for the observed Po halos], the half life is 3.82 days. By contrast, for $^{219}$Rn in the $^{235}$U series, the half life is 3.92 seconds, and for $^{220}$Rn in the $^{232}$Th series, the half life is 54.5 seconds. **The $^{222}$Rn has about 84,000 or 6,000 times as much time to enter the biotite as does $^{219}$Rn or $^{220}$Rn!**

Second, the $^{216}$Po, $^{215}$Po, $^{212}$Po, and $^{211}$Po daughter isotopes have half lives that are measured in fractions of seconds rather than the 140 days for $^{210}$Po and 3.05 minutes for $^{218}$Po. **The $^{210}$Po has over three million times the longevity of the sister series equivalents!**

Third, the relative abundance of released $^{222}$Rn gas is proportionately much greater in most terranes than the abundances of released $^{219}$Rn and $^{220}$Rn gas.

Therefore, the combination of extremely short half lives of $^{219}$Rn and $^{220}$Rn gas [and their daughter polonium isotopes] and the relatively small quantities generated make the formation of $^{216}$Po, $^{215}$Po, $^{212}$Po, and $^{211}$Po halos impossible. These isotopes of radon and polonium, which could produce the missing Po halos, convert to Pb isotopes so quickly that their radon gas can never travel far from its source before decaying. Polonium from these isotopes can never migrate to and accumulate in distant biotite or fluorite in sufficient quantities to produce any halos.

## The Buckhorn Pegmatite, A Test Case

Let us try these deductions on a suitable terrane. In 1988 Richard Wakefield and I examined several uranium-bearing and Po-halo-bearing pegmatites in an area of coexisting myrmekite in pegmatites, gneiss, and granite. The latter lithologies comprise the wall rocks adjacent to the pegmatites.

One particular pegmatite that we studied is the Buckhorn pegmatite that occurs along Highway 36, 19 km southwest of Bancroft, Ontario. Thick quartz vermicules characterize the myrmekite in this body, suggesting that both the pegmatite and granite result from replacement of an older hornblende-biotite diorite or gabbro gneiss at temperatures below the melting interval of granite. The myrmekite in wall rock and pegmatite is similar, with medium to coarse quartz vermicules.(Haynes 1986)[76]

These vermicules in myrmekite-bearing granite and Buckhorn pegmatite are diagnostic of a non-magmatic origin by replacement of diorite and gabbro rather than from a magma. The compositional relationship depends on the calcium content of the plagioclase in the original rock from which the granite and pegmatite were derived.[77]

Island remnants of diorite and gabbro occur in the granite to attest to this history, and they support the hypothesis that the precursor rocks were diorite or gabbro. Examples of these island remnants can be observed in roadcuts at the intersection of Highways 36 and 507; and their wide distribution supports the hypothesis that replacement occurred on a pluton-wide scale.

[76] *Geologic Setting of the Buckhorn Pegmatite: The uranium- and thorium-bearing Buckhorn pegmatite is a terrane of hornblende-plagioclase (diorite) gneiss, quartzo-feldspathic (diorite) gneiss, and pink, coarse-grained, massive, biotite granite. The hornblende-plagioclase gneiss in the wall rock of the pegmatite consists dominantly of polygonal hornblende, biotite, and normally zoned plagioclase. Quartz, microcline, magnetite-ilmenite, sphene, and calcite are minor constituents. The microcline fills interstices and locally encloses the other minerals. The quartzo-feldspathic gneiss is similar but contains less biotite and hornblende and more quartz and feldspar.*

*The pegmatite consists of coarse crystals of pink, perthitic microcline, biotite, albite, and quartz. Broken fragments of plagioclase are commonly surrounded by microcline. Myrmekite with relatively coarse vermicules occurs occasionally along the borders of the large pink microcline crystals. Albite lamellae in the perthitic microcline are not uniformly distributed as in granites that have crystallized from a melt, but are irregularly scattered through the crystals. Much of the albite in the microcline occurs as veins, which in some places have the same optical orientation as the plagioclase in adjacent myrmekite. Some veins also contain quartz blebs or vermicules. Po halos occur sparsely in large biotite books, two to four cm in diameter.*

*In the quartzo-feldspathic gneisses adjacent to the Buckhorn pegmatite, microcline, albite, quartz, and biotite are the dominant minerals; hornblende is generally absent and presumably has been replaced by quartz. Magnetite, calcite, and sphene are common accessories. Rounded sphene granules are the same dimensions as in the adjacent hornblende gneiss. Myrmekite with medium to coarse quartz vermicules borders the microcline grains. Chlorite containing iron oxide dust is a common alteration product of the biotite. Some biotite crystals have ragged edges and quartz sieve textures. Zircons with halos are rare. Po halos are absent.*

[77] *Generally, in diorite or gabbro that contains hornblende and pyroxene, the composition of the plagioclase tends to be calcic but not as calcic as in diorite or gabbro in which biotite is the sole ferromagnesian silicate. Hornblende and pyroxene incorporate calcium into their lattice structures. Thus, when these minerals are abundant, less calcium is available to go into plagioclase. In contrast, because biotite does not accommodate calcium in its lattice structure, most of the available calcium in that kind of diorite or gabbro goes into the plagioclase. This high-calcium plagioclase has the effect, when biotite gabbro is replaced by granite, of producing quartz vermicules in associated myrmekite that are coarser than where hornblende or pyroxene gabbro is replaced by granite.*

*Regionally, granite in and near the Buckhorn pegmatite contains myrmekite with coarse quartz vermicules. This suggests that the granite adjacent to the Buckhorn pegmatite was originally a biotite gabbro or calcic diorite.*

## Uranium Mines of the Bancroft District, Ontario

Three uranium mines [Silver Crater, Fission, and Faraday] are situated in the same terrane as the Buckhorn pegmatite.(Hewitt 1957; Bedell 1985)  Local Po halos are found in biotite and fluorite in these mines.  Several rock types including gabbro, diorite, monzodiorite, monzonite, and syenite comprise the country rock in the region surrounding these mines, all of them essentially free of myrmekite.  Beyond the Bancroft area in the Anstruther area, however, granite is present instead of syenite, and it exhibits myrmekite with coarse quartz vermicules.

The absence of myrmekite in the region where syenite occurs has a logical explanation as excessively sodic replacement. Cataclastically sheared diorite and gabbro bodies are replaced by fluids carrying $Na^{+1}$, $K^{+1}$, and $Ca^{+2}$.[78]  These ions have converted the diorite and gabbro to monzodiorite, monzonite, and syenite, but only locally into granite.[79]

The fluids that created this regional evolution from mafic to siliceous rock also brought in uranium, which became concentrated in localized sites.  The biotite and fluorite in the uranium ores in the three mines contain Po halos formed essentially at the time the diorite and gabbro were recrystallized and replaced [as monzodiorite, syenite, and granite].

It should be noted, however, that not all biotite "books" and samples of fluorite in the pegmatites and calcite veins in these modified rocks contain uranium halos or Po halos.  The local absence of the Po halos suggests that only in certain places were concentrations of radon sufficient to produce visible Po halos.

---

[78]  *Lumbers, personal communication, 1987.*

[79]  *In the former diorite and gabbro, relatively calcic plagioclase crystals have become sodic (albite to oligoclase) instead of being relatively calcic (andesine and labradorite). In these modified rocks calcite is found filling the interstices between the broken silicate grains or concentrated in veins.  K-feldspar locally is introduced, but the coexisting plagioclase is so sodic in the syenite and/or granite that myrmekite is not formed (Collins 1988a).  Precursor hornblende and pyroxenes in the diorite and gabbro bodies have also recrystallized (as biotite and/or sodic amphiboles).*

## Polonium Halos Explained

The formation of granite by replacement of solid rocks means that Gentry's theory is no longer tenable. He can no longer legitimately say that Po-halo-bearing granites must form by supernatural means.

Solid diorite and gabbro rock, which had previously crystallized from magma, has been subjected to repeated cataclasis and recrystallization. This has happened without melting; and the cataclasis provided openings for the introduction of uranium-bearing fluids and for the modification of these rocks to granite by silication and cation deletion.

In uranium ore-fields the extra uranium provides an abundant source of inert radon gas; and it is this gas that diffuses in ambient fluids so that incipient biotite and fluorite crystallization is exposed to it. Radon ($^{222}$Rn) decays and Po isotopes nucleate in the rapidly growing biotite (and fluorite) crystals whence they are positioned to produce the Po halos.

**The whole process of Po halo formation can be accomplished without calling on a Creator to do it. The serendipity that has emerged from these observations implies that Po halos in myrmekite-bearing granite indicate a non-magmatic origin for the granite. The argument comes full circle when it meets Gentry's initial [*truthfully made*] observation that magmatically-derived granites cannot contain Po halos: the half-lives of the Po isotopes are simply too short.**

Chapter V

# SILANE SYSTEMATICS: INTERPRETATIONS OF GRANITIZATION IN SITU

## PETROLOGIES AND DISCONTINUITIES

**L. G. COLLINS & C. WARREN HUNT**

If granite forms *in situ* by upward migration of silanes that replace mafic minerals, major changes must follow and affect all rock suites of the outer geospheres. Among the profound consequences of this process are the mineral transformations that occur at the 670-km, 400-km, and shallower upper-Mantle seismic discontinuities. These are the crustal sites, where volcanic rock with sodium-bearing plagioclase forms, and where the entire spectrum of metal cation mobilization, displacement, and upward transfer is started. The effects include (1) rock volume that results from the substitution of high-volume phases containing additional $SiO_4$ in place of low-volume mafic phases, (2) isotope fractionation that produces oxygen-isotope composition anomalies (between volcanic and plutonic igneous rocks), (3) rare-earth-element fractional redistributions, (4) zonation of plutons, and (5) isotope fractionation that gives anomalous $^{87}Sr/^{86}Sr$ ratios in lamprophyres and mafic volcanic lavas, and (6) discordant zircon populations relative to uranium-lead dates.

## Magmatism vs. Silane Replacement

Conventional wisdom favors a magmatic interpretation for all these features. Within a cooling magmatic plume the minerals crystallize in predictable order. For example, in a melt with the components of clinopyroxene, orthopyroxene, and olivine, the olivine crystallizes first, then orthopyroxene, and finally clinopyroxene. Where plagioclase is present, as it is higher in the Mantle and in the lower Crust, it generally will crystallize after these ferromagnesian silicates. Dry rock crystallizes at much higher temperatures than wet; and as little water as 0.04 % is found to lower crystallization temperature significantly, the magmatic plume is imagined to congeal in the aforesaid order.

**Magmatism** is perceived, in addition, to result in downward accumulation of more mafic components (heavy ferromagnesian silicate and oxide minerals) by selective settling during progressive crystallization. Although clearly occurring on a local scale, this process, as a mechanism for planetary differentiation, is declared by Tarling to be impossible,[80] a position also taken by Gottfried.

In any case, fractionation into more mafic and less mafic levels of a pluton does not entail volume change. Where it lightens upper levels of plutons, their isostatic rise can be imagined as a diapiric displacement of the cover rocks.

**Granite formation *in situ* comprises an alternative process, wherein displaced mafic components are entrained in hydrous fluids and flushed *upward*.** Volume changes are inevitable with this process. To generate granite in the Crust involves the addition of bulky oxygen, which must be present either as ambient water or as an extracted element during conversion of relatively-dense to less-dense silicate minerals. In the rare case

---

[80]   *Editor's note: The disarray in geology described by Skobelin in his introduction is shown well by the D. H. Tarling quotation (p23) pointing out the impossibility for gravitational segregation of crystals from a magma.*

of the process happening near Earth's surface, the oxygen source may be atmospheric gas dissolved in groundwater in rock fractures.

In anhydrous granitic rocks of the Crust, where silanes replace ferromagnesian silicates to form quartz, and where silanes plus sodium ($Na^{1+}$) replace calcic plagioclase to form sodic plagioclase, little to no volume change occurs. This conversion is a volume-for-volume replacement. as will be explained in coming paragraphs.

## Silane Behavior in the Transition Zone

Where silanes replace minerals in the upper Mantle and Crust, the availability of oxygen controls the end result. The deductions I had drawn, first at Temecula and then in Colorado pertaining to open systems, crystal expansion, and the impropriety of balanced mass-for-mass equations, are fully applicable in the upper Mantle and Crust, where minerals in the solid rock are not free to expand.

Silane gas, under pressure and electrically neutral, is able to pervade microcracks in deformed crystals and to react with water (Aston, 1983) or with oxygen released from dense minerals by replacement. Oxygen ions ($O^{2+}$) are huge relative to metal ions in silicates, occupying on average 93.8 % of mineral volume. Thus, oxygen is preferentially in contact with invasive silane. Silica is provided in this way for volume-for-volume exchanges and for supplementation of total rock volume by generation *in situ* of less dense rock.

In silane reactions at mantle depths, where water is absent and the solid rock harbors no free oxygen or water, oxygen is drawn from the abundant oxygen in oxide or silicate minerals. Most oxygen is bound to metal cations in ferromagnesian silicates and oxides. This is the case in all geospheres, the upper and lower Mantle, the Transition Zone, and the Crust. Silanes mainly bypass this oxygen and combust only with available water in the lower Crust and with free oxygen that is released where granitic rocks replace mafic rocks in the middle or upper Crust.

This paper will now advance the concept that major physical/chemical changes stemming from silane emanation are prominent at three geosphere levels:

(1) the mid-mantle 670-km seismic discontinuity,

(2) upper mantle sites of sodic plagioclase growth, and

(3) water-permeated upper mantle and lower- and mid-crustal depths. The last two of these include the Benioff zones, which will be discussed in Chapter VI.

## Experimental Work

Interpretations of how silanes react with minerals at various depths in the Mantle rely heavily on many years of experimental work (Ringwood 1991). Some of this work is summarized briefly in **Table V-1**, which lists rock zones, discontinuities, mineralogical compositions, and approximate density changes that occur in mantle rocks at increasing depths in the Earth. Using mixtures of natural minerals or of elemental compounds found in Mantle rocks and subjecting them to increasing pressures and temperatures in closed systems to determine phase changes, research has given a good approximation of what can be expected if rocks of these compositions occur at various depths. The validity of this experimental work is not in doubt, and its results do not conflict with the silane-replacement model.

Notwithstanding, it is still moot whether these data truly represent conditions at 700 km in the Mantle. Do the minerals convert to less-dense phases in the manner predicted from the experiments? Do the mantle rocks below 700 km have the compositions as shown in **Table V-1**? Perhaps.

**TABLE SHOWING POSSIBLE MINERAL ASSEMBLAGES OF UPPER MANTLE, TRANSITION ZONE, LOWER MANTLE, AND DISCONTIUNITIES.**

| Zone | Depth Range (km) | Mineralogy | | Density |
|------|------------------|------------|---|---------|
| Crust (oceanic) | 0 to 10 | | | |
| (continental) | 0 up to 80 | | | |
| Mohorovicic Discontinuite | below Crust | Olivine $(Mg,Fe)_2SiO_4$<br>Orthopyroxene $(Mg,Fe)SiO_3$ | | 3.2 |
| Plagioclase Peridotite (Harzburgite) | 11 to 25 (oceanic) | Olivine $(Mg,Fe)_2SiO_4$<br>Orthopyroxene $(Mg,Fe) SiO_3$<br>[Clinopyr. $(Ca,Mg,Fe)_2(Na,Al)(Si_2O_6)$]<br>Plagioclase $CaAl_2Si_2O_8NaAlSi_3O_8$ | | 3.23 |
| Spinel Peridotite (Lherzolite) | 25 to 75 (oceanic) | Olivine $(Mg,Fe)_2SiO_4$<br>Orthopyroxene $(Mg,Fe)SiO_3$<br>Clinopyr. $(Ca,Mg,Fe)_2(Na,Al)(Si_2O_6)$<br>Spinel $(Mg,Fe)Al_2O_4$ | | 3.3 |
| Garnet Peridotite (Lherzolite) | 75 to 350 (oceanic) | Olivine $(Mg,Fe)_2SiO_4$<br>Orthopyroxene $(Mg,Fe)SiO_3$<br>Clinopyr $(Ca,Mg,Fe)_2(Na,Al)(Si_2O_6)$<br>Garnet $(Mg,Fe,Ca)_3(Al,Cr)_2Si3)_{12}$ | 58%<br>17%<br>12%<br>14% | 3.4 |
| 400-km Discontinuity | 350 to 400<br>400 to 430 | Orthopyroxene to garnet<br>Olivine to B-Phase $(Mg,Fe)_2SiO_4$ | | 3.59 |
| Transition Zone | 400 to 550 | ß-Phase<br>$(Mg,Fe)$ garnet<br>Sodic Clinopyroxene<br>(Ca) garnet to (ca) perovskite | 57%<br>40%<br>3% | |
| | 500 to 550 | ß-Phase to spinel structure | | 3.68 |
| | 550 to 650 | Spinel structure<br>$(Mg,Fe)$ garnet<br>(Ca) perovskite<br>Sodic Clinopyroxene | 57%<br>38%<br>2%<br>3% | 3.71 |
| 670-km Discontinuity | 650 to 700 | Spinel str. to perovsk. str.<br>(Mg,FeO garnet to perovskite structure<br>Sodic clinpoyr. to Ca-perovskite | | |
| Lower Mantle | 700 to 1050 | Perovskite structure<br>Magnesiowüstite<br>$(Mg,Fe)$ garnet to perovskite structure<br>Ca-perovskite | 73%<br>19%<br><br>8% | 4.16 |

**Table V-1**
**POSSIBLE MINERAL ASSEMBLAGES OF THE MANTLE**

*Estimated weight percent of each mineral after Ringwood (1975) and Liu (1979). Adapted from information in Ringwood (1991), Kostopoulos (1991), and Condie (1982).*

In the silane-replacement model, the rocks of the lower Mantle are taken to be slightly deficient in silica relative to rocks in the overlying Transition zone and the upper Mantle (far above the 670-km discontinuity). The results of silane replacement under these conditions are described in the following sections.

## The 670-km Discontinuity:

### The Perovskite To Spinel, Ca-perovskite To Sodic Clinopyroxene, And Perovskite To Garnet Transitions

A broad zone occurs in the Mantle between 650- and 700-km depth (hereafter called the 670-km discontinuity) in which earthquake waves change speed because of changes in rock density (**Table V-1**). In current theoretical models these changes in density are thought to result from "phase changes," in which denser minerals below convert to less-dense minerals above. The phases are here regarded as internal to the crystals, atom transfers that result in volume changes. No provision is made for elements to move in or out of the rock fabric at the discontinuity.(Ringwood 1975; Liu 1979)

Earlier experimental work had suggested that at prevailing temperatures and pressures, at the 670-km discontinuity, mineral conversions occur between perovskite structure and spinel structure, between ilmenite structure and garnet, and between Ca-ferrite structure and sodic clinopyroxene (not shown on **Table V-1**).(Ringwood 1975; Liu 1979) This earlier work and later studies that supersede it are described in the following discussion.

In the Crust at the Earth's surface perovskite is $CaTiO_3$, and spinel is $(Mg,Fe)Al_2O_4$. At the 670-km discontinuity these minerals do not have these compositions, and certain

ferromagnesian silicates assume identical atomic, "perovskite," and "spinel" structures.  The following formulas compare these crustal and Mantle ferromagnesian silicate mineral analogs (ß-phase,[81] and olivine, $Mg_2SiO_4$), which are stable at higher levels in the upper Mantle.

**670-km Discontinuity**                    **Transition Zone**

**perovskite    →    spinel    →    ß-phase**

$(Ca)TiO_3$          $(Mg,Fe)Al_2O_4$

$(Mg,Fe)[(Al)Al]O_4$

**perovskite str.    →    spinel structure    →    ß-phase    →    olivine**

$(Mg,Fe)SiO_3$    →    $(Mg,Fe)[(Mg,Fe)Si]O_4$    →    $(Mg,Fe)_2SiO_4$    →    $(Mg,Fe)_2SiO_4$

Let us set all these formulae to 12 oxygens and compare their metal (cation) contents:

**perovskite str. →      spinel str.      →    ß-phase    →      olivine**

$(Mg,Fe)_4Si_4O_{12}$   $(Mg,Fe)_3[(Mg,Fe)_3Si_3]O_{12}$   $(Mg,Fe)_6Si_3O_{12}$   $(Mg,Fe)_6Si_3O_{12}$

8 metal cat.      9 metal cat.              9 metal cat.      9 metal cat.

In this consideration, masses do not balance for converting perovskite structure to spinel structure, but masses balance for converting spinel structure to the ß-phase and then to olivine. In order to make masses balance in the conversion of perovskite structure to spinel structure, existing theory combines

---

[81]   *"ß-phase" is an intermediate-phase mineral form in which a slight shift of atom positions, and no change in composition has occurred.*

perovskite structure with magnesiowüstite (Mg,Fe)O (or periclase MgO) as follows:

**perovskite str. +   magnesiowüstite →   spinel structure**

$$(Mg,Fe)SiO_3 \quad + \quad (Mg,Fe)O \quad \rightarrow \quad (Mg,Fe)(Mg,Fe)SiO_4$$

This equation is predicted from experimental work, using laboratory "bombs." The silane-replacement model provides an alternative explanation. In the conversion, one additional metal cation for every twelve oxygen atoms enters from an outside source to bring the total to nine metal cations instead of eight metal cations. That addition lowers rock density by mineral volume change and, predictably, should occur in higher geospheres. The added cation may, in fact is most likely to be, silicon from silane. The reaction may then be written as follows:

**perovskite str. + silane →   spinel structure +   hydrogen**

$$(Mg,Fe)_4Si_4O_{12} + SiH_4 \quad \rightarrow \quad (Mg,Fe)_3[(Mg,Fe)Si_5]O_{12} + 4H$$

$$8 \text{ metal cat.} \quad + 1 \text{ metal cat.} \rightarrow \quad 9 \text{ metal cat.}$$

However, if this is true, this variety of spinel structure does not convert directly to the ß-phase and thence to olivine by a phase change, because the formula: $(Mg,Fe)_3[(Mg,Fe)Si5]O_{12}$, has too much silica and not enough magnesium and iron to be able to produce olivine $(Mg,Fe)_6Si_3O_{12}$.

The alternative silane replacement model does much better, as it has eight metal cations in the perovskite structure, of the form: $(Mg,Fe)_4[(Mg,Fe)_2Si_2]O_{12}$ rather than the form: $(Mg,Fe)_4Si_4O_{12}$. The silane replacement formula better fits with existing theory of the deep Mantle, because toward the Core, the Mantle probably contains more iron and magnesium and

less silica in silicate form (Ringwood 1991) and because magnesium and iron ions fit into the atomic coordination positions of the perovskite structure more easily than does the silicon ion.

Enhanced silicon levels characterize higher geospheres, where additional seismic discontinuities are recognized. If we take the liberty of presuming that a magnesian perovskite structure, $Mg_4[Mg_2Si_2]O_{12}$, has a permissible density of 3.8 (in contrast with a lower density of 3.5 for a magnesian spinel, $Mg_3[Mg_3Si_3]O_{12}$), then the following formulation follows (ignoring hydrogen which may enter and depart with little effect on volume).

| silane | + | perovskite str. | → | spinel str. | +hydrogen |
|--------|---|-----------------|---|-------------|-----------|
| $SiH_4$ | + | $Mg_4[Mg_2S_{12}]O_{12}$ | → | $Mg_3[Mg_3Si_3]_{12}$ + | 4H |
| | | (100 cc, 380 g) | | (116.4 cc, 407.4 g) | |

| O | | 186.2 | | 186.2 |
|---|---|-------|---|-------|
| Mg | | 139.5 | | 139.7 |
| Si 27.2 | | 54.3 | | 81.5 |

$$27.2\ g\quad +\quad 380.0\ g\quad \rightarrow\quad 407.4\ g$$

$$407.2\ g\ (total) + 0.2\ g\ Mg\quad \rightarrow\quad 407.4\ g$$

These calculations incorporate all available oxygen, increase volume by 16.4 cc, and add 27.2 g of Si from introduced silanes. The remaining 0.2 g of additional Mg needed to balance the reaction may be derived from other simultaneous mineral transformations, or it may be an artifact of the assumed densities. In fact, the reaction may be perfectly balanced without external additions other than the aforesaid silicon from silanes.

In the event of water being present and reacting with silanes, extra oxygen would be present above that required for these transformations. This extra oxygen would be expected to be immobilized by incorporation into new spinel structures, its large ionic radius precluding easy movement through solid rock even with substantial pressure differentials. Nevertheless, should some of this available oxygen – perhaps as little as the 0.04 % mentioned at the opening of this discussion – combine with released hydrogen to form water, such water would facilitate partial melting because of its effect on mineral melting temperatures.

Some water, albeit very little, is contained in lava that emerges from volcanoes overlying hot spots, such as the Hawaiian Islands, or along mid-ocean ridges. Magmatic models postulate a small continuous supply of it from the Mantle to account for the evolution of volcanic rock. The silane-replacement model manufactures water at whatever stage the hydrogen is oxidized. This becomes more likely in higher geospheres.

The feasibility of mineral transformation being the explanation for the 670-km discontinuity is supported by the ability of the proposed transformation to produce a spinel structure, $Mg_3[Mg_3Si_3]O_{12}$, which is then capable of direct conversion into the ß–phase spinel, $Mg_6Si_3O_{12}$ at the 500- to 550-km transition and thence into olivine, $Mg_6Si_3O_{12}$ in the 350 to 430-km discontinuity without addition or subtraction of external material.

Perovskite structure at the 670-km discontinuity is unlikely, however, to be a pure magnesian variety with a density near 3.8. Rather more likely, its density is near 4.16 because of the substitution of Fe for some Mg. The 4.16 figure is the density of the lower Mantle derived from geophysics.(Li & Jeanioz 1991a,b)

Recent experimental work is consistent with the aforesaid conversion at the 670-km discontinuity, suggesting that the upper Mantle and Transition Zone have compositions substantially the same but for a silica deficiency in the lower Mantle represented by a ratio of Mg:Si = 1.27 (atomic). Earlier models had proposed a lesser ratio of Mg:Si = 1.05 $g/cm^3$ the average value for the stony meteorites, "chondrites," which are enriched in silica and iron oxide.(Ringwood 1991)

These conversions, then, offer a rational explanation for the heretofore elusive phase changes. They can satisfy known conditions appropriately with minimal volume increase because only the atomic positions are rearranged to provide more open structures.

A second, earlier-proposed, mineral-conversion at the 670-km discontinuity (not shown on **Table V-1**) is one in which ilmenite structure in the lower Mantle changes to garnet in the Transition Zone as follows:

**ilmenite structure** $\rightarrow$ **garnet**

$FeTiO_3$    (ilmenite)

$Fe_4Ti_4O_{12}$

$(Mg_3Al)(AlSi_3)O_{12}$    $\rightarrow$    $Mg_3Al_2Si_3O_{12}$

If this conversion were to occur at the 670-km discontinuity, it would require no addition or subtraction of elemental matter, and only a slight shift in the crystalline configuration of the atoms. Therefore, this phase change occurs without the agency of silane reactivity. However, recent experimental work shows that aluminum can be absorbed into the perovskite structure, and that, where this occurs, the aluminum-bearing perovskite structure will recrystallize as garnet just above the 670-km discontinuity. Therefore, the conversion of ilmenite structure to garnet should not be regarded as a "necessary phase change." Instead a slight change in atomic positions changes the ilmenite structure to the perovskite structure, which then converts to garnet with an additional slight change in atomic positions.(Ringwood 1991)

**perovskite structure** $\rightarrow$ **garnet**

$CaTiO_3$ (perovskite)
$(Ca)_4Ti_4O_{12}$

$(Mg_3Al)(AlSi_3)O_{12}$ $\rightarrow$ $Mg_3Al_2Si_3O_{12}$

A third mineral conversion, once considered necessary but now superseded, was a proposed change of Ca-ferrite structure (hercynite-spinel structure) in the lower Mantle to sodic clinopyroxene in the Transition Zone (not shown in **Table V-1**). This phase change was originally proposed in order to provide for a sodic mineral in the lower Mantle to match sodic clinopyroxene in the Transition Zone. This relationship is illustrated in the following equation, which requires the addition of silica. It should be noted that hercynite spinel has twelve cations in its structure, while only eleven fit the Ca-ferrite structure (an effect of ionic charge-differences when sodium is included).

**Ca-ferrite structure + quartz (stishovite) → sodic clinopyroxene**

$FeAl_2O_4$ hercynite spinel

$(Fe_4)(Al_8)O_{16}$

12 metal cat.

$[Ca_2(Mg,Fe)_2](NaAlSi_5)O_{16} + SiO_2 → [Ca_2(Mg,Fe)_2]NaAlSi_6O_{18}$

11 metal cat.   +   1 metal cat.   →   12 metal cat.

Recent experimental work shows that sodium can enter Ca-perovskite structure instead of the Ca-ferrite structure.(Ringwood 1991) Ca-perovskite combined with less calcic (Mg,Fe)-perovskite, sodic clinopyroxene can form, as follows:

**Ca-perovskite str. + (Mg-Fe)-perovsk. str. - sodic clinopyroxene**

$CaTiO_3$ (perovsk.) + $CaTiO_3$ (perovsk.)

$(Ca)_7Ti_7O_{21}$     +   $Ca(Mg,Fe)_6Si_7O_{21}$

$NaAlSi_2O_6 \cdot Ca_5Si_5O_{15} + Ca(Mg,Fe)_6Si_7O_{21} → [Ca_6(Mg,Fe)_6]NaAlSi_{14}O_{42}$

$→ 6[Ca(Mg,Fe)Si_2O_6 \cdot 1NaAlSi_2O_6$

14 metal cat.   +   14 metal cat.   →   28 metal cat.

Note that in this process, silica in the form of stishovite need not form. The total amount of $Na_2O$ in the upper and lower Mantle and Transition Zone is small (0.3 to 0.4 percent). Proportional amounts of Na and Ca in the Ca-perovskite structure and sodic clinopyroxene in the foregoing formulae need to be adjusted, thus increasing the calcium component, to reflect this small $Na_2O$ percentage. In this conversion model, the simultaneous recrystallization in two perovskite forms can occur without the addition of silanes. In any case, the volume of Ca-perovskite is proportionately small (about 8 %, **Table V-1**), and therefore, its conversion to clinopyroxene has little effect on the total volume of mantle rock that is transformed.

On the basis of these calculations and assumptions in the silane-replacement model, the lower Mantle below the 670-km discontinuity is interpreted to be essentially composed of calcic and magnesium-rich perovskite structures containing some iron but lacking magnesiowüstite and stishovite. This is unlike that shown on **Table V-1**, which includes magnesiowüstite.

The upward movement of silanes from the lower Mantle through the Transition Zone catalyzes the recrystallization of these mineral phases and adds silica to the magnesium-rich perovskite phase.

## The Transition Zone and Upper Mantle

### Depletion of Transition Zone and Upper Mantle of Lithophile Elements

Above the 670-km discontinuity, in the Transition Zone and upper Mantle, there is a general deficiency of large-ion elements, the so-called lithophile "rock-loving" elements, such as potassium, rubidium, strontium, barium, thorium, zirconium, and the light rare-earth elements. This contrasts with the condition below the discontinuity, where the lithophile elements, although present, are not prominent (**Table V-I**)

Kimberlites are generally attributed to lower mantle provenance, despite the confliction created by their high silica content. The occurrence of silica and other lithophile elements in kimberlites, as well as in volcanic emissions from isolated volcanic islands in oceanic terranes is then taken to show that they exist in the lower Mantle, a tenuously defensible form of circular logic.

## Depletion Of Lithophile Elements from the Mantle

A possible explanation for depletion of the lithophile elements from the upper Mantle and Transition Zone is implied by their sequential melting behavior. When minerals melt, the elements tend to fractionate, the lithophile types tending to move out in the melt, and the siderophile, more refractory elements (nickel, chromium, etc.), tending to stay behind. Thus mobilized, lithophile elements depart in coalesced and advancing melts, while unmelted residues, their fractionated lithophile elements depleted, are left behind. In this way, **the upper Mantle and Transition Zone overlying the 670-km discontinuity in oceanic terranes,** *because of a prior history of partial melting,* **are depleted in the lithophile elements.**

This explanation leads to a dilemma, however: the lack of a heat source for partial melting. Silane reactions offer a ready solution to this dilemma. Silanes provide fuel for melting from the inception of their activity in the 670-km discontinuity, persisting with enhanced activity (including melting) at higher levels, and ultimately evolving into granitic rocks in the Crust, alkali-rich continental flood basalts, and the volcanics of oceanic terranes.

Let us look briefly at the current explanations. The prevalent idea is that, when the Earth's continental Crust first formed on a liquid geosphere, lithophile elements had floated up and solidified as the Crust. After the primary Crust was formed, huge convection cells, rock fountains in the ductile Mantle, split the continental Crust and circulated underlying depleted rock to the surface. After cooling, the increased density of the cool rock, in contrast to the hot upwelling rock, cause it to recirculate down again in "subduction zones," whence, another fountain cycle is set to commence. The heat for remelting in this system is primordial radiant heat from the Core, and the convection process is thought to have gone through many cycles during 3.5 billion years of Earth history. The upper Mantle has, thusly, become depleted in lithophile elements by successive fractionations. Subducted basaltic crustal plates, now depleted, are imagined to accumulate above the 670-km discontinuity, the work of aeons.

One problem with this model is that at the points of emergence of the Mantle plumes, the "spreading centers," where melting brings basalt and gabbro to the surface, the process should have depleted the underlying Mantle in all lithophile elements. The fact is that the same rock types with lithophile elements incorporated keep coming up, a deficiency that cannot readily be explained. A solution has been suggested whereby convection cells bring the basaltic lithospheric plates down, all the way, in fact, to the 670-km discontinuity, where their lesser density relative to the lower Mantle causes them to glide laterally. In this way the Transition Zone and upper Mantle are "fertilized" again below spreading centers by new "plates" containing lithophile elements.

## An Alternative Scenario

The silane-replacement model offers a less-contrived alternative explanation. In this model huge convection cells above the 670-km discontinuity are unnecessary. The flushing out of lithophile elements from rock-forming minerals occurred in early stages of the Earth's formation, when its diameter was much smaller. Silane replacements of perovskite structures to form spinel structure, garnet, and sodic clinopyroxene were the deepest levels of this process, and led to partial melting and upward migration of evolved magmas. Diapiric rise with entrained unreacted silanes resulted in subsequent partial melting and catalyzed the further displacement of lithophile elements into the original continental Crust.

When the process compounded by expansion of the Earth to its present diameter, space was created between the continents

in which today's oceans are accommodated. The Mantle necessarily expanded by plastic flowage. Its most depleted and most ductile – least siliceous – portions flowed toward ocean basins from continental centers. Its least-plastic and least-depleted – most siliceous – portions remained as continental underpinnings. This global expansion and mass redistribution produced the present heterogeneous upper Mantle and Crust.

Huge convection cells are not required to produce a depleted Mantle underlying oceanic terranes. Instead of "fertilization" of the Mantle beneath spreading centers with recycled basaltic components, these components are resupplied steadily from the lower Mantle into upper geospheres by escaping silanes. An adequate supply of necessary basaltic components is available for continued crustal expansion *because* the upper Mantle and lower Transition Zone have not gone through multiple convection cycles.

## The 400-km Discontinuity:

### Garnet to Pyroxene
### and ß-phase to Olivine Transitions

A broad zone occurs between 350 and 430 km that resembles the broad recrystallization zone of the 650- to 700-km discontinuity. This is generally called the 400-km discontinuity. In it, garnet and olivine are converted either to orthopyroxene and spinel or to orthopyroxene, clinopyroxene, and spinel. At the same time, the ß-phase converts to olivine. These reactions result in a spinel peridotite at higher levels in the Mantle above the garnet-bearing peridotite. On the basis of experimental work on closed systems, the equation shown first below is usually the one that is used to illustrate the change that occurs; but the second equation is also possible.

**Garnet-bearing Peridotite     Spinel Peridotite**

$$\text{garnet} \quad + \quad \text{olivine} \quad \rightarrow \text{orthopyroxene} \; + \quad \text{spinel}$$
$$Mg_3Al_2Si_3O_{12} \; + \; Mg_2SiO_4 \; \rightarrow \quad 4MgSiO_3 \quad + \quad MgAl_2O_4$$

Or:

$$\text{garnet} \quad + \quad \text{olivine} \quad \rightarrow \quad \text{clinopyr.} \; + \; \text{orthopyr.} \; + \; \text{spinel}$$
$$CaMg_2Al_2Si_3O_{12} \; + \; Mg_2SiO_4 \; \rightarrow \; CaMgSi_2O_6 \; + \; 2MgSiO_3 \; + \; MgAl_2O_4$$

If these equations and the ß-phase change to olivine represent the only conversions at the 400-km discontinuity, the transformation should be accomplished without a change in mass. If that is true, no excess oxygen is available to react with silanes at this level; and any silanes moving upward through this zone should continue upward toward the Crust. However, if sodium has been mobilized and rises with silanes from the lower Mantle in the manner of the lithophile elements to this discontinuity, then potentially both sodium and silicon (from silanes) could combine with aluminum released from garnet to produce additional sodic clinopyroxene. In that case not all released aluminum would go into spinel.

## Spinel Peridotite to Plagioclase Peridotite

Current experimental work shows that higher in the Mantle (above the 400-km discontinuity) spinel peridotite recrystallizes to form plagioclase peridotite (**Table V-1**). Generalized equations that illustrate this transformation are usually written as follows:

**Spinel Peridotite          Plagioclase Peridotite**

$$\text{orthopyr.} \; + \; \text{clinopyr.} \; + \; \text{spinel} \; \rightarrow \; \text{olivine} \; + \; \text{plagioclase}$$

$$2MgSiO_3 \; + CaMgSi_2O_6 \; + \; MgAl_2O_4 \rightarrow 2Mg_2SiO_4 + CaAl_2Si_2O_8$$

Written in this way, the equation shows that the conversion is simply a metamorphic reconstruction of minerals at lower pressures near the Earth's surface without addition or deletion of new material. If that were true, no reactions with silanes would be possible. However, plagioclase is rarely of pure calcic nature, and usually has some sodic content. Under that constraint, additional silica is required because three silicon atoms per formula are needed for sodic plagioclase [$NaAlSi_3O_8$] instead of two as in calcic plagioclase [$CaAl_2Si_2O_8$]. Two kinds of mass-for-mass equations illustrating this requirement are shown below.

### Iron-rich Garnet-bearing Peridotite

**soda-pyroxene + garnet + quartz (stishovite) + rutile =**

$$NaCa_2MgFe^{2+}AlSi_6O_{18} + Ca_2Mg_3Fe^{2+}{}_4Fe^{3+}Al_5Si_9O_{36} + SiO_2 + 2TiO_2 =$$

#### Plagioclase Peridotite
And:
augite           +           plagioclase    + ilmenite

$$Ca_2Mg_2Fe^{2+}Fe^{3+}AlSi_5O_{18}{\cdot}Mg_2Fe^{2+}Si_4O_{12} + NaAlSi_3O_8{\cdot}Ca_2Al_4Si_4O_{16} + 2FeTiO_3$$

#### Magnesium-rich Garnet-bearing Peridotite

**clinopyroxene   +     garnet    +    quartz (stishovite)**

$$2NaAlSi_2O_6{\cdot}nCaMgSi_2O_6 + 3CaMg_2Al_2Si_3O_{12} + 2SiO_2$$

#### = Plagioclase Peridotite

= olivine   +   diopside   +     plagioclase

$$= 3Mg_2SiO_4 + nCaMgSi_2O_6 + 3CaAl_2Si_2O_8{\cdot}2NaAlSi_3O_8$$

It is illogical that the needed $SiO_2$ in these equations should be attributed to deep sources in the Mantle or that it already exists *in situ* because (1) the mineral transformations at lower discontinuities consume it; none produces it, (2) Mantle is characterized as silica-deficient, (3) olivine, which is abundant in the upper Mantle, reacts with quartz to produce orthopyroxene and, therefore, would eliminate any available quartz, (4) vertical, upward-directed motion characterizes magmas at spreading centers, so that a counter movement, downward from some unknown, upper, quartz-rich source for the needed $SiO_2$ is unlikely, (5) even if primary $SiO_2$ were present, quartz being the first mineral to melt during partial melting, should have been flushed out in the earliest magmatic events, and (6) any quartz in supposedly re-cycled basaltic rock that could be invoked to "fertilize" the Mantle should have been subtracted by partial melting during deep burial of the oceanic plate, and, hence, never reach the roots of a spreading center.

The continuing formation of basalts containing soda-bearing plagioclase in spreading centers requires a supplemental provenance for the extra silica. One possible source of extra silica, as mentioned above, would be the breakdown of orthopyroxene to produce olivine plus quartz.

| orthopyroxene | $\rightarrow$ | olivine | + | quartz |
|---|---|---|---|---|
| $2MgSiO_3$ | | $Mg_2SiO_4$ | + | $SiO_2$ |

Experimentally, however, it has been shown that orthopyroxene does not begin to melt until all clinopyroxene is gone. Furthermore, it is possible to argue that the emerging basalt at spreading centers commonly contains abundant orthopyroxene (i.e. it is "tholeiitic"), and, thus, that the breakdown of orthopyroxene is not occurring. In summary, although unlikely as the source of extra silica, partial melting of orthopyroxene cannot be completely ruled out.

Therefore, silanes must be considered an alternative source of silica. Instead of quartz being introduced or being present as a primary mineral, silane influx can supply the silicon. If the $Na_2O$ content in lower Mantle rocks is 0.3 % and transformed basaltic magmas emerge with 2.0 % $Na_2O$, the least-contrived source for extra silica is one that relates it to volatile influx. Only as silanes can silicon be mobilized in that fashion.

In the silane-replacement model, density differences between primary ferromagnesian silicates and newly-formed plagioclase provide enough oxygen per unit volume to allow the production of soda-bearing plagioclase. The following volume-for-volume calculations for the above equation are more complex than previous calculations because more than one mineral is involved in the silane-replacement process. The results, however, show that plagioclase will form when oxygen is kept constant.

The comparison is made between 107 cc of reactants (garnet, soda-pyroxene, rutile) in the proportions that occur in balanced mass-for-mass reactions, and 123.8 cc of the products (augite, plagioclase, ilmenite) in the proportions in which these reactants produce them (see previously shown formula). Hydrogen is not included; and excess metal cations are presumed carried upward with as dissolved gas, hydroxides, or oxides in water.

| | Silane + | Garnet + | Soda-Pyroxene + | Rutile | Sum |
|---|---|---|---|---|---|
| | | D = 3.91 | D = 3.40 | D = 4.18 | |
| | | 64.6 cc | 36.4 cc | 7.0 cc | 107.0 cc |
| | 2.3 g | 252.6 g | 123.8 g | 29.3 g | 408.0 g |
| 0 | | 104.3 | 53.5 | 11.7 | 169.5 |
| Si | 2.3 | 45.5 | 31.2 | 0.0 | 79.0 |
| Al | | 24.5 | 5.1 | 0.0 | 29.6 |
| $Fe^{3+}$ | | 10.1 | 0.0 | 0.0 | 10.1 |
| $Fe^{2+}$ | | 40.7 | 10.4 | 0.0 | 51.1 |
| Mg | | 13.1 | 4.4 | 0.0 | 17.5 |
| Ti | | | 0.0 | 0.0 | 17.6 |
| Ca | | 14.4 | 14.9 | 0.0 | 29.3 |
| Na | | 0.0 | 4.3 | 0.0 | 4.3 |
| | 2.3 g | 252.6 g | 123.8 g | 29.3 g | 408.0 g |

|        | = Augite + | Plagioclase + | Ilmenite | Sum | Loss |
|--------|-----------|---------------|----------|-----|------|
|        | D = 3.45  | D = 2.72      | D = 4.70 |     |      |
|        | 59.3 cc   | 53.0 cc       | 11.5 cc  | 123.8 cc | |
|        | 204.7 g + | 144.1 g +     | 53.8 g = | 402.6 g | |
| O      | 84.9      | 67.6          | 17.0     | 169.5 | 0.0 |
| Si     | 44.4      | 34.6          | 0.0      | 79.0  | 0.0 |
| Al     | 4.7       | 23.8          | 0.0      | 28.5  | 1.1 |
| $Fe^{3+}$ | 9.9    | 0.0           | 0.0      | 9.9   | 0.2 |
| $Fe^{2+}$ | 29.6   | 0.0           | 19.8     | 49.4  | 1.7 |
| Mg     | 17.0      | 0.0           | 0.0      | 17.0  | 0.5 |
| Ti     | 0.0       | 0.0           | 17.0     | 17.0  | 0.6 |
| Ca     | 14.2      | 14.1          | 0.0      | 28.0  | 1.0 |
| Na     | 0.0       | 4.0           | 0.0      | 4.0   | 0.3 |
|        | 204.7 g   | + 144.1 g     | + 53.8 g = | 402.6 g | 5.4 g |

A similar calculation, which uses the compositions of magnesium-rich garnet and clinopyroxene in a garnet-bearing, olivine-rich peridotite and plagioclase, olivine, and clinopyroxene (diopside) in plagioclase peridotite, causes a volume increase of 19.5 cc and similar losses of metal cations.

## Conversion of Eclogite to Plagioclase Peridotite

Eclogites result from metamorphism of basalt at depth; and the progressive conversion is established by the recognition of intermediate rock types (**Table V-1**). This conversion is illustrated by the following generalized formula.

**Basaltic Crust** ⟶

augite + plagioclase + ilmenite

$$Ca_2Mg_2Fe^{2+}Fe^{3+}AlSi_5O_{18} \cdot Mg_2Fe^{2+}{}_2Si_4O_{12} + NaAlSi_3O_8 \cdot Ca_2Al_4Si_4O_{16} + 2FeTiO_3$$

⟶ **Eclogite**

quartz

= clinopyr. (omphacite) + garnet + rutile + (stishovite)

$$NaCa_2MgFe^{2+}AlSi_6O_{18} + Ca_2Mg_3Fe^2{}^+4Fe^{3+}Al_5Si_9O_{36} + 2TiO_2 + 3SiO_2$$

It is notable that this recrystallization to form eclogite also produces quartz (stishovite). Quartz-bearing eclogites resulting from this conversion often occur in association with deeply-buried oceanic basalts. Eclogites not associated with such transitions lack quartz, an absence that may stem from partial melting and quartz extraction. In any case, not all eclogites appear to be formed through basalt metamorphism. Quartz-free eclogites, underlying continental regions (**Table V-1**), could be remnants of relatively silica-deficient upper Mantle.

If the above equation were reversed to show the conversion of eclogite to plagioclase peridotite (in the same manner as garnet-bearing peridotite is changed to plagioclase as shown in the previous section), then silanes in an expanding-Earth model could be the source of silicon to produce the needed extra silica and recrystallize the eclogites in the form of soda-bearing plagioclase in plagioclase peridotite. For the extra oxygen, we must look to the dense garnets.

Oceanic mantle rocks have little garnet, less than 5 % in garnet peridotite, for example.  Continental mantle rocks often contain more garnet than 15 % (e.g. in the peridotite that is here called eclogite) and sodic-clinopyroxene, omphacite.[82] The average composition of garnet in eclogite is 19.5 % ferric (almandine), 41.1 % calcic (grossular), and 39.4 % magnesian (pyrope).(Alderman 1936)[83]  Aluminous clinopyroxene (omphacite) in eclogite has been shown experimentally to be able to contain up to 8 % $Na_2O$, but usually contains no more than 4 %.

## CONSEQUENCES OF PHASE CHANGES AT DISCONTINUITIES AND THE FORMATION OF MELTS

As discussed in previous paragraphs, the Transition Zone and upper Mantle are layered geospheres separated by discontinuities.  The mineral menu of each varies according to lithostatic pressure, temperature, infusion of volatile elements from below, and the upward displacement and fugacity of immanent elements (**Table V-1**).  Advancing from layer to layer toward the Earth's surface, mineral oxygen and silicon levels tend to rise.

Where melting led to the mobilization of magmas, sills propagate into hydraulically "fracked" horizontal sheets, and magmatic plumes stope their way through the "Moho" before crystallizing in the Crust as batholiths and lesser plutons.

Skobelin's innovative theory of the Mohorovicic discontinuity, presented in Chapter II, depicts it as a primordial solid surface, which through geological time, has been overlaid

---

[82]  *Omphacite, $2NaAlSi_2O_6 \cdot nCaMgSi_2O_6$; or, $NaCa_2MgFe^{2+}AlSi_6O_{18}$*

[83]  *Almandine garnet $Fe_3Al_2(SiO_4)_3$; Grossularite garnet: $Ca_3Al_2(SiO_4)_3$; Pyrope garnet: $Mg_3Al_2(SiO_4)_3$*

with the products of endogenic effusion and exogenic infusion. Separating the dense Mantle from less-dense Crust, the seismic picture of the Moho indicates a zone ranging from 0.5-km thickness under stable continental areas to 1.0-km thickness beneath oceans, and even greater thicknesses in tectonically active regions. (Condie 1982) Let us consider the subject in the light of the behavior of melts.

## Formation of Melts

Melts can be produced either in open systems, where elemental materials may enter or depart, or they may be produced in systems closed to introduction and departure of elemental components. **A solid rock may melt if (1) confining lithostatic pressure is lowered, (2) temperature is raised, or (3) water is added, reducing the temperature at which "first melting" can occur. Manifestly, tectonism and hydrodynamic change can bring about melting; silane infusion is not necessary.** Prevalent thought on closed-system experiments has provided these explanations of melting phenomena, which are useful for interpreting some magmatism.

## Transformations: Alternatives to Magmatism

A fourth factor that brings about melting is the exothermia during mineral transformation. Silane permeation of a plutonic body may result in mineral conversions, exothermic oxidation, and release of hydrogen, silica, and water, all of which mitigate partial or total melting. The important point here is that, where silanes are involved, melting ensues after silane combustion reaches advanced levels. Where more pervasive mineral conversions occur, melting may remove the evidence of earlier mineral replacements. Lesser levels of conversion right alongside, so to speak, may be the only surviving evidence that silanes had been involved at all.

The effects of silane are most pronounced where mafic crustal rocks are converted to *anhydrous* granitic rocks. Little or no volume change is entailed as the reaction of silane gas from fractures in the rock converts ferromagnesian silicates [where sodium ($Na^{1+}$) is available] to quartz and calcic plagioclase to sodic plagioclase. The conversion is a volume-for-volume replacement with little volume change, not a "constant volume" conversion, but one with only a small volume increase.[84]

---

[84]  *Footnotes A and B give examples of balanced mass-for-mass equations in which clinopyroxene and calcic plagioclase in anhydrous, K-free, biotite-free gabbro are interpreted as reacting with silane and sodium ion ($Na^{1+}$) to produce quartz and sodic plagioclase in a granitic rock. In this combination of reactions, free oxygen, which is not available at great depths in anhydrous rock, is required. The reaction cannot proceed in the absence of free oxygen, which must be added along with silane gas to convert gabbro (more-dense rock) to felsic (less-dense) granitic rock by replacement.*

*A.  Mass-for-mass equation for clinopyroxene converting to quartz.*

silane + clinopyroxene + oxygen     =     quartz + FeO + MgO + CaO + water

$$SiH_4 + Ca(Fe,Mg)(SiO_3)_2 + 2O_2 = 3SiO_2 + .5FeO + .5MgO + CaO + 2H_2O$$

32 g +     232 g     + 64 g     =     180 g + 36 g + 20 g + 56 g + 36 g-

           328 g total          =          328 g total

*B.  Mass-for-mass equation for calcic plagioclase converting to sodic plagioclase*

silane + calcic plagioclase + oxygen     =     sodic plagioclase + $Ca^{2+}$ + H

$$4SiH_4 + CaAl_2Si_2O_8 + 2Na^{1+} + 4O_2 = 2NaAlSi_3O_8 + Ca^{2+} + 8H_2$$

128 g +     278 g + 46 g + 128 g     =     524 g     + 40 g + 16 g

           580 g total          =          580 g total

*Both reactions require the availability of free oxygen. In anhydrous, potassium-free, solid rock in the Crust, as in the Mantle, the total oxygen contained in the rock is not available for these reactions, because most of it is bound in oxide and silicate minerals. Nevertheless, in volume-for-volume replacement reactions, extra oxygen is not needed because the replacement process releases enough of it to serve the ongoing process. It becomes evident that knowing the oxygen content of each mineral is critical to understanding silane transformations.*

In the following volume-for-volume equations, hydrogen is neglected because of its insignificant size and because it enters combined in silane and goes out as elemental hydrogen (or possibly as metal hydrides or hydroxides). In the first equation, 100 cc of clinopyroxene (340 g; density 3.4) are replaced by 100 cc of quartz (267 g; density 2.67). In the second, 100 cc of calcic plagioclase (278 g, density 2.78) are replaced by 100 cc of sodic plagioclase (262 g; density 2.62). Then, total mass in grams for each element is compared in the same volumes.

### 1. Volume-for-volume reaction for clinopyroxene converting to quartz.

silane + clinopyroxene → quartz + $Fe^{2+}$ + $Mg^{2+}$ + $Ca^{2+}$ + $H^{1+}$

$SiH_4$ + $Ca(Fe,Mg)(SiO_3)_2$ → $SiO_2$ + $Fe^{2+}$ + $Mg^{2+}$ + $Ca^{2+}$ + $H^{1+}$

(100 cc, D = 3.40)      (100 cc, D = 2.67)

|  |  |  |  |
|---|---|---|---|
|  | O | 140.4 | 142.3 |
| 42.4 | Si | 82.3 | 124.7 |
|  | Ca,Mg,Fe | 117.3 | 117.3 |
| 42.4 g | + | 340.0 g | 384.3 g |
|  | 382.4 g (total) | = | 384.3 g (total) |

## 2. Volume-for-volume reaction for calcic plagioclase converting to sodic plagioclase.

silane + calcic plagioclase + $Na^{1+}$ = sodic plagioclase + $Al^{3+}$ + $Ca^{2+}$ + $H^{1+}$

$SiH_4$ + $CaAl_2Si_2O_8$ + $Na^{1+}$ = $NaAlSi_3O_8$ + $Al^{3+}$ + $Ca^{2+}$ + $H^{1+}$

(100 cc, D = 2.78)        (100 cc, D = 2.62)

|       |      |      |      |      |            |
|-------|------|------|------|------|------------|
|       | O    | 128.0 |     | 127.9 |           |
| 28.2  | Si   | 55.9 |      | 84.1 |            |
|       | Al   | 53.9 |      | 27.0 | (+26.9 Al) |
|       | Ca   | 40.2 |      | 0.0  | (+40.2 Ca) |
|       | Na   |      | 23.0 | 23.0 |            |

28.2 g   + 278.0 g + 23.0 g = 262.0 g +  67.1 g

329.2 g  (total)        =      329.1 g  (total)

The first equation shows that, whereas 42.4 g of Si in silane must be added to 100 cc of clinopyroxene to convert it to quartz, the available volume of clinopyroxene has insufficient oxygen to match the oxygen in 100 cc of quartz. Therefore, an additional 1.9 g of oxygen are needed. On that basis, 101.3 cc of clinopyroxene, containing 83.4 g Si and 142.3 g 0, are needed to produce 100 cc of quartz. At this ratio, 41.3 g of Si must be added from introduced silane.

In the second equation, where calcic plagioclase is replaced by sodic plagioclase, 28.2 g of Si must come from introduced silane. A slight excess of oxygen (0.1 g) occurs in the calcic plagioclase, thus making for a nearly perfect volume-for-volume conversion.

In the combined reactions, however, there is no excess free oxygen. Therefore, silane gas molecules cannot combust to produce water, when anhydrous gabbro is converted to more-felsic granitic rock by silane-replacement. Hydrogen that is released in the process may move to higher levels before encountering available oxygen with which to react.

A different result occurs when proportional replacement of 101.3 cc of clinopyroxene by 100 cc of quartz occurs in rocks being converted to granitic compositions. Here, a small 1.3 cc volume loss compares with essentially zero loss where calcic plagioclase is replaced by sodic plagioclase. This tells us that in anhydrous gabbro or diorite under deformational pressure, as quartz replaces some clinopyroxene, unreplaced clinopyroxene is subjected to internal compressive stress that further deforms it. Volume losses in this way cause cracks, which create additional avenues for further introduction of silanes.

Under similar conditions, by contrast, plagioclase holds its shape. For that reason clinopyroxene tends to be replaced by quartz first, before the replacement of calcic plagioclase by sodic plagioclase. Replacements of orthopyroxene, hornblende, and biotite by quartz also follow this same pattern of minor volume loss. Slightly greater losses occur where quartz replaces hornblende (2.7 cc per 100 cc) and biotite (7.1 cc per 100 cc).

In continental areas distant from spreading centers, the formation of gabbro or diorite magmas yields segregations of primary water that are supplemented by water from ongoing silane combustion. Therefore, after magmatism is initiated at high temperature, indigenous supplies of water may be exhausted, and the remelting of newly-solidified anhydrous rocks impeded. Additional water would have to be generated at deeper levels from peridotite and eclogite to sustain magmatism, or it would have to come from other external sources such as hydrous minerals or from silane reaction with atmospheric oxygen in rock fractures. Where no excess oxygen is available,

reactions are dependent upon the presence of residual water left over from previous magmatic phases.

Magma pressure from behind a solidifying pluton may continue, deforming it cataclastically, allowing further permeation by silanes, another round of reactions, and replacements of newly-crystallized ferromagnesian silicates and plagioclase. Remelts of already crystallized, altered gabbro or diorite occur so long as sufficient heat is produced. But sooner or later the heat is insufficient, and conversion proceeds below melting temperatures. This process is mainly one of replacement and exchange of silica and sodium for iron, magnesium, aluminum, calcium, and titanium, which then rise as hydrides or hydroxides, with available water.

In the absence of potassium, formation of sodic plagioclase from calcic plagioclase consumes silica and yields quartz diorite, tonalite, or trondhjemite, but not quartz-rich granitic rock. In this replacement process (perhaps followed by melting and recrystallization) little volume change occurs. The Crust is not expanded to any great degree where silanes and sodium ($Na^{1+}$) are the chief replacing agents.

## FLOOD BASALTS

We have seen that silane reacts with oxides and silicate minerals of *anhydrous gabbro and diorite* terranes in volume-for-volume reactions without the need for an external oxygen source, and that resultant density changes set the stage for emergent scenery, the rise of mountain ranges, and the growth of continents.

A different situation occurs when *peridotite* is deformed and silanes enter its lattices. Volume-for-volume equations involving the relatively heavy minerals of this waterless rock have sufficient oxygen to give the necessary equal-volume results. The ensuing reactions of silanes with dense silicate minerals generate heat and small amounts of water (e.g. ± .04 % in relation to magma) in the process of creating hydrous magmas of gabbro or diorite in the upper Mantle and lower Crust. These magmas expand, lose mass, and may rise diapirically in the Crust or flood out upon the surface as "flood basalts" (**Table V-2**) These are products of silane reactivity generated in some highly-tectonized crustal sites.

## Table V-2

## The Great Continental Basalt Floods of our Planet

Long enigmatic to geologists, these deposits reach into the millions of cubic kilometres:

| Flood | Continent | Age Range | | | Average Age | | Volume |
|---|---|---|---|---|---|---|---|
| Deccan | Asia | 62- | 66 | my | 65 | my | 1 500 000 km³ |
| Karoo | Africa | 120- | 204 | " | 190 | " | 30 000 000? " |
| Siberian | Asia | 600- | 200? | " | 250? | " | 2 500 000? " |
| Parana | South Amer. | 120- | 140 | " | 130 | " | 790 000 " |
| Ethiopian | Africa | 43- | 9 | " | 28 | " | 400 000? " |
| Columbia | North Amer. | 17.5- | 6 | " | 16 | " | 300 000 " |
| North Atlantic | | 52- | 66 | " | 62 | " | ? |

## Mid-Ocean Volcanism

At mid-ocean ridges, where a thin Crust is split by crustal tension, one would expect the reactions of silanes with peridotite to provide a seemingly endless supply of water, heat, and consequent volcanism. If mid-ocean ridges were merely cases of such rifting with a lowering of pressure on the underside of the lithosphere, conducted heat from deeper levels could lead

to melting; but it would not generate water or produce the great blankets of eruptive rock that have accumulated on the world's ocean floors. The conclusion seems self evident, that simple magma formation could not account for the evident volcanic ferocity or even the great mass of the extruded rock. There is just too much basalt on ocean floors for that explanation. Simple melting of the upper Mantle by primordial heat release could not have produced the evidence that confronts us.

The silane model avoids such problems, however. It converts dense minerals of the lavas comprising oceanic Crust to less-dense minerals. Its combustion creates heat and small amounts of water, which facilitate melting and volcanic eruption. All of these processes feed the growth of Earth's surface by relocating mantle rock to the surface or to sites of deposit within the Crust.

## ANATEXIS

Cataclastic deformation of a solidified gabbro or diorite above an active silane plume from the Mantle would be expected to provide fracture porosity and allow entry of additional silanes. Melting of continental rock to make gabbro or diorite magmas flushes out primary water or water newly-produced from silane combustion. Thus, after initial magmatism at high temperature, additional water should not be readily available to aid in remelting of solidified plutons. Nevertheless, such remelting does occur widely; and water must always have been available, perhaps from below, where peridotite is replaced by silanes.

The reactivity of these silanes appears to explain very well the replacement of ferromagnesian silicates and plagioclase and the creation of felsic rocks. Previously-presented equations show, however, that excess oxygen is not available in diorite or gabbro for clinopyroxene and other ferromagnesian silicates to be replaced by quartz or calcic plagioclase by sodic plagioclase.

Therefore, the occurrence of silane reactions that could generate the heat for anatexis is dependent on whether residual water is present from previous magmatism.

Although anatexis may generate felsic plutons in this way, the parallel process of replacement is at least as significant. Primary mafic gabbro and diorite are converted to more silicic (granitic) end-products below melting temperatures by replacement and exchange of silica and sodium for iron, magnesium, aluminum, calcium, and titanium. These released metals rise as hydrides or, if small amounts of water are available, as hydroxides.

Formation of sodic plagioclase from calcic plagioclase consumes silica. Sodic transformation processes neither produce granitic rock nor expanded rock volume. Therefore, where silane and sodium ($Na^{1+}$) are the chief replacing agents, we do not find either the high mountains or the rifts that characterize potassic terranes.

The foregoing scenario adduces an explanation for the zonation that occurs in western North America on a continental scale. This subject will be discussed further in the next section.

Potassium-bearing eclogites in the deep continental Mantle experience similar reactions to those occurring in oceanic regions lacking potassium. Introduction of silanes into zones of plutonic deformation generates steam and converts dense eclogitic rock to magma, which then rises as diapirs and crystallizes as rock of relatively low density (compared to the eclogites). Silane-generated magmatic rocks of this type include biotite-bearing diorite or gabbro plutons and andesite or basalt volcanoes.

After solidification and flushing-out of any primary or newly formed water to the volcanoes, renewed deformation permits the introduction of additional silanes. Then, biotite-bearing diorite and gabbro can be replaced at temperatures below melting, their ferromagnesian silicates first and their plagioclase later being replaced by quartz, as silica is exchanged for calcium, aluminum, iron, magnesium, and titanium.

Volume losses attendant on the replacement of hornblende and biotite by quartz are greater than the volume losses in comparable pyroxene replacement. Such shrinkage facilitates silane entry. Now, however, a new "wrinkle" is added, the release of potassium from biotite. Instead of calcic plagioclase being replaced solely by sodic plagioclase, K-feldspar and myrmekite may be substituted for the sodic interloper.

K-feldspar has a lower density than plagioclase, and thus, does not require all of the oxygen released by a plagioclase crystal to fill its space. This freed oxygen is then available to react with silanes to produce water and heat. Depending upon conditions, some of this water can precipitate in muscovite as tiny sericite-alterations of feldspars or in separate, large crystals in granite. If sufficient heat and water are generated, newly-produced granite (former mafic diorite or gabbro) can partially melt or completely melt to become magma. Being felsic already, it has lower density compared with the higher densities of surrounding mafic wall rocks. This causes it to rise in a diapiric pluton.

Introduction of silanes in deformed mantle rocks east of the aforesaid quartz diorite line creates granite by replacement of former biotite-rich diorite and gabbro, volume-for-volume. These granites need not have been formed by melting of continental metasedimentary rock masses or by differentiation from mafic magma. Space does not need to be made for them by displacement of older, overlying or adjacent, mafic rocks nor do the mafic wall rocks need to sink along pluton margins to replace space vacated at depth.(Roddick 1982) Granites formed by silica/potassium replacements, in contrast to syenites formed by silica/sodium replacements, nevertheless behave in a similar fashion on a plutonic scale.(Hoeve 1978; Kresten 1988)

Confirmation that K-rich granites have formed because of introduction of silanes is the abundant presence in many of them of myrmekite with coarse quartz vermicules that indicate their gabbro parent.

The foregoing explanation for *in situ* transformation of diorite and gabbro into granites, volume-for-volume, through silane pervasion, leads logically into an explanation for zonation on a continental or plutonic scale.

## GRAND CONSEQUENCES OF SILANE SYSTEMATICS

### THE QUARTZ DIORITE LINE: ZONATION ON A CONTINENTAL SCALE

**L.G. COLLINS**

The fact was brought out in Chapter IV (**Fig. IV-5**) that the distribution of peraluminous (two-mica) granites in western North America is confined to the terrain east of a line that bisects the Sierra Nevada Mountains, veers east to Idaho, and thence runs northwesterly again, into Canada, a divide that coincides approximately with Carey's right-lateral "megashear" (discussed in the EV volume, Plate III). Where plutonic rocks to the west of the line are dominantly mafic, consisting of quartz diorite lithologies (on average), east of the line, they comprise mainly quartz monzonite and granitic lithologies. (Moore 1959; Moore et al. 1963)

The "quartz diorite line" is a chemical divide. In the rocks west of this line, the chief replacing agents are silane and sodium ($Na^{1+}$). To the east potassium ($K^{1+}$) prevailed in replacement processes, both of them with silanes. Further discussion of this zonation is now deferred until we can examine how potassic mantle rock behaves as it is modified by silanes.

## Silane Replacement in Hydrous, K-bearing Solid Rock

Cation replacement in hydrous, potassium-bearing gabbro or diorite has a special characteristic, when it occurs in a rock that contains both biotite and hornblende. Hydroxyl ions ($OH^{1-}$), derived from water, are found in biotite and hornblende, rendering these minerals "hydrous." When their crystal fabrics are deformed, silanes are able to enter and replace the ferromagnesian silicates with quartz. But with these minerals, a new twist is introduced.

Potassium is released from replaced biotite. Instead of calcic plagioclase being replaced solely by sodic plagioclase, some crystals are replaced by K-feldspar and myrmekite. K-feldspar, having a lower density than plagioclase, uses less oxygen per unit volume, and this oxygen is released from the space formerly occupied by solid plagioclase crystals, K-feldspar then filling the space.

As hornblende and biotite are replaced by quartz, both oxygen and water are released and available to react with silanes. The oxygen is the more reactive and yields water, silica, and heat. Water reacts, if there is insufficient free oxygen, producing hydrogen, silica and heat. More heat is generated, of course, with elevated levels of hydrogen combustion. As noted earlier, quartz replacement of biotite and hornblende results in greater volume losses than for similar replacement of pyroxenes, and the creation of void space in this fashion promotes still more replacement.

Insofar as advanced hydrogen combustion creates excess water above what can be consumed by ambient silanes, the excess water can precipitate in muscovite as tiny sericite alterations of feldspars or in separate, large crystals in granite. If sufficient heat and water are generated, the former mafic rock, now converted to granite, may partially or completely melt to magma. Lower density (either as a hot, plastic solid or as magma) in comparison to higher densities of surrounding mafic wall rocks, allows the rise of plutonic diapirs.

Introduction of silanes into deformed, biotite-bearing gabbros east of the quartz diorite line creates granite and quartz monzonite, volume-for-volume. These granitic rocks need not have formed by the melting of thick masses of continental metasedimentary rock or by differentiation from mafic magma. There was no need for space to be made for their initial creation by displacement of older, overlying or adjacent, mafic rocks. Neither did mafic wall rocks have to sink along the plutonic margins to fill space vacated by the risen granite plutons.(Roddick 1982)

**This leads us to realize that granites formed by silica- and potassium-replacements are the potassic counterparts of syenites, which are formed by silica- and sodium-replacements.(Hoeve 1978; Kresten 1988) The abundant presence in many of these granites of myrmekite with coarse quartz vermicules is the result of silane introduction, cation removal, and consequent mineral transformation that has converted original gabbro to granite.**

**Many interesting deductions follow from this observation. For example, the lower densities of granites, relative to the greater densities of gabbros, diorites, quartz diorites, and tonalites west of the quartz diorite line cause the east to rise differentially relative to the west. This produces a westward tilt, such as we find in the Sierra Nevada.**

The above explanation of how granites form from diorite or gabbro, volume-for-volume, by silane replacement leads, then, logically into an explanation of flood basalts, mid-ocean volcanism, anatexis, and pluton zonation.

# ZONED PLUTONS

Zoned granitic plutons with felsic cores and mafic rims are common in the Crust. The prevalent explanations for this relationship invoke magmatic processes. One explanation imagines eruption of progressively more mafic volcanic effusives from a magma chamber in which previous magmatic fractionation of mafic components by gravitational settling is thought to have segregated the source into upper levels of the chamber of granitic and lower levels of mafic components.(Hildreth 1979, 1981; Smith 1979) This would set up chemically zoned peripheries to ejection conduits.

In this magmatic model, it logically follows that magma chambers with insufficient volatile components to erupt must crystallize at depth to form coarse-grained plutons in which vertical compositional zoning is preserved. However, plutons with felsic, silica- and alkali-rich tops and relatively mafic bottoms do not establish that the zonation is attributable to gravitational settling of mafic components. The zonation could also result from preferential concentration of silanes at the top of a magma chamber.

In addition, many plutons are also zoned laterally. In some of these, mafic rims result from an abundance of lenticular, mafic xenoliths, which decrease and disappear toward the cores. It is not unusual for such xenoliths to have the same chemical makeup as the wall rocks, which therefore, are the most likely sources for mafic enrichment.(Collins 1989)

In other zoned plutons xenolith compositions may be unrelated to wall-rock compositions, and therefore, presumed, in the magmatic model, to be from great depths. Still other zoned plutons show gradational mineral change from mafic rims to felsic cores but without mafic xenoliths. All zoned plutons, however, are interpreted in the magmatic model to reach their compositional variations by gravitational fractionation, either

by settling of mafic oxide and silicate minerals or settling of mafic xenoliths (whatever their source) or both.(Bateman & Wahrhaftig 1966)

The problem with these interpretations is that no evidence exists to indicate accumulations of dense, heavy, mafic minerals or xenoliths at the base of zoned granitic plutons. Both gravity and seismic data should support the presence of the denser mafic rock below granitic plutons if it were present.

The aforesaid deficiency is demonstrated by observations made on the **Duncan Hill pluton in Washington State**. This pluton is tilted and has a bulbous top, which tapers downward. Erosion has exposed it in cross-section from top to bottom. Granite at the top grades upward and outward into rhyolite. Felsic and lamprophyric dikes, which have been injected into the cover and wall rocks, likely comprised feeders to an overlying volcano. Gradationally downward the pluton grades from granite through a myrmekite-bearing granodiorite and then to tonalite and finally to diorite at the bottom, where it has tapered to a narrow width.(Cater 1982; Hopson et al. 1970)

None of the gravitational segregation anticipated by the magmatic modellers, that is to say, mafic minerals at the bottom in layered cumulates, is to be found. Neither are there streams of settling mafic xenoliths along the margins. **Clearly, gravitational processes in magma fail to explain relationships in the Duncan Hill pluton. Silane-generated transformations do explain observable geology quite well.**

Another example of a zoned pluton occurs in the southern **Snake Range of Nevada**, where biotite granodiorite in the rim grades inward to a two-mica quartz-monzonite granitic core.(Lee & Van Loenen 1971) The mafic rim contains 25 % biotite and > 4.5 % Ca0. Some mafic xenoliths in the rim contain up to 32% biotite. Then, gradationally from rim to core, biotite decreases to less than 0.5 % as Ca0 also decreases to less than

0.5 %. As the biotite disappears, microcline increases, and as the CaO decreases, quartz increases. Thin sections traversing from rim to core of the pluton show that as K-feldspar first appears and then increases, myrmekite also appears and increases. Then, about two-thirds the distance to the core, plagioclase becomes very sodic (albite), and myrmekite disappears.

These textural changes suggest that early magmatic, mafic, biotite granodiorite, that solidified adjacent to wall rocks, was modified later by hydrous fluids containing silanes so that biotite and relatively calcic plagioclase were replaced to form quartz, K-feldspar, and myrmekite. Subsequently, where the rock composition in the core became granitic and where combustion of silanes produced sufficient heat and steam, the solidified, but modified rocks remelted to form a magmatic granitic core.

**Thus, once again gravitational processes in magma are seen to be wholly unable to explain relationships, whereas silane systematics provides a simple and elegant explanation.**

Other examples abound. In the myrmekite-bearing **Donegal granites of northwest Ireland** (Pitcher et al. 1953, 1959, 1972, 1982), for example, hornblende tonalite in the rim of the Thorr pluton grades through a myrmekite-bearing biotite-hornblende granodiorite to myrmekite-bearing biotite granite (the "Gola" granite) in the core.[85]

In addition to the examples shown in the footnote, a spectrum of zoned, alkali-rich plutons is prominent in many places in the **White-Inyo Ranges of east-central California. These include the Eureka Valley, Marble Canyon, Joshua Flat, Beer Creek, Cottonwood Creek, Sage Hen Flat, Birch Creek, and Papoose Flat plutons.**(Nelson & Sylvester 1971; Sylvester 1978; Sylvester et al 1978; Idiz 1981; Crowder & Ross

---

[85] *In outer portions of the pluton, biotite and hornblende are replaced by quartz and epidote, as plagioclase is replaced by K-feldspar and myrmekite. Cross-cutting, granitic, felsic dikes show advanced stages of these replacements.*

1973)[86]  Early magmatic intrusions are quite mafic; later replacement has created myrmekite-bearing granitic cores.

Structural evidence suggests that some of the older, mafic, magmatic plutons in the  western United States made space for themselves by forceful diapirism through the overlying sedimentary rocks.  This shows well in the **Sierra Nevada and Klamath Mountains, California** (as described in the EV volume), where the deformed and metamorphosed remnants of former cover rocks are now disrupted, breached, and isolated as roof pendants.

[85] *cont.   Locally, in the Gola granite, orbicular segregations of granite are rimmed by mafic overgrowths.(Pitcher et al. 1987)  These suggest that during an early stage of granite formation, steam generated by silane reaction was vented chaotically around blocks of granite ("horses" in miners' terms).  Rapid pressure reduction caused fugaceous cations to reprecipitate in radial patterns as new mafic mineral deposits on the horses.*

*The Ardara pluton adjacent to the Thorr pluton follows the same zonal pattern, although its core does not become as granitic as the Gola granite.  Nearby is the myrmekite-bearing, felsic, biotite Rosses granite pluton, which penetrates the Thorr pluton and grades from rim to core through four granitic pulses; each successive pulse is more felsic toward the core.  Each granite shell, successively inward, as it was replaced by silanes, may have reached melting temperatures, then solidified, and in later stages been replaced again by silane, thus forming additional K-feldspar and myrmekite.  During these replacements stages, the solidified masses likely moved up plastically through the center of the pluton, deforming minerals with microfractures through which rising silanes could move and modify the rock composition still further.*

*The Main Donegal Granite that also borders the Thorr pluton does not have a centrally zoned felsic core, but nevertheless, is compositionally zoned in parallel sheets of younger felsic layers and older mafic layers.  Each relatively more-felsic layer shows progressively more deformation and replacement by K-feldspar, myrmekite, and quartz.  These replacements converted older, more-mafic layers to more-felsic compositions.*

[86] *Limestone-granite contacts surrounding the myrmekite-bearing granitic Papoose Flat pluton have a marble zone less than two metres wide.  On that basis, following formation of the granite from older mafic rocks by silane replacement, newly created felsic portions were squeezed upward, plastically entering cracks and fractures at relatively low temperatures. These temperatures would be below the melting interval for granite and too low to cause significant metamorphism.  In the process, the same silane-bearing fluids that replaced the mafic rocks of the pluton also replaced plagioclase in the mafic wall-rock gneisses to form K-feldspar metacrysts (bordered by myrmekite) whose composition, size, and appearance match those of metacrysts (supposed-phenocrysts) of K-feldspar (bordered by myrmekite) in the granite pluton.*

*The Skidoo and Hall Canyon two-mica granites in east-central California must also have been plastically introduced as low-temperature granite "mushes" because dikes, extending from these plutons into surrounding limestones and shales, show essentially no contact metamorphism [Griffis 1987; Hunt & Mabey 1966].  Some of these granitic dikes have myrmekite with very coarse quartz vermicules, indicating a former source from a gabbro.*

*Examples of terranes in which biotite-rich diorites and gabbros were theoretically replaced by silanes to form granite, but subsequently melted, include the Kern Knob pluton near Lone Pine, California, (Griffis 1987) and several granite plutons in Arizona (Reynolds 1980).  The myrmekite that would have been the clue to prior replacement has been destroyed by the remelting stage in development of two-mica granites.*

The opposite is the case with many younger granitic plutons, which show no evidence of having forcefully shoved aside the mafic plutons to make space for themselves. For example, gradational contacts (1 to 2 km wide) occur between the myrmekite-bearing Sacatar quartz diorite and the Summit gabbro of the Isabella region of the southern Sierra Nevada.(Taylor, B.E. et al 1986)[87]

## UPWARD MOVEMENTS OF MAFIC COMPONENTS

In the formation of granitic rocks *in situ*, the primary direction of movement of mafic components must be upward. In early magmatic phases in which gabbro and diorite magma are formed, elements extracted from replaced Mantle rocks are carried upward in magmas that may extend to the surface in basalt and andesite volcanoes. In these magmatic phases some differentiation by crystal settling of heavy minerals is possible so that volcanic rocks derived from tops of magma chambers may be slightly more silicic and felsic than compositions of underlying plutons. Continued upward movements of silanes, potassium, and sodium in these magmas, however, could also create replacements of early formed crystals so that the tops become silicic and felsic. In that way, metacrysts of K-feldspar can replace early plagioclase phenocrysts or enclose and replace broken fragments of early formed crystals.

Where continued upward movements of silanes enter earlier, partly- or completely-crystallized rocks in the top of a pluton, elements displaced to hydrous fluids would continue upward. Moreover, even if crystal settling of dense minerals should occur in very early phases, these minerals would also be replaced later by upward moving silanes. Therefore, the net motion of mafic components is vertically upward in spite of local, possible, gravitational settling of some mafic crystals.

[87] *My analysis of 45 thin sections collected across these gradational contacts establishes that K-feldspar gradually appears along with myrmekite with coarse quartz vermicules. The concomitant substitution of mafic silicates with quartz indicate that the Sacatar quartz diorite and monzodiorite, next to the Isabella pluton, California, were derived from the Summit gabbro (Fig. IV-7).*

Evidence for upward movements of mafic components comes from the association of mafic lamprophyre dikes with myrmekite-bearing granitic rocks, from basalts that have anomalously high $^{87}Sr/^{86}Sr$ ratios, from rare-earth compositions of volcanic rocks, and from distributions of oxygen isotopes in plutonic and volcanic rocks.  Each of these items is discussed separately below.

## Evidence from Lamprophyres

Ultramafic composition, which characterizes lamprophyric dikes, contrasts with felsic composition of granite, and inclines geologists to interpret lamprophyres as Mantle derivatives,(Rock et al. 1991) despite the observable association of many lamprophyres with granitic rocks.

For example, at the top of the **Duncan Hill pluton**, numerous dikes, some of them lamprophyric, extend into wall rocks adjacent to the granite-rhyolite pluton.  Yet, lamprophyre dikes are notably absent in the basement rocks underlying the pluton, in the narrowed root, in the midsection, or in the bulbous granite cupola.

These dikes should be represented, if their mafic components ascended from a Mantle source, in all these places.  Their exclusive association as emanations from the top of the pluton implies that they represent mafic components extracted from the original diorite of the pluton, when it was converted to granite.  These mafic components, displaced during silane replacements, would have moved upward in hydrous fluids, ahead of the active replacement front, whence they would be on hand to be deposited within lamprophyre dikes.  Later, the rise of the granitic diapir would overtake and partly resorb them.

A similar situation occurs in the **Spanish Peaks of Colorado.** (Jahn 1973; Jahn et al. 1979; Johnson 1961, 1968)  Here, numerous mafic lamprophyre dikes radiate from a central stock like spokes of a wheel. The natural expectation in a magmatic model, that the stock should also be mafic, is unfulfilled: it is granite. This makes no sense unless the lamprophyres represent derivative mafic components, which have been displaced by silane transformations, when the original mafic rocks of the central pluton were converted to granite *in situ*.

A relationship may be deduced between lamprophyre dikes and some basaltic lava flows in western United States wherein both can be recognized as derivatives of the same mafic rocks converted to granite *in situ*.[88]

## Comparison of lamprophyre dikes with Cenozoic basalts

Isotopic age-dating of lamprophyre dikes from Cenozoic basalts throughout the western United States often results in anomalously old Rb/Sr age dates, some greater than the Earth's age! The Rb/Sr ratios are too low to account for the generation of strontium in the observed $^{87}Sr/^{86}Sr$ ratios. At the same time the strontium ratios, unexpectedly, are above permissible levels (0.705 to 0.710) for Mantle-derived basalts, which have $^{87}Sr/^{86}Sr$ initial ratios of 0.702 to 0.704.(Peterman et al. 1970; Leeman 1970, 1974; Hedge & Noble 1971; Kudo et al. 1971; Laughlin et al. 1972; Pushkar & Condie 1973; Collins 1988a).

The conventional explanation for these anomalous data is variation in mantle levels of Rb and $^{87}Sr$. Contamination is ruled out to explain the low values on the grounds that there are such low Rb/Sr ratios in the basalts, that they could not have

---

[88]  *In the outer 200 metres of the Sacatar quartz diorite (Fig. IV-7), there are numerous black, basaltic lamprophyric dikes (4 cm to 1.5 m wide). These dikes contain about 72 % hornblende, 5 % biotite, 23 % plagioclase, and a trace of epidote. They are lamprophyric, with hornblende phenocrysts rather than plagioclase. The groundmass consists of fine-grained plagioclase and hornblende. Chemical analyses of fine- and relatively coarse-grained portions of the dikes indicating enrichment in $MgO$, $Fe_2O_3$ (total Fe), $CaO$, Mn, Co, Ni, Cu, Pb, and Zn, but depletion of $Al_2O_3$, $K_2O$, $Na_2O$, Rb, and Ba in comparison with replaced portions of the Kernville and Sacatar plutons; $TiO_2$, $P_2O_5$, and Mo are about the same. As there is enrichment of these elements and oxides in the dikes; and as the same elements and oxides are subtracted from the myrmekite-bearing plutons; the dike chemistry, logically, must incorporate these relocated components.*

assimilated radiogenic material during ascent through the Crust.(Hedge & Noble 1971)

Comparison of the Cenozoic basalt data and lamprophyre data from the Kern River (Isabella) area, where I have a collection of chemical and petrological data suggests different interpretations. The following anomalies, which apply to both the basalt terranes and the Kern River terrane, are cogent to the problem:

(1) The age dates are scattered.[89]

(2) Plotted data of Rb/Sr ratios versus $^{87}Sr/^{86}Sr$ ratios have unusual slopes.[90]

(3) $^{87}Sr/^{86}Sr$ ratios are variable in the same rocks.[91]

(4) $^{87}Sr/^{86}Sr$ ratios are high relative to Sr concentrations.[92]

[89] *Data point scatter: Rb and Sr isotopic data obtained from Cenozoic basalts have a broad scatter rather than plotting along isochrons. (e.g., Hedge & Noble 1971). Therefore, determining $^{87}Sr/^{86}Sr$ initial ratios is difficult or impossible.*

*If an initial $^{87}Sr/^{86}Sr$ ratio of 0.703 for a supposed Mantle source of the basalts is assumed, then a theoretical isochron of 1.5 by can be drawn as a reference line on the assumption that the preCambrian source rocks are 1.5 by old. Also, for comparison, a theoretical isochron of 4.6 by can be drawn to represent the age of the Earth. Scatter of age dates for the Cenozoic basalts leads to some ages greater than the age of the Earth. Therefore, some process other than what is normally found in magmas has disrupted the normal distribution of Rb and Sr isotopes, or a very unusual distribution of Rb and Sr isotopes occurs in the supposed Mantle source. Introduction of silanes could do this.*

[90] *Plotted data of Rb/Sr ratios versus $^{87}Sr/^{86}Sr$ ratios have unusual slopes. Where basalts have low Rb/Sr ratios, they also have relatively high $^{87}Sr/^{86}Sr$ ratios. Furthermore, where basalts have high Rb/Sr ratios, they also have relatively low $^{87}Sr/^{86}Sr$ ratios (Hedge & Noble 1971; Pushkar & Condie 1973). This trend is in opposition to Rb-Sr isochron data generally used for measuring age dates.*

[91] *$^{87}Sr/^{86}Sr$ ratios are variable in the same rocks. In some volcanic terranes $^{87}Sr/^{86}Sr$ ratios in lava flows, erupted at different times from the same volcano, are markedly different, ranging from 0.704 to 0.708 (Laughlin et al. 1972). Moreover, in some lava flows $^{87}Sr/^{86}Sr$ ratios vary laterally and vertically within the same flow.*

[92] *$^{87}Sr/^{86}Sr$ ratios are high relative to Sr concentrations. In many Cenozoic basalt terranes high $^{87}Sr/^{86}Sr$ ratios are associated with low strontium concentrations, and low $^{87}Sr/^{86}Sr$ ratios are associated with high strontium concentrations (Kudo et al. 1971; Leeman 1974). In some basalt terranes, however, high $^{87}Sr/^{86}Sr$ ratios are also associated with high strontium concentrations (Hedge & Noble 1971).*

If these four unusual trends are compared with the trends of isotopic data for lamprophyre dikes that cut the Sacatar monzodiorite of the Kern River terrane, similar relationships produce a scatter that precludes determination of an initial ratio or age date. Similar lamprophyre samples from the rim and core of the same dike collected less than one metre apart are found to have marked differences in $^{87}Sr/^{86}Sr$ ratios (e.g. 0.70522 vs 0.70600) and samples with higher strontium concentration (e.g. 563 ppm) exhibit lower $^{87}Sr/^{86}Sr$ ratio (e.g. 0.70522) while samples with lower strontium concentration (e.g. 507 ppm) have higher $^{87}Sr/^{86}Sr$ ratio (e.g. 0.70600). Moreover, the particular lamprophyre samples in this example have $^{87}Sr/^{86}Sr$ ratios that are high relative to $^{87}Sr/^{86}Sr$ ratios in rocks having similar strontium concentrations from other terranes.

The explanation for the strange Rb-Sr isotopic data is as follows. The original biotite in unmodified portions of the original diorite or gabbro would have contained most of the rubidium and proportionately only tiny amounts of strontium. In contrast, the original plagioclase would have contained most of the common strontium and little rubidium. Considering that $^{87}Rb$ decays to form $^{87}Sr$ and that only tiny amounts of common strontium crystallize in biotite, unmodified biotite in Sacatar quartz diorite would have contained high $^{87}Sr/^{86}Sr$ ratios. In contrast, $^{87}Sr/^{86}Sr$ ratios in original plagioclase would have been near the initial ratio and relatively low because small amounts of $^{87}Rb$ in this case would contribute only minor amounts of its daughter $^{87}Sr$ to raise the $^{87}Sr/^{86}Sr$ ratio.

These relationships are important in understanding how rubidium and strontium would have moved, and in what proportions during introduction of silanes. Biotite is replaced early by quartz during introduction of silanes and prior to replacement of plagioclase by K-feldspar and myrmekite (Collins, 1988a). Therefore, the small volumes of strontium in biotite with high $^{87}Sr/^{86}Sr$ ratios would be released from altered Sacatar quartz diorite before common strontium with low $^{87}Sr/^{86}Sr$ ratios is released from the plagioclase.

Rb$^{1+}$ is also released during replacement of biotite by quartz, but this Rb$^{1+}$, whose chemistry is similar to K$^{1+}$, tends to remain behind in residual biotite or to follow K$^{1+}$ into nearby K-feldspar, whereas released Sr$^{2+}$, whose chemistry is similar to Ca$^{2+}$, tends to follow Ca$^{2+}$ as it is carried out of the system. The released Sr$^{2+}$ moves in hydrous fluids that also carry the released ferromagnesian elements from replaced biotite. The logical site of deposition for this material is in nearby lamprophyre dikes or in basaltic flows at the Earth's surface. Little silica accompanies these fluids, because most of it replaces biotite to form quartz, and therefore, the resulting lamprophyre dikes or basaltic flows are silica deficient (e.g., Kudo et al. 1971).

If the patient reader has been able to absorb the intricacies of the foregoing, he should be willing to concede that the concept of magmatism cannot explain the observed conditions, and, on the other hand, that **the data are satisfied if both lamprophyres and basalts are derived by conversion of diorite and gabbro to granite *in situ*.**

Rb$^{1+}$ released during replacement of biotite by quartz is either retained in residual, recrystallized biotite or transferred with released K$^{1+}$ to K-feldspar, which replaces plagioclase. Less than normal amounts of Rb$^{1+}$ are normally transferred in hydrous fluids feeding lamprophyre dikes or basaltic flows. Therefore, in deformed and replaced portions of the Kernville and Sacatar plutons during early stages of their replacement by silanes, a differential movement of $^{87}$Sr relative to $^{87}$Rb must have occurred.

This differential movement is natural fractionation, which results in extraction from biotite of small volumes of strontium with high $^{87}$Sr/$^{86}$Sr ratios and at the same time even smaller amounts of rubidium. Therefore, Rb/Sr ratios in fluids moving out of modified rocks are very low (0.02 to 0.10). The system is open where replacement by silanes occurs, and that permits variations of $^{87}$Sr/$^{86}$Sr ratios, Rb/Sr ratios, and absolute strontium and rubidium values to occur in response to

subsurface hydrodynamics through time and from place to place in the lamprophyre dikes and basalt flows.

**In any case, the introduction of silanes eliminates the need for any unusual Mantle magma source with low rubidium and high $^{87}Sr/^{86}Sr$ ratios. It also provides the first comprehensive explanation for anomalous data reported here on Cenozoic basalts and lamprophyres from many parts of western United States and the world.**(Rock et al. 1991)

## Evidence from Rare-Earth Studies

Evidence for upward transfer of mafic components from underlying rocks is circumstantial insofar as it is derived from rare-earth studies. This is a result of the inhibition to direct examination because of the enormous scale of rock masses that are involved in the transfers of elements. Where samples of volcanic rocks can be obtained, generally it is impossible to determine magma sources. The assumption is usually made for a magmatic model that basalt derives from partial or complete melting of rocks of equivalent or more mafic composition. Thus, andesite would be obtained from diorite or more mafic magma. Obtaining mafic volcanic rock from more granitic source rocks would be illogical in a magmatic model.

In mafic volcanic and plutonic rocks, heavy and light rare-earth elements are primarily concentrated in pyroxenes, hornblende, sphene, and apatite and less abundantly in plagioclase. In contrast, in granitic volcanic and plutonic rocks, light rare-earth elements are enriched, and heavy rare-earth elements are diminished in biotite, apatite, K-feldspar, and sodic plagioclase. Heavy rare-earth elements are concentrated, however, in zircon and garnet. Therefore, processes that change abundances of these minerals affect amounts of rare-earth elements in the whole rock.

The observed contrasting distribution patterns between mafic and felsic end-members of plutonic and volcanic rock suites are usually explained in magmatic models by subtracting heavy rare-earth elements during crystal fractionation of pyroxenes, hornblende, sphene, and calcic plagioclase and by upward migration of light rare-earth elements in residual fluids.

An alternative method that is used is to fractionate the elements by partial melting, retaining heavy rare-earth elements in a residue of pyroxenes, hornblende, sphene, and calcic plagioclase while light rare-earth elements move up in a melt. If either magmatic process occurs, then volcanic rocks, which are supposed to be derived from the top of a magma chamber, should, on average, contain the same or more light rare-earth component and the same or lesser heavy rare-earth component than the magmatic source.

When average rare-earth compositions for plutonic and volcanic rocks are compared, however, volcanic rocks of equivalent compositions have higher amounts of heavy rare-earth elements than predicted by fractionation. Ultramafic lamprophyre dikes are, in particular, enriched in light and heavy rare-earth elements.(Rock et al. 1991) The logical explanation for this anomaly is that **silanes, in replacing ferromagnesian silicates by quartz, have released both their heavy and light rare-earth components, which have then been carried upward in hydrous fluids to concentrate in the lamprophyre dikes and overlying mafic volcanic rocks.**(Collins, 1988a)

## Evidence from Oxygen Isotope Studies

The relationships between the isotopes of oxygen[93] help to explain why myrmekite-bearing granite that is formed below 600°C has high $^{18}O/^{16}O$ ratios. Above 600°C rocks partially or completely melt to become magmas. At these high melting temperatures fractionation of oxygen isotopes between aqueous

fluids and crystals is so small that differences between $^{18}O/^{16}O$ ratios are insignificant. This means that simple fractional crystallization should produce no differences in oxygen isotope ratios between late-stage magmas and primary magmas and that plutonic and volcanic rocks of equivalent compositions should have the same oxygen isotope ratios. This, of course, is not what is observed.

Fractional crystallization by gravitational settling of iron oxides (magnetite) or ferromagnesian silicates, as proposed in magmatic hypotheses for the origin of granitic rocks, produces the wrong effect to explain the observed partitioning of oxygen isotopes between volcanic and plutonic rocks. Gravitationally-settled mafic minerals contain less $^{18}O$ than the residual liquid, and therefore, the top of a magma chamber that is enriched in silicic and felsic components should also be enriched in $^{18}O$ relative to $^{16}O$.

Fractional melting produces the same wrong result: volcanic rocks that are derived from the top of a magma chamber should have higher $^{18}O/^{16}O$ ratios than the residual plutonic rocks remaining at depth. As this is not the case magmatic fractional melting does not solve the problem any better than fractional crystallization!

---

[93] *Partitioning of the heavy oxygen isotope, $^{18}O$, and the light oxygen isotope, $^{16}O$, between crystallizing minerals and available water (the water reservoir) is controlled by temperature. This partitioning is generally expressed as an $^{18}O/^{16}O$ ratio and compared to a fixed value, which is called the "standard mean ocean water" (SMOW), (Taylor & Epstein 1962). Deviations from the standard value are expressed in parts per thousand ($^o/_{oo}$) and yield values of +5.4 to +5.8 $^o/_{oo}$ in Mantle-derived gabbro. For rocks of increasing granitic compositions, values range up to +10 $^o/_{oo}$ in granite and +13 $^o/_{oo}$ in pegmatite or peraluminous granite. By comparison, volcanic rocks have a lower range, extending from +5.4 $^o/_{oo}$ to only +9 $^o/_{oo}$ in obsidian or rhyolite.*

*The lower range in volcanic rocks is unexpected on the basis of magmatic hypotheses. Increases in $^{18}O/^{16}O$ ratios generally mean that $^{16}O$ has been lost from the whole rock, because the lighter $^{16}O$ isotope diffuses faster in aqueous fluids than the heavier $^{18}O$ isotope and because the heavier $^{18}O$ isotope forms a stronger crystalline bond than its lighter sister. Fractionation of the two isotopes in exchanges between mineral and water is greater at low temperatures (0 - 600°C) than at higher temperatures. (Garlick 1966; Taylor 1968)*

Only rocks affected by silane replacement at temperatures below 600°C should produce significant fractionation of $^{18}O$ and $^{16}O$ between plutonic and volcanic rocks. Introduction of silanes brings in silica that forms a stronger bond with $^{18}O$ than with $^{16}O$. This bonding concentrates $^{18}O$ in quartz that replaces the ferromagnesian silicates. Moreover, conversion of silanes to quartz plus water provides aqueous fluids which carry out the mafic components. The greater rate of diffusion of $^{16}O$ in these aqueous fluids relative to $^{18}O$ means that greater amounts of $^{16}O$ relative to $^{18}O$ are released from ferromagnesian silicates and accompany displaced mafic components that move upward to overlying volcanic rocks. In that process a lower $^{18}O/^{16}O$ ratio is created in volcanic rocks than in plutonic rocks. Voilà! The problem is solved!

## Evidence from Discordant Zircon Populations

Uranium-lead (U-Pb) isotopic age dates are commonly determined from uranium-bearing minerals, such as zircon, monazite, uraninite, orthite, allanite, apatite, or sphene. The age dating is based on radioactive decay schemes in which the isotope $^{238}U$ (half life: 4.468 by), converts to $^{206}Pb$, and the isotope $^{235}U$ (half life: .7038 by), converts to $^{207}Pb$. If ratios of $^{206}Pb/^{238}U$ and $^{207}Pb/^{235}U$ in any of these minerals are measured, the age of a rock can be estimated. The estimation is obtained from a graph on which a curve is plotted, called the "concordia." The concordia consists of the loci of all data points for which $^{206}Pb/^{238}U$ ages are equal to $^{207}Pb/^{235}U$ ages (**Fig. V-2**; Wetherill 1956) This is shown in the diagrams on page 194.

**Case 1:** Each aligned set of points represents an age of crystallization. Uranium contribution by overgrowth is small compared to that in mineral cores. Uranium in rims has a $^{238}U/^{235}U$ age of 2.5 by, but lacks lead. This combination creates a fictitious apparent age. Progressive uranium addition to the rims creates increased divergence of discordant lines from the concordia.

**Case 2:** Each terrane had the same primary age of 2.5 by (upper intercept). The time of replacement is indicated as the present (at zero years, lower intercept).

**In case 1** a primary rock has aged to 2.5 by. and its Pb/U isotopic data in zircon populations all plot as a single point on the concordia at 2.5 by. At that time the rock was sheared, and uranium and zirconium (and thorium) were released from ferromagnesian silicates, where they were replaced by quartz. They deposited then as zircon overgrowths on existing larger zircon and uraninite crystals.

**In case 2** the primary rock had only trace amounts of zircon and uraninite scattered as tiny crystals throughout. Unlike case 1, proportionally more abundant ionic Zr and U are locked in lattice structures of the primary ferromagnesian silicates.

As in case 1, the Pb/U isotopic ratios in zircon and uraninite and in the ferromagnesian silicates all fall on the concordia at a single point representative of 2.5 by. When ferromagnesian silicates in this rock are sheared and replaced by silanes, most released zirconium and uranium deposit as new zircon crystals devoid of older zircon or uraninite cores. For these crystals the geologic clock is reset to zero, and their Pb/U ratio plots on the concordia at the zero point.

Measured ages are accurate (1) if $^{238}U$ or $^{235}U$ and their intermediate daughter products have not escaped or entered the crystals, (2) if corrections are made for included lead isotopes, and (3) if, also, no radiogenic daughter products are included at the time of the original crystallization of the rock.(Rankama 1963)

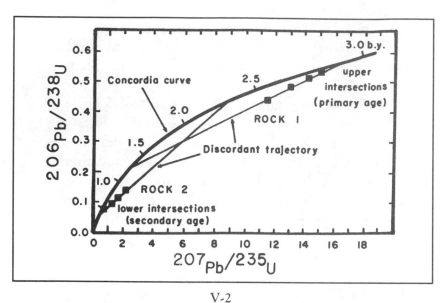

V-2

CONCORDIA DIAGRAM SHOWING DATA POINTS OF
DISCORDANT ZIRCON POPULATIONS FROM ROCKS OF TWO
DIFFERENT AGES.

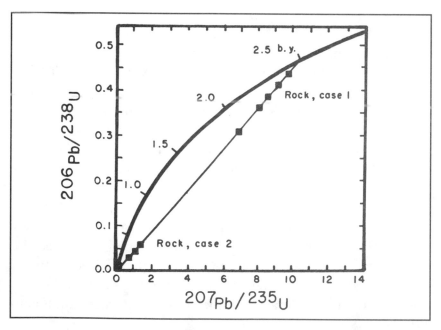

V-3

CONCORDIA DIAGRAM FOR DISCORDANT ZIRCON
POPULATIONS FROM TWO TERRANES IN WHICH SILANES
HAVE EFFECTED REPLACEMENTS AND RADIOMETRIC
ZIRCON DATA POINTS HAVE BEEN RESET.

Zircon is commonly used for U-Pb age measurements because of its crystal stability and because uranium itself readily fits into the zircon lattice, whereas its decay products (including lead) do not. In some granitic rocks all zircon dates (which may come from a diverse population of many different sizes and magnetic properties) conform to the concordia. In others the Pb/U isotopic ratios vary from the concordia, generally falling on a "discordant" straight trajectory below the concordia (**Fig. V-3**).

The trajectory has for each of two different rocks on **Fig V-3** two intercepts on the Concordia. The upper line represents the older rock; and its upper intercept is interpreted as giving the age of the original rock, while its lower intercept is interpreted to have no geological significance.(Faure, 1986) Many different models have been formulated to explain these discordant data points.

When data points for discordant Pb/U isotopic ages fall near the origin on the concordia graph (**Fig. V-3**), the lower trajectory, if assigned any significance, is taken to be the age of the intrusive source of discordant zircon. The upper trajectory is interpreted as the age of inherited zircon cores.

The possibility that zircon crystals retain the same apparent ages after uranium addition is considered inadmissible. There is usually no apparent uranium source in the host rock to provide a uranium infusion to permeate the whole magmatic rock body. Discordance of zircon populations, therefore, in magmatic models is ascribed to local causes of unknown nature.

This contrasts with attribution of zircon to crystallization from magma. There, the positions of zircon populations along the discordant line represent *averaged ages of older (primary) zircon crystals with varying rim overgrowths of younger zircon*. In all magmatic interpretations discordant zircon populations are assumed to be totally magmatic in origin, whether they are modified by lead diffusion or result from secondary overgrowths.

The secondary overgrowths are interpreted to be crystallized from a newly-formed magma that has incorporated zircons from an older, solidified, magmatic rock.

## Inadequacies of Magmatic Models

This interpretation, however, raises the first inadequacy of magmatic models by demanding the astringent qualification that the older zircon cores would necessarily have been sufficiently refractory not to have melted as they were assimilated by intruding magma.

Other arguments show that these magmatic models are inadequate to explain discordant zircon populations. This is true whether the data points fall on the upper or lower trajectories. Let us look at the inadequacies of lead-diffusion models first.

Diffusion of lead out of zircon is an unproven assumption and illogical for the following reasons: First, if lead diffusion were a significant factor at low temperatures, then no preCambrian rocks should ever contain zircon crystal populations that plot on the concordia. The long time interval in the preCambrian should have allowed lead to diffuse out of the crystals, so that all Pb/U isotopic data from preCambrian rocks would be discordant. This is not the case.(Gebauer & Grüenfelder 1973, 1977, 1979; Grauert et al. 1974)

Second, if diffusion of lead were the cause of discordant data, then the slightly heavier $^{207}$Pb isotope would have to diffuse faster than the lighter $^{206}$Pb isotope in order to produce the observed patterns. ***Faster diffusion of a heavier isotope is contrary to known isotope migration behavior.***

Now, let us look at the problems that result if older zircon crystals were to be incorporated into younger granitic magma. There is the physical problem of first extracting zircon crystals

from an older rock and then of mixing them uniformly into younger magma. If the younger magmatic rock arrives as a separate intrusion that assimilates an older rock, there ought to be regions in the younger rock where uncontaminated magma produced new zircons without any older cores. These should have Pb/U isotopic ages that would plot only on the concordia. No such place has ever been found. **Zircon populations in younger granites that intrude older granites are always discordant.** If the assimilation model is correct, thorough mixing between the concordant and discordant zircon populations is implied, a coexistence that is illogical and improbable. The physical problem of segregating whole, tiny unmelted zircon crystals from the matrix of a solid igneous rock body, while large crystals of major minerals of similar melting temperatures (pyroxenes, hornblende, calcic plagioclase) are completely melted, would seem insuperable.[94]

Another way to view this is to consider whether $Zr^{4+}$ ions would be dissolved in a melt of introduced magma, while older zircons in the invaded rock are not soluble in this same magma. In order for the introduced magma to contain abundant $Zr^{4+}$ ions that are eventually deposited as overgrowths, hot fluids in a former zircon-bearing rock had to dissolve or melt zircons from some source when temperatures became hot enough to convert the rock to magma. It would be impossible for older zircons to be preserved under such conditions. Obviously the magmatic model could not give the observable zircon distribution.

---

[94] *Magnesium-rich ferromagnesian silicates and calcium-rich plagioclase melt above 700°C; and zircon is not stable above 700°C.(Frondel & Collette 1957) Therefore, if these high-temperature minerals are assimilated completely, the tiny zircon crystals should also break down and mix their components into the melt, or at least show evidence of partial solution or melting to produce corroded edges or embayments. The simultaneous complete disappearance of larger, high-temperature minerals and absence of any corrosion and embayments on the zircons suggest that incorporation of zircons has not occurred.*

## The Silane-replacement Model

The silane-replacement model provides a logical explanation that eliminates the problems engendered by magmatic models. The explanation is based on the following rationale. The zirconium ion is common in pyroxenes, hornblende, and biotite (12 to 125 parts per million), and in sphene (1 to 2 percent).(Wedepohl 1978)

These same primary minerals also contain trace amounts of uranium.[95] At the time of crystallization this uranium had a certain $^{238}U/^{235}U$ isotopic ratio that is common for, and uniform throughout that particular magma. Therefore, when ferromagnesian silicates and sphene were replaced by silanes to produce quartz, the zirconium ion, set free to move in hot hydrous fluids, preferentially should nucleate on unmelted residual zircon or uraninite crystals in the primary ferromagnesian silicates.

With continued solidification, the dispersed uranium isotopes then became reconcentrated with the original ratio in the residual uraninite and zircon[96] and also other places scattered through other primary silicate, oxide, and phosphate minerals wherever uranium occurs in trace amounts.

In time, both $^{235}U$ and $^{238}U$, whether concentrated or scattered, decay to produce lead isotopes. When the primary rock is sheared, however, so that silanes can enter and replace pyroxenes, hornblende, biotite, and sphene with quartz, the remaining, undecayed, uranium isotopes and their daughter isotopes, including lead, are all released along with trace zirconium. **Lead and other daughter isotopes are not transferred to nearby zircon crystals, as part of its overgrowths.** Rather, they move out of the system in escaping

---

[95] *Thorium is also present and behaves chemically in the same way as uranium.*

[96] *In thorite, or thorium-bearing zircon as well.*

hydrous fluids or into feldspars, sulfides, or other accommodating crystals. Both the released uranium isotopes and zirconium,[97] however, can now nucleate on primary zircon or uraninite crystals to form zircon overgrowths. This uranium, stripped of its decay products is *"rejuvenated uranium."*

If original zircon and uraninite crystals are relatively large and modified by only a small amount of zircon overgrowth, then Pb/U isotopic data-points plot mostly near the upper intersection of the discordant line with the concordia. If original zircon and uraninite crystals are tiny and overgrown by thick rims of secondary zircon, the data points plot mostly near the lower intersection of the discordant line with the concordia. **These are the relationships that produce the discordant zircon populations.**

Basically, the same silane replacement process that results in discordant zircon populations produces myrmekite as well. Where greater replacement has taken place, and where the replaced rock is relatively old, the discordant line tends to deviate far from the concordia. Where diorite and gabbro are minimally replaced by silanes and lacking in myrmekite, discordant zircon populations may still occur, but with less deviation of the discordant line from the concordia, and with most data-plots falling near the upper intercept.

Episodic replacements by silanes that produce episodic zircon overgrowths can also add complications to the way in which Pb/U isotopic data plot on the concordia diagrams.

In the silane-replacement model all discordant zircon crystal populations are truly inherited features that have been produced during replacement *in situ* of mafic magmatic rocks by silanes to form younger felsic rocks. After replacement, such transformed rocks may be mobilized by isostasy, and rise plastically as diapirs into the Crust. There is no need for mixing of "refractory" zircons from older magmatic rock into a newly-formed intrusive magma to get overgrowths of young mineral on older mineral.

---

[97]   *Thorium also.*

## Behavior of Corollary Minerals

Discordant zircon formation suggests why other minerals of magmatic rocks produce Pb/U isotopic ratios that also plot near the upper intercept with the concordia.(Faure, 1986) It is noteworthy that sphene, monazite, and apatite in most granitic rocks formed by silane replacements have little to no secondary overgrowths on their primary uranium and zirconium.

The lack of overgrowths is consequential upon calcium in the mineral fabric. Minerals, with trace uranium and calcium content, tend to be replaced rather than accept peripheral overgrowth. In contrast, zircon (a non-calcium-bearing mineral) is generally stable in alkali-rich rocks that contain K-feldspar and sodic plagioclase. $Zr^{4+}$ ions precipitate from dilute solution (as little as 100 ppm).(Watson 1979) Thus, **zircon overgrowths are promoted on minerals which are preserved during silane replacements.**

One last bit of circumstantial evidence supportive of the silane replacement model is the common association of zircons with quartz and magnetite rather than with feldspars.(Moorhouse 1956; R. L. Harris 1977) **This association reflects the fact that quartz and magnetite are produced in the process of silane replacement of ferromagnesian silicates.**

## Significant Facts in the Silane-replacement Model

*The age represented by the upper intercept of the discordant line with the concordia is the age of a primary rock. The age represented by the lower intercept is the age in which silane replacements of this rock occurred. These are true ages, whether most data plot near the upper intercept or most near the lower intercept (Fig. V-4).*

In magmatic models, if data is plotted only near the upper intercept, as already shown by **Fig. V-4**, the lower intercept is usually interpreted as meaningless.(Gebauer & Grüenfelder 1979)

*By contrast, in the silane model, both intercepts have geologic significance.*  In both of cases 1 and 2, as the younger rock ages, after silane replacement and recrystallization from older rock, the slope of its discordant line changes, and the "zero point" moves up the concordia.  *Both intercepts of the discordant line with the concordia give significant dates, and should be taken seriously.*

Silane replacement may not explain all variations in Pb/U distribution; but successive episodes of it in the same rock do appear to explain some of them.  Weathering effects, acid leaching, introduction of U-bearing fluids of different $^{238}U/^{235}U$ ratios, etc., may differentially redistribute isotopes, and account for others [Faure (1986)].  **Nevertheless, the silane-replacement model offers a logical explanation for discordant zircon populations, especially where other evidence of replacement such as the presence of myrmekite can be demonstrated.**

## Chapter VI

# EARTHQUAKES AND VOLCANISM

## THEORIES AND FACTS

### THE TECTONIC NATURE OF EARTHQUAKES

**E. A. SKOBELIN**

> "The forces that generate faults and earthquakes are the same forces that result in mountain building; and the nature and cause of mountain building are the central problem of geological history."
>
> **C. F. RICHTER**

This author, being not a specialist in seismology, justifies his incursion into this branch of Earth science by quoting an eminent modern seismologist.

For many years the author has studied trapp rocks, the basaltic and doleritic sills of the Siberian platform in connection with petroleum exploration activities. Fortuitously, the facts revealed by this work have led to simple solutions of many geological problems. Among the resolutions of major geological problems, a new non-traditional understanding of the tectonic nature of earthquakes has been developed.

Key to the link between the tectonics and earthquakes are the trapp rocks of the Siberian platform. This kind of rock is widely but unevenly distributed in the upper lithosphere. Within some terranes it may, in an individual series of flows, be 200 to 600 m and up to 1 000 m in thickness. Trapp is found in

stacks of 5 to 10 flows in some places while sparse or absent in adjoining terranes.

Correlation of individual sills from one place to another is usually difficult due to monotony and composition. Correlation is not a problem in the case of the south Siberian platform, where the Usolsky sill (first named by C.D. Feoktistov 1976) occurs widely. With a lateral extent of 1 200 km (Ryzhov & Mordovskaya 1979), the extent of this feature has been established by the drilling of hundreds of deep wells.

Significant potential is implied that additional, areally-extensive sills are present. Many previously-uncorrelated sill intersections are yet to be defined on the north Siberian platform.

F.U. Levinson-Lessing (1949) suggested that sill injection occurred passively during undersill subsidence. This view was supported by P.E. Offman (1959, 1981) and continues to be upheld to this day (Staroseltsev 1981). Many data contradict the concept, however, especially those of the Jujumba field, Podkamennaya, Tunguska. In that important region profiles 50 to 90 km in length interconnecting drillhole intersections show that the undersill suite of Cambrian rocks retains its horizontal attitude without variation of more than 40 m. However, in the oversill suite, folded and faulted structures are contoured; and they show amplitudes up to 150 to 250 m and, generally as much as 550 to 600 m on the Bugariksky horst.

These data are interpreted as proving that the oversill structure (excepting for 40 m or so of amplitude) is consequential upon the disturbance caused by sill injection and by sill thickness. Therefore, **sill injection is always accompanied by lifting of the oversill terrane to the height of sill thickness** (as shown by Feoktistov 1976).

The foregoing information shows that sill injection must have been accompanied by earthquakes. Furthermore, the great

dimensions of the Siberian platform imply that **earthquakes connected with their intrusion could have exceeded the scale of known earthquake events.**

The Alaskan occurrence affected an area of 150 to 250 km by 700 to 800 km; and it lifted the surface 10 m in some places and depressed it two m in others (Rikitake 1979). The Kanto event compares with the Siberian ones in that the bottom of Sagami Bay was reported as changing by subsidence of 110 to 200 m in some places and 400 m in others, while concomitant uplift of 250 m occurred in still other places. "Response to this report varied from complete rejection as 'obviously false' to full and unconditional acceptance." (Richter 1963)

A process of isostatic net uplift in oversill topography covering an area of 150 to 250 km by 700 to 800 km above a sill of 10-m thickness (or, perhaps, much more) thickness in the Alaska quake would compare with a similar volume change in submarine Sagami Bay that resulted from a 400-m sill injection in the Kanto quake.

The link between tectonism, magmatism, and earthquakes implies uplift and subsidence of Earth's surface. According to the author's geological concept, this process is likely to be a function of the external gravitational field. The greatest and fastest such changes are the lunar and solar tides, both of which reach maximum values at the zenith and nadir of Moon and Sun.

Tides in solid Earth are measurable in dozens of centimetres; on the ocean surface by metres; in the atmosphere by larger but unknown amounts. It follows that the gravitational field influence on earth substances depends on their viscosity as well as on the intensity and duration of the field of influence.

The large uplift here envisaged would certainly affect the entire lithosphere and result in complexities in local faulting and magmatism. Where the largest bowing of the lithosphere

occurs there would, doubtless, be fractures – lithosphere-penetrating faults.

Let us consider such faults that may be connected with *subsidence*. These should be produced in the deepest parts of geosynclines during subsidence. Their counterparts that should be produced during dilation should be found in the foredeeps. Such faults at the depths of geosynclines and in foredeeps are the signatures of subsidence. Their lower reaches should be *"zones of gape,"* widening downward. Their upward continuations should be *"zones of compression."* In fact, what

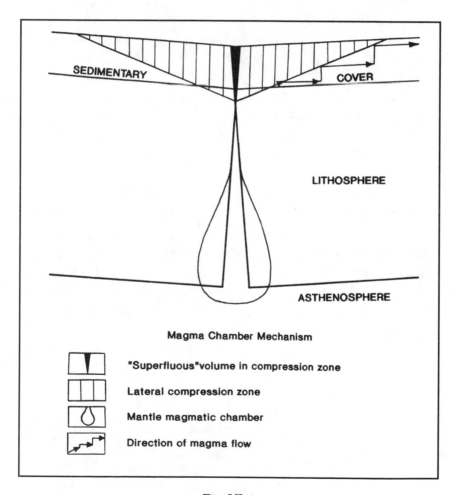

Fig. VI-1

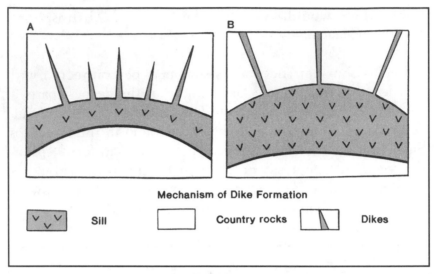

Mechanism of Dike Formation

Fig. VI-2

likely occurs in the higher zones is a *mechanism of lateral compression* (Yoder 1979). Melting in a zone of gape, which may occur as a response to pressure decrease, immediately creates a mantle magma chamber.(**Fig. VI-1**)

These tectonic movements never retain their initial directions over any long period and are always accompanied by inversions - movements of contrary sign and lesser magnitude. Thus, during inversion, a mantle chamber is compressed and the magma forced toward Earth's surface. As it leaves the *gape zone* and enters the *zone of lateral compression,* magma is forced to stream aside to escape compression. In *homogeneous rocks* its path follows an isobaric path at the base of the compression, whether parallel to the induced lithosphere curvature or not. In *sedimentary rocks* it tends to be injected along bedding planes.(**Fig. VI-2**)

The foregoing mechanism probably explains the injection of Mantle magma as sills. Intrusive injectites in the majority of mafic and ultramafic situations are generated in this manner. The widely-held idea of ultramafic rock emplacement by "cold

injection" is refuted entirely insofar as sill emplacement is concerned. In many occurrences this is obvious even from geological maps such as those of the Zond and Philippine Islands (Mishina 1978) and Hokaido Island (Markovsky 1975).

Magma injected from a Mantle chamber first establishes a thick sill with comparatively small lateral extension from the lithosphere fault. Oversill rock-resilience precludes fast spreading. Only with long-term isostatic equilibrium does the magma become dispersed over a broad area. With the passage of time, uneven magma pressure results in locally weakened oversill conditions. This allows oversill collapse under its own weight, squeezing magma out and causing an earthquake. In this context, concomitant isostatic levelling and sill injection are, of necessity, intermittent.

If sill injection occurs among aquifers, steam will enter the sill chambers in enormous volumes, especially at deeper levels. Steam entrapment in a structure will expel its magma. The evacuated magma chamber then, held open by steam (a **"steam pillow"**) may collapse very quickly, provoking a particularly energetic earthquake shock. The steam-filled chamber may raise the ground surface ("seismic dilation"), as happened to a 40-m diameter circular crater above the 3-km-depth epicenter of the Djirgatal earthquake in 1984. In that event 5 000 to 6 000 m³ of rock was ejected over an area of 100 to 180 m (Belousov 1987).

Tectonic earthquakes are generally thought to be associated with sills of mafic and ultramafic magmas, those of most fluid nature. Magmas of other compositions are more viscous and, despite being subject to the same isostatic effects, are considered unlikely to result in earthquakes.

Other interesting facts connected with earthquakes were collected by Byerly (1987) and generalized in the following manner [our translation]:

"Omory (1905) drew attention to the fact that earthquakes from any one region always produce the same type of first motion, compression or tension, at any given station. Labozetta (1916), studying Italian earthquakes with seismograms obtained from Italian stations, discovered that the only straight line on his map separated a region where all first motions were tension from another region where all first motions were compression. Garcy (1923) noticed at the Ci-Ka-Vey station that, if the earthquake is west of the station, the first arrival is a tensional impulse; and, if the earthquake is to the east, a compressive impulse. Somwill (1925) published an epicenters map on which every epicenter was marked by black or red, depending on whether the P-wave for it at the UKK1 station is the compressive impulse or the tension impulse. The distribution of the black and red epicenters was not random; they were grouped geographically. For instance, the earthquakes of the Aleutian region, Alaska to Kamchatka, always produced the compressive impulse at the UKK1 station, the western USA stations, the tensional impulse."

These earthquake data agree well with the concept of sill injection - collapse of an oversill complex and the squeezing of magma from such areas produce an initial tensional impulse. This leads to the surge of magma into neighboring areas, where the first impulse is compressive.

Japanese scientists were quite near such conclusions. Thus, Sida (reference unavailable) considered that earthquakes could be a falling-in of rocks under the influence of gravity; and Ishimoto (1932) suggested that magmatic intrusion was the source of earthquakes. This concept of roving magma was very popular in Japan. (Byerly, 1987, p97)

Of course, the seismic effects of a sill-injection event involving redistribution of magma can have very different results. For example, the isostatic equilibration after sill injection in the Jirgatalsky earthquake of 1884 was accompanied by very powerful shocks and "seismic dilation," whereas in Japan this process is splendidly quiet. In the words of Weda (1989, p143):

> "One other remarkable fact should be mentioned. It is known that very great earthquakes in Japan cause uplift of its coast of a few metres. But between such quakes, the coast slowly subsides."

The signature of magma injection from a "lithosphere fault" is bimodal: The first is general dilation of the Earth surface in the quake area without compensating subsidence in adjoining regions. The second is a proximal area of structural depression, beneath which is the active "lithosphere fault."

The spreading of a sill is not apt to be simultaneous over its length along a lithosphere fault. Thus it is that the powerful consecutive earthquakes in the region between the Kurile Island arc and the trench of the same name occur successively in areas of previous recent inactivity (Weda 1980). This regularity is displayed so clearly that the zones of "seismic quiet" are considered probable areas for the next powerful earthquake.

Evidence for sill injection from a lithosphere fault over a mantle magma chamber is clearly given by the presence of a deep-sea trench and a negative gravity anomaly aligned parallel to it. The sill flowed first in separate tongues toward the island arc, later infilling the areas between the tongues.

Most earthquakes affecting large areas (for example, the Alaska earthquake of 1964) are accompanied by uplift of some areas and subsidence of others. These are likely to be the results of ordinary isostatic equilibration. Wave guides are usually

associated with earthquake sources in seismically-active regions. As the preferred paths for propagation of seismic energy, they also usually exhibit increased electrical conductivity, qualities that are probably effects of unconsolidified sills.

This suggested model for tectonic earthquakes accommodates well many exotic phenomena that find no explanation in conventional theories. For example, luminescence may be observed in large fissures that open during earthquakes. This fact may be explained as magma itself or as the burning of hydrocarbons, hydrogen sulphide, or hydrogen. The gases may be the product of country rock saturated with organic substances being heated by sill magma.

A convenient fact, corollary to this explanation, is the ability of the suggested model to resolve the otherwise-unsolvable problem of primary migration of hydrocarbons from impermeable source rock. It also can explain the presence of hydrogen in hydrocarbon gas as a product of thermal destruction of methane and hydrogen sulphide. Hydrogen cannot be formed from organic substances by "maturation."

The model also explains the often-reported "waves" on Earth's surface, which are usually ascribed to observers' imaginations "earthquake excitement." These are tsunamis on "magma oceans."

The great areal expanse and great depths of the sills on Earth's platforms imply much greater sill spreading and seismicity in the geological past than today. The presence of these features on such aseismic and amagmatic regions as the Russian platform of West Siberia imply that no part of Earth is exempt from such activity. There is no "safe" place.

The model advanced here limits and focuses the search and methods of prediction of tectonic earthquakes. To solve the problem will require reinterpretations of seismic data and drilling in areas of shallow-focus earthquakes of the recent past.

It can be emphasized that the adduced opinions are a direct result of general geological concepts developed by the author to provide simple and interconnected solutions to major problems, some of which have been briefly examined.

## PHREATIC EARTHQUAKE MECHANISMS

### C. WARREN HUNT

There is a common feature between the foregoing Skobelin explanation of earthquakes and my own (discussed briefly in Chapter III). Where I interpret explosive rupture by injection of volatile matter, Skobelin evokes "steam pillow" collapse. Both, manifestly, negate the idea of rock masses rubbing together. And both are events of earthquake energy release into the lithosphere by sudden volume change, whether of collapse or growth.

The Skobelin contribution also brings up the interesting concept that "first motion" waves in some entire seismic terranes are exclusively either compressive or tensile. On the west coast of the USA earthquakes deliver first motion of tensile sense, whereas in Aleutian earthquakes compressional first motion prevails. Although Skobelin offers no resolution for this dichotomy, the concept is extremely interesting and deserves further research. It would not seem unlikely, as Skobelin implies, that steam chamber collapse could release tensile first motion, whereas injection of sills (Skobelin's preference) or volatile matter (my preference) could propagate compressive first motion.

Let us look at some more evidence that bears on the question of the stress-release versus phreatic earthquake mechanisms. Although speculation on the nature of earthquakes has been with us since antiquity, let us start with the nineteenth century, when geology began to be taken as a serious science. The idea emerged in the period 1878-1893 that earthquakes resulted from fault movement. First off was Rudolf Hoernes, who related earthquakes to "mountain building" in 1878. This was a position that was endorsed by G. K. Gilbert in 1883.[98] Meanwhile John Milne and Andrew Ramsay in 1878[99] had ascribed Japanese earthquakes to effects of "crustal splitting," a contrarian view.

Whereas both mountain building and crustal splitting are generally associated with faults, the first person to link earthquakes and fault movement was the Japanese, Bunjiro Koto,[100] who referred in 1893 to "fault displacement, which causes the land to move both horizontally and vertically." This pleased the mechanistic mind of the turn of the century. Great blocks of rock lurching past one another were just the right image of imperfect Earth for a society convinced that nature was a machine running on Newtonian physical principles [even subject to repair when man understood it sufficiently].

My own recollection of popular perceptions of the earthquake mechanism goes back to 1933, when the subject came up after the Long Beach earthquake. Along with such ideas as "earthquake weather," I recall the frequent expression, "earthquake fault" being used to mean the surface trace of the planar structure upon which terrain displacement occurred consequential upon earthquake. The "uninformed public," since the time of Hoernes, Gilbert, Milne, and Ramsay has regarded the tremor as primary, the ground breaching, secondary. **Fault movement resulted from earthquake, not the other way around, as the scientific community believed.**

---

[98] *Erdbeben-Studien, in Geologisches Jahrbuch, v28: p387-448. Anon. 1884 "THEORY OF THE EARTHQUAKES OF THE GREAT BASIN WITH A PRACTICAL APPLICATION" Salt Lake Tribune, Sept. 30, also Am. Jour. Sci., 3rd Ser., v27: p49-53, 1884.*

[99] *Report of the Committee Appointed for the Purpose of Investigating the Earthquake Phenomena of Japan, in British Assoc. for Adv. of Sci. Rpt., 200-204.*

[100] *On the Cause of the Great Earthquake in Central Japan, 1891, in Tokyo University College of Sci. Jour., v5:297-253.*

The "experts" have persisted, however, with this view right up to our own day. The predisposition to the Koto view, that "fault displacement causes land to move ..." while earthquakes are bursts of energy released as blocks of rock slip past one another, is an idea that has advanced to the category of "established fact," while the Hoernes, Gilbert, Milne and Ramsey view has been largely ignored.

"Earthquake fault" is still a term that is heard in public usage, although not by "better informed" people who shun it as redundant. **They "know" that,** *since fault movement causes earthquakes, all faults are actually or potentially, earthquake faults.*

The evidence I will now present shows that the view of the "experts" is wanting and public intuition, once again [incredibly] right. The esoteric "expertise" of the scientific community on the nature of earthquakes is clearly untenable nonsense.

To focus on the evidence the following subjects will be taken up:

**Characteristics of the San Andreas fault ("SAF")**
    **Parkfield Sector Characteristics**
    **Exotic rock bodies**
    **Vertical displacement on the SAF**

**The Loma Prieta earthquake**
    **Aftershock scatter: heterogeneity of "stress"**
    **Aseismic creep: crustal accretion**

**The Cajon Pass drillhole**

**Benioff Zones**
    **Epicenter scatter: confusion as to what it shows**
    **Japanese earthquakes**

**Electromagnetic Effects Associated with Earthquakes**

**Global Distribution of Deep-focus Earthquakes**

**Great McQuarrie ridge: epicenter of confusion**

## Characteristics of the San Andreas Fault

### Parkfield Sector Characteristics

The Parkfield sector of the SAF has been seismically quiet in recent years, a fact which has led researchers to conclude that Parkfield is a prime site for a quake in the near future. In anticipation of release of built-up stress and earthquake energy with a new slip, extensive instrumentation has been installed and prior geophysical surveying carried out.

Mooney et al. (1986; 1987) depict the low-velocity, reflection seismic characteristics on the Calaveras splay of the SAF north of Parkfield and the low density found by gravity traverses of the SAF at Parkfield. These studies are combined to provide an interpretation of the **fault as an upward-tapering, funnel-shaped zone of fractured rock and fault gouge extending to 10-km depth with porosity of 12%.**

**Fig. VI-3**

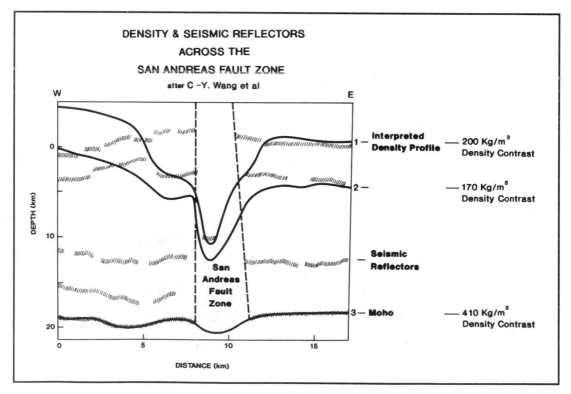

The research indicates that the downward extension of the fault zone penetrates the Moho with several km of width. It is noteworthy that similar research on a transcurrent fault in the Nagssugtoqidian region of western Greenland likewise resulted in an interpretation of widening with depth.(Bak et al. 1975)

**Nothing in any of this work on the Parkfield sector of the SAF supports the idea that built-up stress even exists in the SAF zone, much less that it produces earthquakes upon being released.**

---

## Exotic Rock Bodies

---

Two hornblende quartz-gabbro bodies of closely similar petrology and Jurassic age (137 to 159 my) intrude basement rocks on opposite sides of the SAF and spaced about 315 km apart. They have been interpreted as separate segments of one original body.(Sims, J. D. 1989) Deposited over the gabbros are Eocene to Miocene conglomerates. The Miocene strata are interfingered with lavas and pyroclastic deposits that are exposed on opposite sides of the SAF at widely-separated occurrences. Dated early Miocene (22 to 24 my), the volcanics have been correlated (Weigand & Thomas 1989) so as to confirm the displacement interpreted for the gabbros. Late Cenozoic right-lateral strike offset of 315-km magnitude is implied.

These age and lithological correlations cannot reasonably be doubted. The inference of late Cenozoic faulting, however, is not proven. A third gabbro exposure, the Gold Hill occurrence (Sims, J. D. 1989) about midway between the distal gabbro exposures confuses the situation by implying displacement only one-half the 315-km figure. Sims contrives a complex scenario of fault motions to explain the Gold Hill "exotic block," in which block movement is first one way, then the other, as the main fault slippage surface skips over the block, from one side to the other.

Alternatively, the entire terrane between the gabbro and volcanic outcrops may have been paved with gabbro and other basement rocks during the 114 to 134 my of elapsed time from the date of gabbro crystallization until the Miocene volcanism. **Exhumation of that surface could have provided a floor for the Cenozoic conglomerates and Miocene volcanics. Laid discontinuously but contemporaneously over the entire region from a series of volcanic vents along a rift that was to become [and is today] the SAF, the volcanic assemblages could create the modern topology. Horizontal fault displacement on a large scale, while an attractive explanation, is not essential to explain the present geology.**

Long-distance offset being neither proved nor required to explain the age correlations of these basement and volcanic rock distributions, it may be appropriate to note that interspersed gabbro, granodiorite, and metamorphic basement is known to underlie the sedimentary cover in many places adjacent to the SAF; and effusion of Miocene volcanics was common elsewhere in California as well as North America and the world.

In view of these facts, the injections and ejections of igneous rock would seem best taken as an indicator of a lithosphere-penetrating rift, a precursor of the narrow, tensile, trench-like feature we see as the SAF today. Vertical fault movement may predominate over strike motion on various sectors and in various periods of activity on the SAF rift.

Of course, proving that strike faulting is unnecessary to interpreting the SAF is negative evidence, which neither discredits nor negates the large body of indisputable evidence for right lateral strike-slip offsets on some sectors of the SAF as well as other northwesterly-trending faults of western California. There surely is good and sufficient evidence that *significant right-lateral strike slip motion **has** occurred in many places. Let us look at some of them.*

## Vertical Displacement on the San Andreas Fault

Opposition to major right-lateral movement on the north end of the SAF was my vigorously expressed view in the EV volume (Plate II). The idea of the fault turning nearly 90º west to merge with the Mendocino escarpment was taken as structurally impossible. The escarpment was not apparently disrupted by the junction, an impossibility if SAF motion of 300 km had occurred, most of it in Pliocene and Pleistocene time, as conventional wisdom would have it. There is no field evidence that such motion could have been accommodated by the subduction of an "oceanic plate" beneath the Klamath Mountains. Such a construct is pure speculation aimed to prove a dubious theory.

I went to some trouble to point out that the Klamath Mountains are a westward-translated salient. The Mendocino escarpment, which coincides with the south edge of the Klamaths, would have a left-lateral sense of motion, not right-lateral. I drew attention to the undisturbed sedimentation [e.g. Pliocene strata on-shore between Eureka and Ferndale] immediately north of the Mendocino escarpment to illustrate the absurdity of projecting the SAF couple into this area.

Bruce D. Martin (1991) expresses similar objections, in his paper on the constraints to major right-lateral movement of the SAF. He points out that **"arguments for major right-lateral movement ... are weakest at Cape Mendocino, precisely where they should be strongest."** Observing that to the south in the Tehachapi area the fault changes direction by 20° or so and exhibits major adjacent tectonism to accommodate that shift, Martin notes the contrasting situation in the Mendocino area. He notes delicate sedimentary structures on the ocean floor at the base of the Mendocino escarpment (estimated age: one million years). **"Tectonic activity associated with rotation should have ... destroyed these delicate sedimentary features."**

Martin proceeds with his second constraint. **In the central California Coast Ranges normal faulting is essential to account for the topology of the environs of the SAF: (1) granites west but not east of the SAF, (2) a buried escarpment of 240-km length, on which 2 790 m relief [in the Santa Cruz Mountains] at the north end of the escarpment and 930 m [at King City] at the south end express vertical motion, (3) 2 170 m of vertical offset but no horizontal offset affecting the Oligocene-to-Pleistocene seaway that drained the Great Valley through Pajaro Canyon into Monterey Submarine Canyon [across the trace of the SAF], and (4) the implication of depression by major *normal* faulting in Monterey Bay, where a marine, sediment fill of 5 720-m thickness is present.**

Martin makes the points that in Eocene time the seaway from the Great Valley of California to the Pacific Ocean had occupied Valecitos Canyon, but from Oligocene to mid-Pleistocene times, all sediment-laden drainage from the Great Valley to the Pacific Ocean passed through the 2 170-m gorge of Pajaro Canyon. **Right-lateral offsets on the SAF would have blocked both the major Vallecitos and Pajaro drainage channels from the Great Valley. Since we know the seaways must have remained open, Martin avers that right-lateral offsetting must have been minimal.** *All of Martin's evidence implies a supervention to major strike motion on the SAF north of the Parkfield sector.*

## The Loma Prieta Earthquake

In the EV volume I pointed out that the actual movement in the Loma Prieta earthquake was not either strike or dip motion, but a hybrid: the epicentral area moved north **parallel to Earth's rotational axis,** the direction in which conservation of momentum could best be achieved. The epicenter was 6.5 km west of the most prominent surface rupture in the San Andreas zone. I suggested that the motion illustrated an explosive character suggestive of the upward bursting of volatile exhalites into chambers[101] created ahead of them by their own pressure dynamics. This interpretation was set forth as the probable nature of earthquake energy release.

Whether, on the one hand, silanes and hydrocarbons hydraulically fracture the solid Mantle or Crust and erupt into the lithosphere in the manner of underground volcanos, or whether they seep in, permeating mineral fabrics and transforming minerals, the mechanism is one of oxidation and heat evolution. It is capable of melting rock, and creating the sills and steam chambers of the Skobelin theory.

The chambers I had in mind in the EV volume were *upward* ruptures. The idea of natural horizontal hydraulic fracturing as a major process in geology is a variation that is developed in this chapter and further in Chapter IX. Skobelin deserves the credit for the idea of sill emplacement by hydraulic overpressure. However, the oil industry practice of creating a horizontal fracture by hydraulically fracturing reservoir rock has been well-established for many years as a technique for inducing fluid inflow to wells. Bailey (1990), a geophysicist, has been studying the concept to explain behavior of fluid up-flow in the solid Crust.

The mechanism of hydraulic fracturing deserves the focus of our attention, because it is the only mechanism that can explain earthquakes in the plastic lower Crust and Mantle. A ductile medium can be fractured if it is struck suddenly so that inertia precludes plastic response. Skipping a rock on water invokes this principal: the water cannot deform quickly enough to envelop the rock, and acts as a solid shield. Incisive cleavage of the atmosphere by lightning is another example, the extreme speed of impact precluding atmospheric deformation, and a

---

[101] *The word, "chamber" was used in the EV volume to describe the space occupied by volatile materials. Its connotation of empty space is unfortunate and inappropriate because the space may be no more open than under-pressured inter-granular passages or paper-thin fracture surfaces. Better terms are sill, dike, intrusion chamber, or sand-filled crypt. Porosity can be implicit in words like sand or cataclasite.*

brittle split opening ahead of the electrical charge. The rupturing and subsequent collapse of the split generate a radiation of sonic energy known as a peal of thunder.

*The explanation for earthquakes in ductile Crust or Mantle is a similar explosive rupture/collapse, from which underground sonic energy radiates. A thunderclap in the solid Earth only differs from its atmospheric counterpart by the different vibration characteristics of rock and air.*

*The initial earthquake is an hydraulic-pressure phenomenon powered by liquefied highly-compressed volatile components, mainly silane, hydrogen, hydrocarbons, carbon dioxide, carbon monoxide, nitrogen, water, and various polymers and free radicals. The sudden volume changes are gas-driven, phreatic, that is to say.*

*Magma surges and secondary gas releases after an earthquake make up a field of aftershocks, which, together with the initial rupture, all bring about geodetic terrain adjustment. Both the initial rupture and the aftershocks express volume changes which show up as aseismic, creeping, terrain distortion, creeping fault movement, surface dilation, or surface depression.* In the next sections aftershock distributions, the Cajon Pass drillhole, and aseismic creep are described.

## Aftershocks of the Loma Prieta Earthquake

In the Loma Prieta earthquake there was no recognized fault displacement, although seismic analysis of the radiated sonic energy indicated specific movement in the epicentral area of a minute (<2-m) translation. This tiny distortion of a huge terrain is taken to mean that great blocks of the whole terrain have moved relative to one another. Realistically, aseismic after-response may as well as not have augmented, restored, or even reversed the trivial interpreted net motion.

No causal connection is shown between the earthquake phenomenon and fault displacement by the Loma Prieta tremor; the entire earthquake mechanism remains enigmatic; and all explanations that have been advanced up to now are dismal failures in accounting for observed facts. Analyses of the aftershock pattern illustrate the confusion.

A transverse plot of the Loma Prieta aftershocks **(Fig. VI-4)** shows a distribution scatter of 35 km or so on both sides of the surface expression of the SAF. These have been "migrated" in from 10 km to the line of the profile and represent a sampling of the much larger aftershock area, which is remarkably circular in plan view and not strongly elongate along the SAF or oriented to any single "fault plane." To summarize the situation, the author of the paper from which the figure was copied, D.H. Oppenheimer (1990), opines that "a highly variable residual stress field ... is required to explain the variety of mechanisms." This is an arcane way of saying the data cannot be interpreted to illustrate fault behavior consistent with the anticipated stress-release mechanism.

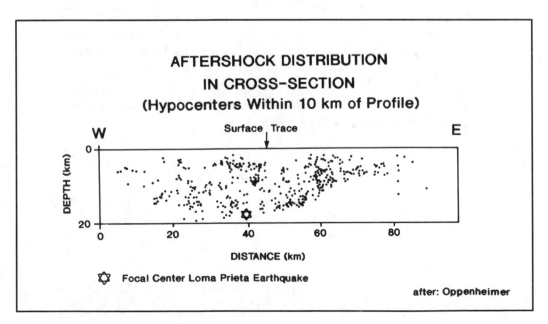

VI-4
**LOMA PRIETA EARTHQUAKE AFTERSHOCK DISTRIBUTION
PATTERN ACROSS THE SAF ZONE**

Noting that the circular area of aftershock seismicity centers on the epicenter and is not the narrow trail of activity that a planar, stress-releasing rupture should produce, my explanation as the aftereffects of gaseous, diapiric intrusion above a lithosphere rift is a better resolution of the SAF enigma than stress release by rupture.

The nature of an earthquake, then, is sudden volume change within a plume area. As Oppenheimer puts it, "most aftershocks near the rupture have focal mechanisms that disagree with the *pre-stress force field*." (emphasis added) Michael et al. (1990) determine that the earthquake caused a "*total stress drop*." These investigators are perplexed by their own results, because they seem unable to recognize that built-up stress between brittle rock masses as a cause of earthquake is discredited by their own work, which fails to support its existence.

## The Cajon Pass Drillhole

In the mid-1980s the United States government financed drilling in the San Andreas fault zone near Cajon Pass, in southern California. The drilling was terminated, when funds ran out, before the intended 16,000-foot total depth was reached. Notwithstanding early termination, the well made an enormous contribution to knowledge.

It was intended to measure the stress buildup, which plate tectonicists confidently expected to find in the SAF. Such stress buildup is essential if the "Pacific plate" is, in fact, impacting against the "North American plate." Indirectly then, the Cajon Pass well was expected to give confirmation of the plate tectonics theory ("PT").

The Cajon Pass drillhole proved nothing that was expected of it. Instead it proved beyond doubt that **there is no stress** built up at that location within the San Andreas fault zone and, hence, that SAF movement must not cause earthquakes by frictional stress release. Indirectly, it proved that there is no subduction of "Pacific plate" in progress under the continent at that site, nor any evidence for the corollary, that PT drives the SAF.

Absolutely nothing was found to support the preconceptions of plate tectonics. Has there been any recognition that PT is a myth? Has anyone even asked: "Could the theory be deficient?" The answer is "No!" so far as I have been able to discover. Quite to the contrary, the order of the day has been "business as usual" with no wavering on the important matter of faith in PT!

Fifty or more scientists worked on the drillhole results and published papers on it, and some are still coming out. Despite the enormous contribution to knowledge, not one of the published papers has questioned publicly the theories that are certainly called into question by the extensive data. Some quotations from this volume [Geophysical Research Letters, v15, #9, Aug. '88] are instructive:

R. J. Weldon, and J. E. Springer report that **"stress orientation measured in the well is not consistent with right-lateral slip on the SAF."**.

J. H. Healy and M. D. Zoback report that **"the stress data indicate that there is no right-lateral shear stress acting on planes parallel to the SAF."**

G. Shamir, M. D. Zobach, and C. A. Barlon report that **"All measured stress orientations result in left-lateral shear stress ... contrary [to SAF] slip [sense]."**

P. A. Pezard and S. M. Luthy report that the **"borehole electrical image analysis suggests that the observed present-day stress regime [of the SAF] might have remained constant over the past few million years."**

E. W. James and L. T. Silver report abundant zeolites[102] permeating the host rock up to and including the late Cenozoic Cajon formation. **These minerals required low pore pressures and open space to have been able to enter the rock.** The previously-mentioned gravity survey on the Parkfield sector of the SAF established 12 % porosity within a deep slot through the Crust; and these authors assert the necessity for such conditions in order for zeolites to have entered.

A. H. Lachenbruch and J. H. Sass, reporting on thermal and stress conditions in the well, say that **"no anomaly is observed from over 100 heat-flow determinations along 1 000 km of the SAF,"** and **"stress [in the well] is near normal."** The drillhole does, in fact, show a 30 % positive heat anomaly, but only in its *upper* levels. As the well encountered some radioactive minerals at the basement contact, these may explain the shallow-zone heat anomaly. Alternatively, these authors think it may be explained by the high rates of uplift and erosion common historically for the Plio-Pleistocene Transverse Ranges south of the well.

The foregoing excerpts express the extraordinary discoveries of the Cajon Pass drillhole, a "wealth of data," but more than data, insights to new understanding. Three main conclusions can be drawn:

First, *there is no significant stress buildup transverse to the San Andreas fault.*

Second, *there is no right lateral shear stress in the San Andreas fault zone.*

Third, *there is no heat flow indicative of prior stress release at the Cajon Pass location.*

---

[102] *Minerals characteristic of volcanic effusion, low PT conditions. Further discussion of zeolites will be found in Chapter VII.*

*Chapter* VI

These facts make a shambles not only of fault theory but of plate tectonics concepts, which depend heavily on fault theory. It can now be fairly said that:

**1. The concept of stress buildup as a precursor to fault motion in which earthquake energy is released is wholly disproven; there is no stress buildup at all.** This long-held theory, treated as fact, a "given" in geology, must now be treated with skepticism.

**2. The expectation of compression across the SAF was based on the idea that the Pacific Ocean "plate" is underthrusting itself beneath the continent. Absence of compression disproves impacting and underriding. It completely discredits the "plate theory," relegating it to the category of figment of imagination.**

**3. The above can be considered applicable to the last two or three million years at the Cajon site, enough duration to establish that similar conditions in all likelihood prevailed over the 20 millions or more years during which the SAF is thought to have been active.**

Laundering the plate tectonics idea in this acid bath has at least bleached it badly, if not entirely destroyed it! One cannot but wonder why not one of the geoscientists involved with the Cajon Pass drilling project has expressed these evident and inescapable conclusions?

## ASEISMIC CREEP

It is only in recent years that creepmeters have replaced surveying for routine measurement of small amounts of ground distortion. Measurements in many places have shown that in most places studied some motion is recognizable. Most creep cannot be sensibly related to earthquakes: it is, thus,

225

"aseismic."(Behr et al. 1990) For example, at two sites traversing the SAF where readings were taken after the Loma Prieta earthquake, the creepmeters found "significant creep, 10 mm net in four months".(Savage et al. 1990) SAF movement is, thus, not a series of lurches, but a creeping effect with occasional lurches.

In another example, this one in eastern California, a creeping right-lateral distortion of about eight mm/yr is found to occur on a N 35°W trend extending from the western Mojave desert near the SAF to northern Owens Valley (the eastern flank of the Sierra Nevada Range).(Savage et al. 1990) The zone of distortion transects the pan-like Mojave Desert and the southeast flank of the Sierras. Northerly, it converges at a low angle on Owens Valley, the graben that marks the eastern edge of the Sierra Nevada. There is no identifiable surface fault unique to the trend of the distortion.

This latter creep analysis was based on measurements at four arrays, which, unfortunately, are almost on a straight line. Thus, interpreting the results as indicative of a **simple** *shear across a vertical northwest-trending plane* separating two moving blocks is unjustified. Regional distortion in other directions may also be in progress without being recognized by the available array of sensors.

Geodetic measurements show without doubt that gradual differential movement, "*creep*," changes the separation of points **across active faults**; *and* **within individual "fault blocks."** The term "block," implying rigid behavior is, then, a misnomer, and should not be used in connection with faults or earthquakes. The perception that fault movement releases earthquake energy is evidently quite incorrect.

A proper view of terrain distortion is one of creeping more than sudden offsetting. Visible fault offsets may gradually develop from creeping movement as well as from sudden rupture.

The release of energy in an earthquake is an associated phenomenon of chaotic dissipation of (1) previously-latent chemical energy, (2) energy of compressed gas, and (3) heat generated within non-rigid members of the terrain. Rarely, sufficient violence is released to activate an existent fault surface and create a sudden surface offset that supplements the normal creeping behavior.

Thus, the stress-buildup concept on which the Savage analysis was based is fallacious. Rather, surface rock undergoes a sort of stately seething process, analogous to the churning lava in a caldera, but much slower. The plastic lower Crust appears to reshape itself constantly, like glacial ice, by crystal deformation and recrystallization. The near-surface rock adjusts on myriad micro- and macro-fractures, joints, and faults, stress buildup being precluded by the relative inelasticity of brittle rock and by micro-fractures.

Another apt analogy to aseismic creep is the seething of water coming to a boil. Taking the process in very slow motion, various places are moving at different rates, some hardly moving. The analogy is apt because silane, hydrocarbons, and their oxidized gaseous products, as they rise through the Crust, are analogous to the steam bubbles in nearly-boiling water.

Seething in an active melange, such as one finds in some California terranes, may variously be attributed to shifting subcrustal magmas, to active mineral transformations, to hydrological volume transfers, or to isostatic adjustments due to off-loading of overburden by erosion. Silane systematics may inspissate a magma or deposit a barrier to its migration.

Micro-quakes and rock bursts in mines are manifestations of local adjustments to these phenomena, although entrapped gas from earlier silicification may sometimes produce the latter. Local stress, the cause of minor quakes, is an aspect of oxidation in which silanes and hydrocarbons from a "mother gas," in the process of replacing the original metal cations in rock-forming minerals, produce water, silica, and effluent gas. Terrane seething is brought on by volume and density changes, and by heat release inherent in silane and hydrocarbon oxidation processes.

As no large-scale stress buildup such as might release earthquake energy of a magnitude 6-8 earthquake can build up across the San Andreas or any other fault, it must result from hydraulic fracturing after overpressure buildup from a complex of mineral transformation, rock density loss, silicification, inspissation, and expansion.

This brings us back to the San Andreas. Creek offsets at numerous places along the surface trace of this great fault and large earthquakes near its surface trace are convincing evidence of two things: small-scale right lateral displacement and release of seismic energy. Beyond that we cannot claim to "know" very much. Long-distance horizontal offsetting is an attractive but tenuously held idea. The great length of the fault encourages the image of great blocks of the Crust interacting disruptively. However, motion all at once along the great length *never* occurs; and that fact alone discredits the "great blocks" image.

What then *is* the relationship between vertical and horizontal motion at various places along the SAF? How can the absence of strike slip at the Mendocino Escarpment and, perhaps, of small strike slip in the Santa Cruz sector be reconciled with apparent major strike slip in the length of fault between Parkfield and the Garlock splay 150 km to the south?

A "seething" motion should characterize terrain adjacent to the fault, especially the western side. This image is a stark contrast to the popular image of great blocks of rigid rock grinding past one another. **The seething produces very different net motions at different places along the length of the SAF, emphatically not the same displacement at all points.**

Motion on one sector that is in excess of that on another sector is dispersed by micro-faulting in horsetail fashion, splaying away from the main fault direction and dissipating the net motion. In addition to giving motion dissimilarities along the SAF, the seething readily accommodates block rotation, and splay faults that disperse net throw on both vertical and horizontal surfaces. *Net right-lateral offset of significant magnitudes on some sectors of the SAF is not inconsistent with small to non-existent net strike movement at others.* This is the image that fits all the facts.

## DEEP-FOCUS EARTHQUAKES

Notwithstanding "a persistent dialogue for the past 60 years between seismologists and experimental geophysicists, the origin of deep-focus earthquakes remains unidentified."

**C. FROLICH (1989)**

Most seismic activity worldwide occurs in and close to zones where the Crust is deforming. Crustal rock slips laterally past other crustal rock, encroaches glacier-like out upon neighboring terrain, or emerges buoyantly as mountains. The most active zones are the margins of the Pacific Ocean, a belt stretching from Indochina to Spain, and a few well-defined mid-ocean ridges. Most earthquakes occur on or near these sites and are largely shallow. This seismicity is, thus, a phenomenon of the Crust.

The deepest earthquakes occur at depths of 300 to 710 km in a few well-defined places. The most active is a northerly-trending submarine area known as the Tonga-Fiji trench (north of New Zealand), with a westward branch that reaches almost to southern Japan. Other deep earthquakes occur under western South America as far east as the eastern foothills of the Andes.

Earthquakes in the 100- to 200-km depth range occur in greatest numbers also in the Tonga-Fiji region, but with belts of activity westward from the Tonga-Fiji region to Papua New Guinea, along the west coast of South America and central America, and, to a lesser degree, on the eastern rim of Asia, the Aleutians, Indonesia, and Lake Baikal. The topology of quakes in the 200 to 300-km depth range is intermediate between the 300 to 700 and 100 to 200-km ranges. Thus, although the worldwide distribution of seismic foci (Barzangi & Dorman 1969) is quite different for the deep and the shallow quakes, there is a continuous transition from the deep to shallow quakes, a fact that implies a single process.

The energy radiated from deep-focus earthquakes has a dual signature of both explosive and implosive sense, not what one would expect from steam-chamber collapse. The signature is, however, compatible with hydraulic fracturing. A rupture opens the fracture explosively and is followed by its collapse. As there is unlikely to be free water in the Mantle, the "frack" fluid can best be envisioned as emerging gases, the silanes, hydrocarbons, and inert gas plus any gaseous hydridic derivatives. Sudden, phreatic opening and immediate resealing of a major fracture or myriad micro-fractures should produce the observed explosive/implosive energy radiation.

## BENIOFF ZONES

In Benioff zones "... slip vectors show that the quake motion is not generated by movement of the slab through the asthenosphere but is internal, within the slab, showing that it retains the crustal property of deforming by fracture far down into the asthenosphere. Ultimately it breaks up by internal stress, like an iceberg."

**LESTER C. KING (1983, P63)**

Swarms of earthquakes, including both shallow and deep, at some ocean basin margins define zones of active seismicity with tabular forms that slope away from the bordering ocean and under the bordering landmass. These are the "Benioff zones." They penetrate deep into the Mantle with dips that vary from 35° (as beneath Japan **Fig. VI-5**) to near vertical (as below parts of the Mariannas trench).

Geophysicists attribute lower temperatures to the Benioff zones, because they exhibit a negative gravity anomaly, increased density at depth, and low attenuation of seismic waves. From these tidbits of information, the PT enthusiasts assert that cooler oceanic crust must be "diving" into the Mantle and creating earthquakes as it deforms. However, the density differential between cool oceanic mafic rock and hot ultramafic mantle rock is unlikely to be sufficient for a "plate" to penetrate the solid Mantle even without the energy loss inherent in the seismicity of the process.

On the other hand, the prospect is significant that isostatic rise of granitic rock, transformed from mafic rock, may explain the negative gravity anomaly and the seismicity. The cooler temperatures at depth could result from gas expansion as rising hydrocarbon, silane, hydrogen, and inert gas enter the fractured Benioff zone. The fracturing would explain the seismic wave attenuation. The subduction interpretation is superfluous.

Let us examine these typical Benioff zones further. **Figs. VI-5 & VI-6** show the scatter of epicenters beneath two prominent ones, Japan and the Aleutians.(Mooney & Ginsburg 1986; Toksoz 1975) The epicenters do not fall exactly on the profiles, but are "migrated" on the approximate strike of the dipping zone to the plane of the profiles.

A first point of noteworthiness is the width of the scatter zone, about 25 km on both profiles. The second point to note is that the dip is very different on the two sites, 60° for the

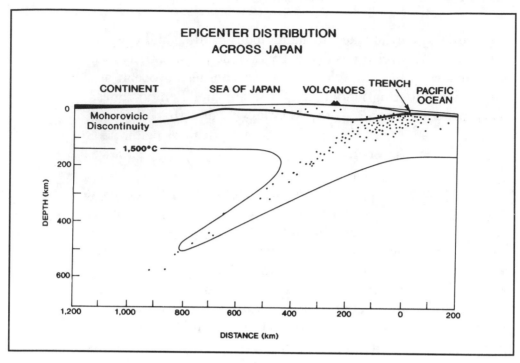

VI-5

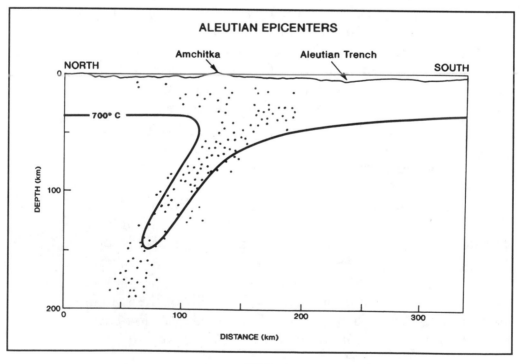

VI-6

Aleutian site and 35° for the Japan site. The plane attributed to "plate subduction" off southern California, a surface determined by seismic reflection, was a mere 0° to 16°!

Indeed, the Aleutian and Japanese profiles are not known from seismic imagery but deduced from gravity survey, seismic wave travel times, seismic wave attenuation, and the preconception of subduction. The premises should be reexamined. **Siliceous lithofacies and ambient gas evolution can produce the observed geophysical conditions. Silane influx may be the cause of the seismicity of Benioff zones.**

## Japan

The landmass of Japan is a creation of endogenic processes in Tertiary times. **There is a thin Crust over the Sea of Japan** *[where effusion has created little new rock].* **A thickened Crust under the Japanese Islands has resulted from the silane-linked crustal addition that has built the islands.**

As explained already, geophysicists interpret low-temperature rock in the 25-km-wide Benioff zone to represent a "diving slab" of oceanic Crust. The area where the Benioff zone approaches the surface centers on eastern Japan, the most active volcanic and magmatic region. That is the area where silanes have the most access to water and free oxygen. **Thus, the epicenter topology coincides with the locus of phreatic behavior.**

The terrane of western Japan, which is situated above thin Crust of the Sea of Japan, is an area that has extensive deposits called "Green Tuff", of Miocene age. These contain resources of natural gas in felsic volcanics that are noteworthy for their high $^3He/^4He$ ratios (up to $9.37 \times 10^{-6}$). These are the signatures of helium from the Mantle, and impossible to ascribe to helium entrapped through biogenic processes on the surface. Thus, the gas in Japan's Green Tuff formation is Mantle-derived. Its occurrence is above the active epicenters of about 100-km depths and, thus, is related to submarine volcanism in the Sea of Japan. Wakita et al. (1990; 1983; 1987) set out the systematics of these accumulations with great thoroughness.

## Aleutians

The deep Benioff zone in the Mantle beneath the Aleutians has a 60° dip and no continental mass behind it in the Aleutians. The Aleutian epicenter distribution is similar to that of Japan but for greater depth and a steeper dip. If it also is the consequence of silane migration in a zone of petrologic transformation, why the steeper dip? The answer, I would suggest, is in the orientation:

The direction of Earth rotation is only 12 to 26° from the strike of the Aleutian Benioff zone, whereas it is 65° or so from the Japan Benioff zone. Effusion moves mass away from Earth's rotational axis. This creates a strike shear on the Aleutian Benioff zone as its hanging wall absorbs 90 to 98 % of the inertial force generated by effusion [upward from Earth's axis]. On the Japan zone the strike component of the shear takes up only 42 % of this inertial force.

These orientations have the effect of diverting the inertial energy of effusion. Only 2 to 10 % of inertial force resulting from effusion can act parallel to dip on the Aleutian Benioff zone as opposed to 58 % on the Japan zone. Underground resolution of the dip component of inertial force released by effusion impels horizontal underthrust shear. These horizontal-component moments of inertia should account for some, perhaps all, of the difference between the dips of the Aleutian and Japan Benioff zones. Actual Benioff dips will be vectoral combinations of the underthrust effects and the unpredictable effects of differential silane effusion, inhomogeneities in the Mantle, and crustal loading (isostasy).

## Coastal California

The case of coastal California that was discussed in Chapter III is quite a different matter, and was dismissed for lack of a case favoring subduction.(**Fig. III-1**) Meltzer and Levander elegantly show continental mass extending far out over the oceanic Crust, its basal, aseismic contact undulating gently, and precluding an interpretation of faulting or Benioff zone. Rather, the scene is easily explained as a case of the loading of continental material over adjacent oceanic Crust, not one of subduction.

Linkage between the California depositional structure and the onshore, vertical San Andreas Fault, which parallels the shoreline, is unwarranted and irrational. Scholz (1969) summarized the California onshore earthquake picture with prescient cogency 23 years ago:

> "In California no earthquakes have been observed below about 15-20 km. This remarkably non-uniform geographical distribution implies that the occurrence of earthquakes requires very special conditions – the presence of very acidic rock.

> Earthquakes are characteristic of the rheology of only certain rock types. The silica bond has been found to play a vital role in the rheological behavior of silicates, for it is this strong bond which gives these minerals their generally-high strength. ... cleavage [of silicate minerals] almost always occurs in planes which do not cut Si-0 bonds."

**Thus, to close this discussion, it is reasonable to conclude that no serious problem exists in apprehending Benioff zones as fractured zones in which silanes and hydrocarbons rise, "fields of migration," so to speak, in which diapiric breccias, energized by compressed gas, progress upward in bursts of explosive volcanicity, which explain earthquakes.**

# ELECTROMAGNETIC EFFECTS
## ASSOCIATED WITH EARTHQUAKES

In Chapter IV Collins established firmly the interrelationship between mineral transformation and deformation of rock bodies. In Chapter II I pointed out that geomagnetic fluctuations are known to induce very low frequency (VLF) electromagnetic (100 to 1 000-hertz) and ELF (0 to 100-hertz) fields. I alluded to the researches of R. O. Becker, a medical doctor, whose research implies that these fields effect biomedical responses. This could be due to subtle chemical responses of the inherently unstable elements, carbon and silicon. If that be true, geomagnetism through induction may effect chemical responses within transforming minerals; and geomagnetic fluctuations could result in deformation of Earth's surface.

In these paragraphs the point is made that some geomagnetic fluctuation originates with the same forces that cause fault movement, volcanism, and earthquakes, thus coming full circle. Mineral transformation causes diapirism, magmatism, and phreatic volcanism; these, in turn cause geomagnetic fluctuation, earthquakes, surface deformation, and fault movement; these open the rock cataclastically, thus allowing access to more silanes, mineral transformation; and so on in cyclical fashion.

The two citations from the geophysical literature (Adams 1990; Parrot 1990), which were given in Chapter II, showed that recent earthquakes in western California have occurred during periods of greatly enhanced VLF and ELF electromagnetic activity. They also show, however, that periods of such activity occur without associated earthquakes.

In eastern California the Long Valley caldera is known for its highly active volcanism, seismicity, geodesic instability, and geomagnetic activity. "Changing magnetic fields are often observed associated with eruptions from active volcanoes ... and result from magnetohydrodynamic and electrokinetic effects.(Mueller et al. 1991)    A connection between magmatism, swelling of the surface, incipient volcanism, and geomagnetic fluctuation is clearly established, although neither the order of cause and effect nor the physical nature of any of the phenomena is known or likely to be defined precisely from geophysical data alone.

Our understanding is hardly satiated by being informed that the enhanced electromagnetic fields are due to "magnetohydrodynamic and electrokinetic effects."  What else might cause them?  Let us take a look at an oceanic site that receives much attention due to its seismicity.

## GREAT MCQUARRIE RIDGE:
### Epicenter of Confusion and Consternation

Perhaps the area of the most active shallow seismicity on Earth, the Great MacQuarrie ridge lies beneath the ocean southwest of New Zealand.  Historically the largest strike-slip motion ever instrumentally recorded happened there on May 23, 1989.

Several authors interpret the Australia plate to be in the process of being subducted beneath the Pacific plate(Ekstrom & Romanowicz 1990; Kenji & Kanamori 1990; Braunmiller & Nabelek 1990; Ruff et al. 1989), reversed subduction, as it were, a continental plate sliding under an oceanic plate.  Not impertinent is the fact that Australia is perhaps 1 500-km distant, and that very deep water flanks the ridge both east and west.  Aftershocks are distributed over a broad area asymmetric to the ridge and centering west of the focal point.  The aftershock distribution occurs over an area almost as wide as it is long (from the focal point 125 km northeast, 175 km northwest, 225 km southwest, and 125 km south).  This leads one author to deduce the existence of "a 50-km wide by 15-km thick deforming zone."(H. J. Anderson 1990)

Other authors take an opposite view: "The thrust earthquakes are consistent with **westward subduction of the Pacific plate beneath Australia**." (Tichelar & Ruff 1990)

Geography of submarine topography expresses a 2 000-km mountain range, one island peak being the only feature exposed to the atmosphere. The ridge comprises an alignment of summits making a gentle arc concave eastward extending south from New Zealand.

There is no justification for treating the ridge as part of the continent of Australia other than the compulsion to make the facts fit into plate tectonic theory. Even then, the cited authors agree that the fault is vertical (like the San Andreas), and apparently 60 km in depth. The reader is free to exercise his own imagination to find any compelling reason to connect the Great Macquarrie Ridge to plate tectonic theory.

**A much better choice would seem to be that implied by the 300- x 350-km oval area of the aftershock pattern and the palpable evidence for crustal addition given by the mountains themselves. The oval pattern of Great Macquarrie Ridge aftershock epicenters implies an area above a silane plume. Most of the emanation is channelled into a rift zone that strikes along the axis of the mountains. The mountains themselves are crustal additions above the emanating silanes. The rift is the breach in the crest of the cymatogen.**

# FORMATION OF VOLCANIC PIPES

## AND THE GENESIS OF DIAMONDS

### E. A. SKOBELIN

### *Introduction*

Throughout a protracted study of geological structure and trapp magmatism on the Siberian platform, the author was confronted with an array of facts, deliberations upon which became a key to finding a simple solution and to creating a new theory to explain them. The author's new theory, as it applies to diamonds and breccia pipes, is the subject of this contribution.

Mafic and ultramafic magmas are certainly of Mantle origin. The author has impressive evidence that water and other volatile materials are absent from the Mantle and unlikely to emanate from deeper levels of the Earth (subjects that are beyond the scope of this discussion). Mechanisms of magma formation are, consequently, restricted to processes that do not involve such fluids.

## Formation of Sills

In my preceding contribution on tectonic earthquakes the idea was developed that warping of the lithosphere resulted in deep-seated "lithosphere fractures." These fractures are interpreted to occur along foredeeps at platform boundaries, adjacent to fold systems, along oceanic trenches, and, perhaps, in other regions of relative immersion beneath the hydrosphere.

As the lithosphere fracture extended itself downward, its base must have widened. Simultaneously, its upper extent must increasingly have experienced compression. In this way a Mantle chamber in the widening lower part of the fracture would be capped by the increasingly compressed upper part.[103]

However, tectonic movements never take place in one direction for long; inverse movements with contrary sign and lesser amplitudes follow. With reversals of tectonism the magma chamber is compressed and is squeezed upward. When it reaches the zone of lateral compression, magma is squeezed laterally (with inversion, compression is relaxed but does not disappear completely). When the rocks of the containing chamber are consolidated, the magma flows laterally along the base of the compression zone. If the rocks are not consolidated, it is injected, taking advantage of available cracks, bedding planes, and other weak zones.

At lower levels in a sedimentary cover, zones of lateral compression may be significant barriers to upward magma movement because plastic members of the sedimentary column, such as shale or salt, may flow, thereby equalizing lithostatic pressure in all directions. This allows sills to maintain the same stratigraphic positions over great distances.

Thus, it was that the floor of Cambrian salt on the Siberian platform was recognized by Feoktistov[104] to comprise a sill with the enormous lateral extent of about 1 200 km.[105] Similarly, trapp sills with floors comprising clay beds were found by Walker & Poldervaart[106] in South Africa.

Magma intrusion into sedimentary cover may be limited by a zone of lateral compression that may be located either above or below it. Such sills are found to jump to higher stratigraphic

[103] *Yoder 1979*

[104] *Feoktistov 1976, 1978*

[105] *Ryzhov & Mordovskaya 1979*

[106] *Walker & Poldervaart 1950*

horizons on anticlines,[107],[108] where the base of the sedimentary cover is affected by more lateral compression than in adjacent areas.

## A Magmatic Fluid Regime

A sill, thus, intrudes from the Mantle as a dry magma, makes contact with fluid-saturated rocks, and is differentially permeated with ambient fluids, among them water, being the most abundant, but also gaseous hydrocarbons, $H_2S$, N, $H_2$, $CO_2$, $NH_3$, $P_2O_5$, etc. The influence of elevated temperatures on the sedimentary rocks leads to decomposition of their organic components. Heavy [carbon-rich] hydrocarbons are destroyed; methane and free carbon evolve. Ultimately, the methane also decomposes with liberation of hydrogen and deposition of free carbon.

An especially good example of this process is a ring of graphitic rock known as the Ospinsko-Kitoysky massif of the Altai-Sayan fold belt, which "forms an almost continuous ring encircling a swarm of late harzburgites."[109] These graphite deposits surely represent remnants of an ancient hydrocarbon deposit that was destroyed by harzburgite magmatic heat.

The usual reactions between magma and country rock are of the following kinds: $CaCO_3 \rightarrow CaO + CO_2$; $CaSO_4 \rightarrow CaO + SO_3$; $SO_3 + CH_4 \rightarrow H_2S + CO_2 + H_2O$; etc. Sometimes magma is able to melt the country rock (forming granite, gneiss, carbonatite), partly assimilating or mechanically mixing with the melt, and in this way becoming enriched in $Al_2O_3$, $SiO_2$, alkalies, and salts (Na, K, Ca, Mg, B, P, F, Cl, Br, etc.).

Gases and distillates from below are in the best position to mix with magma and to transform some of the components from oxide-silicates to sulphides, native metals, and carbides:

---

[107] *Efimov 1979*

[108] *Frolov, Efimov, Belozerova 1976*

[109] *Kolesnik 1965*

$$Ni_2SiO_4 + 2H_2S \rightarrow 2NiS + SiO_2 + 2H_2O$$
$$Fe_2SiO_4 + CH_4 \rightarrow 2Fe + Si + CO_2 + 2H_2O$$
$$Fe_2SiO_4 + 4H_2S \rightarrow 2FeS + 4H_2O + Si$$
$$CuSi_2O_5 + H_2S \rightarrow CuS + 2SiO_2 + H_2O$$
$$Al_2O_3 + CH_4 \rightarrow 2Al + CO + 2H_2O$$
$$SiO_2 + CH_4 \rightarrow Si + C + 2H_2O$$
$$SiO_2 + CH_4 \rightarrow SiC + 2H_2O$$

Thus, the interaction between magma and hydrocarbon fluids defines an interrelationship between the generation and deposition of native metals, silicon, carbon, and carbides.[110] Possible interactions between magmas and country rocks and their fluids, of course, may be quite varied; and the reactions indicated above are only some of the possible variants.

## Formation of Dikes

The processes of isostatic equilibration of the overburden is protracted after completion of magma injection from Mantle into sedimentary cover. The process requires conformity of the hypsometry and pressure distribution in the chamber to the laws of Archimedes and communicating vessels. Owing to variations in sill thickness, the overlying complex warps. The more frequent the variations in sill thickness, the greater the warping of the oversill complex, and the more numerous the magma-filled chambers become (**Figs. VI-7, VI-8**).

Single dikes from a sill may lead to higher stratigraphic levels, but a proliferation of dikes is always found in proximity to the originating sill and to alterations in its thickness. Contraction of cooling magma at sill contacts may create new voids; and these may fill with residual unconsolidated magma. In the case of thick sills these processes may continue over a lengthy period

---

[110] *Minerals..., 1985; Nuggett metals..., 1985 "NATIVE METALS IN ERUPTED ROCKS" Yakutsk, 124pp*

during which time the composition of the residual magma may change, often from mafic to felsic. This can give the illusion of multiple dike complexes with differing ages and compositions that deviate from "normally functioning deep magma reservoirs."

## Hydraulic Fracture and Diapir Structures

In trapp and other provinces isometric blocks of country rock are found to be enveloped in irregularly-shaped igneous bodies of intrusive and extrusive character. Due to the presence of tuffs and agglomerates, these igneous bodies have usually been interpreted as deposits of paleo-volcanoes. They are widespread on the Siberian platform and well understood from drilling. Elsewhere they are also known, as in the Arthur Mountain complex, South Africa.[106,111] These structures are formed by sill intrusion and by extrusion of lava to the surface, when the oversill complex fails to retain the reservoir pressure.(**Fig. VI-7**) The author has called such structures "frack" structures, drawing on the oil industry short-form for "hydraulic fracturing," which produces horizontal breaking out that is wholly analogous to sill injection.

Many trapp provinces have basalt plateaus with horizontal lava flows of 5- to 12-m thickness (rarely up to 50- to 70-m thickness) and with 800-m breadths.[112] These regions are devoid of volcanoes; and the youngest flows occur atop the highest table mountains.(**Fig. VI-8**)

The formation of these copious floods of lava [the "flood basalts" of the world], required lithospheric rifts for which active modern analogues are not recognized. Notwithstanding lack of a comparative, initiation of such a system could occur where a broad uplift lost its sedimentary cover to erosion and was then

[111] *DuToit 1957*

[112] *Lebedev & Staroseltsev 1972*

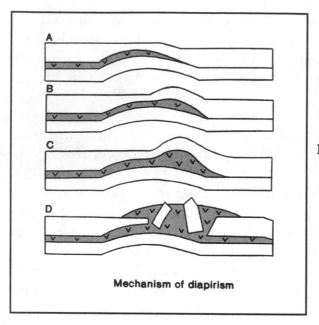

Mechanism of diapirism

**VI-7**

**MECHANISM OF DIAPIRISM**

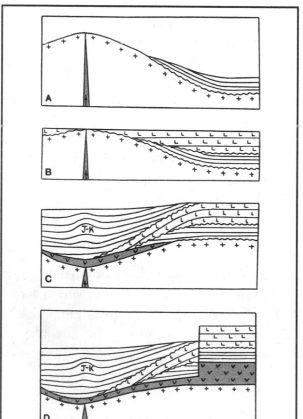

**VI-8**

**PLATEAU BASALT FORMATION**

| | | |
|---|---|---|
| + + | Basement | |
| | Paleozoic | |
| L L | Lava cover | |
| | Mezozoic | |
| | Intrusion of ultramafic sill | |

drawn into "pulsating immersion." Lateral compression above this sort of lithospheric rift would be narrow; and extruded lavas could be accommodated on the surface above and beside it.

In the case of the Siberian plateau platform, basalt flowed from the north out of a lithosphere breach that is now buried under Mesozoic/Cenozoic sediments of the Yenisei-Khatanga depression. Lava flow began when the uplift lost its sedimentary cover and was drawn into pulsating immersion (**Fig. VI-9A**). The flow ceased when the basin ceased to be a positive landform (**Fig. VI-9B**). However, immersion did not stop there, and continued until accumulated deposits reached and even exceeded 4- to 5-km thickness. Continued sedimentation caused a thick zone of lateral compression to develop above the lithosphere rift, which forced magma to be injected laterally as a sill (**Fig. VI-8H,I,J**).

VI-9
**MECHANISM OF VOLCANIC PIPE FORMATION**

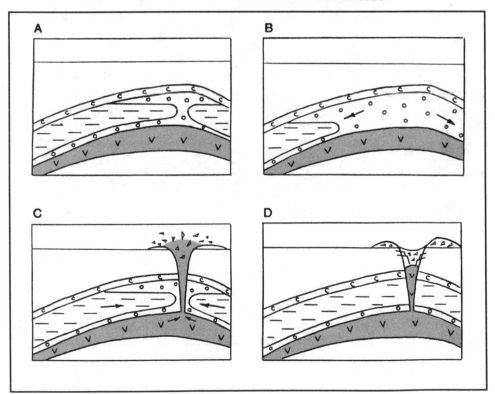

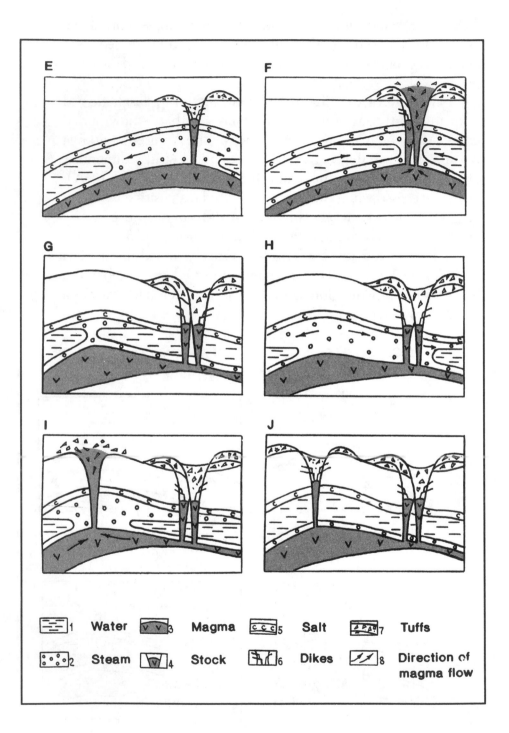

Inasmuch as the basic lava flowed from the upper part of the magma chamber, the subsequent sill injection comprised high-magnesian magma. The ultramafic sill was then emplaced at the base of the sedimentary cover of the Siberian platform with a thickness in places of as much as 1 000 m.

In addition, in accordance with Archimedes Law, the sites of maximal cover thickness should be raised to the highest elevations; and, hence, the youngest flows undoubtedly flowed out into places that were negative landforms previously but are now the highest table mountains (**Fig. VI-8B,D**).

This proposed model is in good agreement with geological and geophysical data on the structure of the Yenisey-Khatanga. The anticipated ultramafic sill reaches the surface as the Gulinskayya intrusive of mainly dunite composition and forms the alkaline ultramafic massifs of the Maymecha-Kotui province. This model conforms well to all geological information on plateau basalts and their habitats. It will be of assistance to us in understanding a kimberlite discovery on the west slope of the Anabar shield.

## ORIGIN OF PIPES

An injection of magma into water-saturated strata of the sedimentary cover leads to immediate steam generation at the contacts. The desiccating contact zone becomes heated, fissured, and highly permeable. Steam, transmitted along the contacts may collect in uplifts and other traps, displacing oil and water as it accumulates.

Sill thickness at axes of uplifts may increase or decrease, and in thickening and thinning, distort and weaken the oversill complex. Such weakening allows the steam to work its way upward. Percolating through micro-pores in the strata, it finally bursts out at the surface.

Only at magma/country rock contacts can magma move rapidly. If it collects in a chamber, bursting out may create a pipe to the surface. At the base of such a pipe the steam moves with maximum velocity, its turbulence pulverizing, disintegrating, and size-sorting entrained rock particles and droplets of magma as they are carried to the surface.

Crystallization within the magma renders it a partially solidified slurry of crystals and melt. This segregation of crystals accounts for the well-faceted crystals that are often found in tuffs. The hot material cools and becomes more abrasive as it rises through cold rocks in the widening conduit. In time this results in the fissure conduit becoming more pipe-like (**Fig. VI-9**) because of wall abrasion by tuff fragments entrained in escaping steam. A conduit with a round cross-section is eventually produced in this way.

After exhaustion of steam resources, magma rushes into the pipe. Its momentum causes it to surge into the entire pipe, after which it settles back, and the pipe walls cave in above it. As this steam/magma cycle is repeated, resurgent steam and magma are compelled to percolate through the existing cooled debris. This additional contact with cool rock causes the effusive steam slurry to cool at progressively deeper levels; and eventually the conduit is plugged.

In a pipe that is incompletely plugged, newly arriving steam may flow unimpeded and explosiveness may not recur. Many Kurile volcanoes have this characteristic, continuing to "smoke," blowing off steam during a quiet period between eruptions.

In the case of virtual plugging of a conduit, when steam is barely able to escape, a new generation of steam must result in creation of a new fissure peripheral to the old pipe or intra-pipe, breaching the debris within it. In any case the new fissure will not match the position of the old one, which is solid with

former magma and its contacts injected and impermeable. (**Fig. VI-9**) These eruptions, then, become intra-pipe or peripheral dikes. The degree of pipe filling may vary in this process practically from zero to 100 %.

Upper levels of the pipe may again be reentered by steam transmitted through a newly created fissure. And once again, the natural pulverizing process and plugging of the new conduit by debris and solidified magma operate after exhaustion of the steam resources.

Emissions from deep sources result in pipes that do not normally exhibit the foregoing behavior because rock plasticity tends to collapse any newly formed conduit. Under these conditions it is appropriate to draw the conclusion that, when

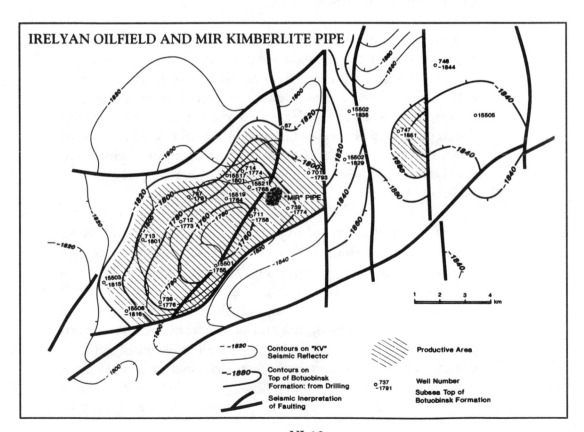

VI-10

viscous magma replaces steam surging through a fissure, the fissure behaves like an hydraulic fracture. Viscosity impedes magma progress, plastic flow of wall rock drives the walls together against the magma, and an hermetic seal is thus formed.

Salt has entered kimberlite- and tuff-filled pipes in many places of the Yakutia province, a behavior not unlike that of surging magma. The Mir structure (**Fig. VI-10**) involves six kimberlite columns, and the Kimberly structure [South Africa], sixteen. Six and sixteen eruptions are indicated respectively. At depth there should be splitting into six and sixteen small tail-like pipes. Such splitting is found in many large kimberlites[113], and also for ironstone (siderite) pipes.[114] The Udachnaya pipe at Yakutiya[115],[116] and the Frank Smith Mine structure in Africa[111] look like two carrots and undoubtedly were formed by two eruptions.

Deposition of extrusive rock near a pipe usually leads to depression of the terrain and squeezing out of underlying magma beneath adjacent terrain. The height of land may shift as a result (**Fig. VI-8**). Then, when a new steam accumulation occurs, it bursts out (**Fig. VI-9**), creating a new pipe. Several dozen pipes may be formed on large uplifts in this way. Usually, an oval shape to a field of pipes typifies a platform uplift, as at the Daldyn, Alakit, and Kharamay kimberlite fields of the Siberian platform.

After completion of explosive and intrusive activity connected with the formation of a pipe, the pipe content undergoes hydrothermal-metasomatic processing by fluids filtering through it. This process proceeds for a long time, until completion of solidification and chilling, at which point the character and scale of metallization will have been set. Both pipe depth and its rate of cooling of metallized fluids are critical in determining metallization character. In shallower pipes there

[113] *Kostrovitsky & Egorov 1982*   [115] *Bobrievich & Sobolev 1957*

[114] *Fon-der Flaass 1980*   [116] *Bratus et al. 1988*

is insufficient cooling time for ore accumulation. Thus, it is a matter of practical interest to explore for endogenous metals only in deep-seated pipes.

If a water-bearing horizon is below a sill, steam will break through the magma in the form of bubbles. A steam accumulation may then occur in the higher part of the sill chamber, where it will displace magma. A "primary" steam eruption commences with bursting upward of steam at near-lithostatic pressure. As the magma follows the steam, it is gas-lifted by steam saturation [in the same fashion that petroleum is gas-lifted by solution gas] to the surface through fractured oversill cover-rocks.

Whereas this mechanism is applicable to any volcanic pipe that pierces a central uplift, local conditions still must be taken into consideration. These include the character of the eruption, its volcanic products, its periodicity, and the power base of the eruption. The model offers some basis for resolving the problems of volcanism and endogenous mineralization. The great volumes of volcanic ash that often occur would be unexplainable without the pulverizing mechanism. Petroleum and natural gas showings found in many pipes with volcanic rock are enigmatic unless one recognizes that both volcanoes and pipes occur on uplifts, the natural structures into which petroleum tends to migrate. An interrelationship between petroleum and volcanoes is clearly implied.

The conventional idea held by many scientists as to the provenance of basement and Mantle rock debris found in pipes is also contradictory to the interpretation advanced here. From the outset it must be recognized that we have little knowledge of basement composition, to say nothing of Mantle composition. Thus, we are unable to counter assertions made by other scientists on this subject. However, in the case of the Mir pipe, where many wells penetrate basement rocks, recognizably contradictory evidence is produced and routinely ignored.

Crystalline argillites in the kimberlite pipes of Yakutiya have been routinely classified as debris of basement rock. Resembling Archean argillites of the Anabar shield, they are, however, rounded to oval in form[111], strange shapes for xenoliths. It is also strange that no such argillites occur in pipes into which erosion has cut deeply and which are situated directly beside the Anabar shield,[117] where basement is so close by.

These data support convincingly an opposite view as to the origin of these crystalline argillite clasts. They derive, indeed, from the Anabar shield, but they entered the pipes from above, not below. They can be attributed to sedimentary deposits that were breached by the pipes. Their oval forms represent erosional shapes of sedimentary clasts that were transported into the area in a period before the pipes burst through the sedimentary cover rocks. The host sedimentary strata contributed some of the clasts to the open pipe throats before erosion removed the formations from the vicinity of the pipes.[118]

Another popular idea of our day is to identify mantle-type xenoliths in pipes and, thusly, to classify the pipe according to its rock type directly. A bold researcher may determine Mantle composition; a timid one may opt for "uncertain" or "heterogeneous" origin. With little known about actual Mantle composition and little likelihood for contradiction or direct observation to verify any determination in the foreseeable future, a monotonous composition of mafic or ultramafic rock is taken for granted.

Kimberlite pipes are related to ultramafic (dunite, harzburgite) sills. Dry magma of this kind at high temperature (>1 900°C) is chemically rather active and able to melt to one extent or another, any country rock, and to mix and react with its components. Just such processes give the wide variety of lithologies found in kimberlites, carbonatites, and massifs of mafic to ultramafic rocks and in magmatic complexes of central

[117] *Ponomarenko et al. 1970*

[118] *Ed. note: The reader will note that this mechanism duplicates that described in the EV volume at the Gros Brukkaros conduit, Namibia.*

uplifts. Such processes generate regional metamorphism and granitization. Thus, there is no need for explaining the occurrence of ultramafic "Mantle debris" in kimberlite pipes.

The other type of "Mantle debris," the eclogites, are formed by metamorphism during high-pressured injection of ultramafic sills into country rock of carbonate-clay lithology. The potential for generation of eclogite lithology by metamorphism of sedimentary rock has been shown experimentally to be feasible.[115,119,121,122,123,124]

As we close our consideration of the pipe-forming problem, it should be emphasized that a variety of special features characterizes specific situations. These special situations are beyond the scope of this discussion. If a water-saturated horizon is beneath a sill, then pipes are formed in the upper levels of breached intrusive complexes. Pipes of carbonatite composition are formed in ultramafic rock.

The first step of this process is steam breaking through the differentiating intrusion. The steam concentrates the most magnesian differentiates in lower levels. Upward, these differentiates give way to more silicic rocks. The second step is melting of undersill carbonate rocks, a process that yields a mobile melt. In the case where pipe formation occurs in unconsolidated rocks that are in the process of achieving isostatic equilibrium, the unconsolidated complex may continue moving upward carrying the pipe along with it.

When a thick ultramafic sill melts country rocks, its heat front may affect water-saturated horizons far from the sill. This interaction can result in the formation of a complex of pipes, their final distribution being dependant upon the compositions of the melted and ruptured rocks. On the other hand, it is possible that pipes form by heating of a water-saturated horizon

[119] *Sobolev 1976*

[121] *Gorokov 1972*

[122] *Lutz 1965*

[123] *Rosen 1972*

[124] *Trusova 1980*

without the melting of any rock.  In this case, the pipes will be filled with breccia.[125]

## THE ORIGIN OF DIAMONDS

The generation of diamonds in nature is limited by the general chemistry of carbon, especially by the opportunities for the formation of the bonds, C-C, C-H, and C-O.  Definitive factors in these relationships are enumerated in the conclusions expressed by A. V. Varshavsky[126]  after experimentation with diamond etching:

"... if potentials of oxygen and hydrogen in the environment of crystallization of any structural form of carbon are significant in any way, a chemosorption of these elements on the surface of the growing crystals of carbon is inescapable.  As oxygen and hydrogen are abundant in nature so, in most cases any newly-started growing particle of carbon takes on the aforesaid elements on its surface, a process that leads to slowing and cessation of growth.  Even a minute concentration of oxygen and hydrogen in the environment bring about deposition of carbon in dispersed form.

... a possibility of generating large, high-quality crystals of carbon is limited to conditions of maximum oxygen-hydrogen sterility ('OHS') of the crystallization environment.

... OHS conditions are rare in nature, a circumstance that explains why primary deposits of diamond are unique and why practically all free carbon in nature is in dispersed form - crystals of graphite are as rare as diamond itself.

... further attempts to synthesize large diamonds under

[125] *Berman 1973*

[126] *Varshavsky 1981*

low-pressure conditions without OHS conditions will be futile.

... OHS conditions may provide a basis for exploring new ways to synthesize diamonds and other structural forms of carbon."

One must agree with A.V. Varshavsky excepting in one respect: OHS conditions in nature are not so rare, and diamond deposits not so unique as they seem to him. His conclusions otherwise must be carefully taken into consideration in devising hypotheses for genesis of natural diamond.

An initial necessary condition for pipe formation is the existence of closed positive structure in sedimentary cover. Availability of hydrocarbon is a second prerequisite, especially its availability in some quantity in the middle and lower parts of the cover, where conditions for hydrocarbon preservation are usually more favorable. As there is insignificant ambient oxygen in proximity to hydrocarbon deposits, only the hydrogen ambience is of concern. Hydrogen is always present with hydrocarbons, usually in small amounts.

The temperature effect of intrusive magma on indigenous hydrocarbons is another matter, however, as it leads to hydrogenation and the following typical reactions:

$C_{10}H_{22} + H_2 = 2C_5H_{12}$ and $C_5H_{12} + H_2 = C_2H_6 + C_3H_8$
In general: $C_{2n}H_{4n+2} + H_2 = 2C_nH_{2n+2}$

After available hydrogen has been exhausted by the agency of magma heat, OHS conditions prevail. Continuing heat in the absence of free hydrogen results in ongoing destruction of hydrocarbons with release of free carbon. OHS conditions favor carbon crystallization in the following reactions:

$$2C_2H_6 = 3CH_4 + C$$
$$3C_3H_8 = 4C_2H_6 + C = 6CH_4 + 3C$$
$$4C_4H_{10} = 5C_3H_8 + C = 10CH_4 + 6C$$

$$nC_nH_{2n+2} = (n+1)C_{(n-1)}H_{2n} + C = \frac{(n^2+n)}{2}CH_4 + \frac{(n^2-n)}{2} \cdot C$$

Phase transitions of liquids (water and oil) to gas at the magma contact with a fluid-saturated stratum provide the mechanism for pipe formation. Owing to destruction of hydrocarbons, the mechanism also allows growth of carbon crystals. Multiple stages in pipe formation allow diamond crystallization to occur in tabular form, a form attributed to "intermittent growth conditions" [127] that are difficult to imagine in the Mantle.

Diamond formation must occur within a fluid-laden horizon that has been in contact with an ultramafic intrusion, which has caused destruction of hydrocarbons. Because hydrogen release by methane destruction at the ultramafic contact would prevent diamond crystallization, the process must occur at a distance from such a contact. This distance is controlled by temperature and migration conditions for hydrocarbons within the stratified rocks. Proximity indicates heat in excess of diamond tolerance; excessive distance provides insufficient heat for hydrocarbon destruction. OHS conditions and diamond crystallization temperatures are possible only within an optimal, "in-between" zone.

The major conclusion is easily recognized: There must be in nature stratiform diamond deposits in which diamonds are sorted as to type, kind, and size in a series of fluid-bearing horizons adjacent to ultramafic intrusives.

Diamonds found in pipes have been transported toward the surface. These should be understood to comprise no more than a minor fraction of the diamonds that are formed. Most

[127] *Bezrukov 1974*

diamonds remain *in situ* after formation. *In-situ* diamonds, unlike diamonds found in kimberlite pipes, have not undergone rough mechanical treatment and sustained severe changes in the temperatures of their habitats. Hence, they have much more chance of being preserved as crystals of maximum size and quality.

## THE GEOLOGICAL CONDITIONS FOR DIAMOND FORMATION

It has already been pointed out that a thick oversill complex and rock plasticity may be insurmountable barriers to steam that is attempting to burst through to the surface. This insurmountability occurs even under conditions of thick, ultramafic-sill intrusion and the presence of copious amounts of fluid in the intruded rocks of a large uplift. Notwithstanding, these conditions are conducive to diamond generation. Time should only improve crystal quality.

The high quality of diamonds found in the Vaal and Orange rivers [South Africa] above the kimberlite pipes of that region[128] is related not only to natural sorting but also to derivation from stratiform deposits to these placers.

Diamond placers within or near fields of kimberlite pipes are not necessarily related to the pipes; and other placer occurrences may be unrelated to recognized sources. However, there is a general association between known placers and mafic to ultramafic massifs at many locations: Southern Rhodesia, the Congo basin, West Africa, Australia, Tasmania, Brazil, Guiana, Canada, Malaysian Archipelago (Kalimantan), Sumatra.[129]

Trofimov[130] considers that "the connection has been discovered and is clear enough between diamonds and

---

[128] *Trofimov 1967*

[129] *Gorokhov 1975*

[130] *Trofimov 1967, p182*

ultramafic rocks in geosynclinal regions, where diamond placers have usually been found near an outcropping of ultramafic rock."

There is no obvious connection between kimberlites and diamond occurrences of the Urals and Ukraine.[135] Notwithstanding, there are numerous diamond discoveries, even mines, in mafic and ultramafic terranes of the Urals, the north Russian platform, eastern Sayan, Kamchatka.[131,132,133,134,135,136,137,138] as well as other areas of the Earth.[130,135,139] As it was shown above, it is clear that diamonds are formed in "exo-contacts" of ultramafic bodies. They may occur within such bodies only accidentally, where they may occur in xenoliths of locally-ingested country rock, or where they are raised from depth by exogenous processes. Thus, such discoveries should be rare; and it is noteworthy that this is confirmed by repeated investigations.

Where an ultramafic sill penetrates carbonate or clay-carbonate rocks, dunite-harzburgite magma causes carbonitization and eclogitization, thus defining the carbonatite and eclogite parageneses of rock that are normally associated with diamonds. This association usually exhibits a direct relationship with magnesium, chromium, nickel, and cobalt and an inverse relationship with iron, titanium, aluminum, and calcium in the kimberlite pipes of Yakutiya.[140,141,142,143,144] An association between diamonds and high-chromium minerals

[131] *Kaminsky 1974*

[132] *Kamchatka -*

[133] *Lukjanova et al.1978*

[134] *Markovsky & Rothman 1981*

[135] *Trofimov 1939*

[136] *Shestopalov 1938*

[137] *Shilo et al. 1978*

[138] *Shilo et al. 1979*

[139] *Trofimov 1980*

[140] *Sobolev 1951*

[141] *Makhotko 1981*

[142] *Ilupin & Nagaeva 1970*

[143] *Marshintsev 1970*

[144] *Nikishov et al. 1979*

would be supported by the presence of knoringite in pyrope and other known features characteristic of kimberlites. A similar possibility for their occurrence in dunite as a separate mineral phase exclusive of knoringite has not yet been discovered.

Diamonds should eventually be found in carbonate "shell" and eclogites including those not associated with kimberlites,[145,146] thus confirming the formation of diamonds in rocks enveloped by ultramafic intrusives.

If an ultramafic intrusive occurs within terrigenous sedimentary rocks, diamond associations should be those characterized by granite and metamorphic parageneses. Occurrences of these kinds are found sparsely in the pipes of Yakutiya.[144] Such parageneses are clearly evident in diamond placers in many parts of Africa, Brazil, India,[147,148,149] and Siberia as well.[150]

We have now considered the clear connection between diamond formation and its proximity to an ultramafic sill. Our theory does not preclude the formation of diamond far from a sill of great thickness nor from diamond association with other magmatic intrusives. Available data imply the widespread existence in nature of diamonds and the advisability of extending exploration to regions not presently considered prospective.

Solonenko,[151] quoting Petraschek,[152] says that in the Kragman coal measures "coal is metamorphosed through the influence of [an intrusion of] andesite to graphite with diamonds." In this occurrence it is probable that the diamonds were formed not directly through the metamorphism of coal but by thermal conversion of organic matter to coal and hydrocarbons, the

[145] *Zorin 1972*

[146] *Zayachkovsky 1972*

[147] *Vernadsky 1959*

[148] *Menyajlov 1963*

[149] *Fersman 1960*

[150] *Meyster 1908*

[151] *Solonenko 1958*

[152] *Petraschek 1931*

latter then being themselves destroyed and diamonds and graphite deposited. Related non-conventional diamond generation has been demonstrated by such distinguished scientists as V.I. Vernadsky[147] and A.E. Fersman[149].

The concept of the involvement of high pressure in the formation of diamond is based on the results of the first successful experiments in artificial production of diamond under pressures of 40 000 $kg/cm^2$ and more. New data indicate the feasibility of diamond generation not only from graphite but also from other materials including methane, ethane, propane, chlormethyl, and carbon dioxide under moderate to low pressures (even atmospheric and vacuum levels down to 0.1 mm of mercury.[124,153,154,155]

What then can be said in support of the requirement for high pressure for the crystallization of diamond? Abundant data exist on the formation of kimberlite and diamond in connection with the MIR pipe in Yakutiya. This pipe is found above an oil and gas field, which sits directly on basement at 2 100-m depth. The pipe is little eroded today. It is believed that lithostatic pressure at the time of its formation was 700 to 800 $kg/cm^2$. The imputed ultramafic sill that initiated kimberlite emplacement has not been found. Furthermore, it is apparent that had the petroleum deposit been breached by the pipe it would have escaped to the atmosphere.

Throughout the Botuobinsk kimberlite fields and the adjoining regions at a level of 30 to 80 m or so above the productive oil and gas formations there is a fractured horizon that causes drilling-fluid loss and to date has not been recovered in cores. Although the lithology is as yet not identified, it is likely an ultramafic sill.

The aforesaid 700 to 800 $kg/cm^2$ lithostatic pressure may not be taken for the pressure of diamond formation at the MIR pipe.

[153] *Bezrukov 1974*

[154] *Bezrukov et al. 1976*

[155] *Shestopalov 1938*

Rather, the pressure in the fluid phase, perhaps a much lesser figure, should be taken. In this respect the observations of A. I.. Botkunov[156] on diamond distribution in the MIR pipe are most interesting. It turns out that diamond distribution has no connection with the location of specific kimberlite chimneys of different ages. This lack of correlation implies late-stage generation, and led Botkunov to interpret a hydrothermal-metasomatic origin for the diamond.

This origin may not apply, of course, to all diamond genesis. Certainly, debris containing diamond crystals has undoubtedly been injected from below in many places. And it is not possible to rule out direct formation of diamonds within some pipes. Under these conditions diamonds must have formed at very low pressures, less than 2 kg/cm$^2$.

Recently there has been great hope that isotope analyses would resolve the question of diamond genesis. Unfortunately, isotopic fractionation during various active natural processes has only served to confuse the matter further.

Other problems as well have become obstacles to resolution of the genesis problem. One has been the age determined for the Yakutian diamonds: 1.5 to 2.0 by. Radioactive decay starts at the "birth" of the rock mass being measured. But what is this date? Is it the time of rising and emplacement at their final resting place of the elemental components of the rock? Is it the moment of crystallization from a melt or hydrothermal solution of the minerals? Is it the time when the latest process of anatexis, metamorphism, diagenesis, or migration occurs in the pore fluids of the rock (the time of "hypergenesis")?

A second problem has been ascertaining that the particular elements analyzed for age determinations have arrived with the newly formed rock and then have remained in place without addition or leakage throughout their long residence-times.

---

[156] *Botkunov 1964*

These are conditions that can hardly be presumed in nature. The full use of knowledge available for isotope analyses is, unfortunately, impossible to apply in geological situations.

A continuous series exists between carbonatites and kimberlites in nature.[111,157] Recognition of "carbonatite-kimberlites" as a variety of diamondiferous rocks[158] of "kimberlite-carbonatites"[159,143] and the association of carbonatites with ultramafic rocks, all imply mutual close interrelationships.

The model here suggested implies zonation of diamond and other minerals (garnet, for example). Such zonation is related to the multistage nature of pipe-forming events. The number of these zones for any given kimberlite chimney should be n-1, where n equals the ordinal number of the chimney. However, as Butkunov[156] surmises, further growth of diamonds in the pipes may not be precluded.

It must be clear that within the framework of our model there are kimberlite chimneys (as, for example, the MIR, Aikhal, and Udachnaya pipes) where "later-arrived kimberlite is richer in diamonds" [than earlier-arrived kimberlite].[160] There occurs a new cycle of diamond crystal growth (including new crystal formation) before every new eruption of kimberlite.

The eruption of kimberlite pipes through granites and other crystalline rocks calls for special attention. Only through the writings of V.S. Trofimov[130,139] has the author been introduced to the preliminary reasoning on this subject, which will require research for its verification.

Granite is created along with gneisses from the thermal effects of mafic and ultramafic intrusives on terrigenous [quartz-rich sedimentary] rocks. Rarely are granites the product of

[157] *Milashev 1972*

[158] *Malkov 1975*

[159] *Lapin, 1984*

[160] *Dobretsov, 1972*

crystal differentiation from mafic magmas. In any case the high-magnesian mafic and ultramafic deposits should be under the granites. Kimberlite pipes within such granites may originate in underlying ultramafic rocks, and in the thermal effects of ultramafic magma acting on a fluid-laden horizon either above or below the intrusive.

Pipes developed in viscous granite magma may follow a distorted course due to isostatic disequilibrium.[161] Their peripheries should incorporate clasts of granite. Pipes lacking these are apt to have developed from deep-seated intrusives that have been injected into unmetamorphosed fluid-saturated rocks.

Stratified diamond deposits that have long been mined in South America are associated with phyllites.[130,139] The phyllites occur either as beds or transverse dikes of small thickness (<100m) and are apt to be altered ultramafic intrusives and dikes. Diamond formation is quite possible in dikes and sills, which are partially congealed but still hot enough for destruction of hydrocarbons that move in contraction fractures that develop as the rock cools. With this type of sill the country rock is not sufficiently heated for its hydrocarbons to be destroyed and diamonds formed in it.

In the case of increased sill thickness, the zone of diamond formation should migrate farther into the country rock, which then may comprise the principal source for diamonds in placers in many diverse terranes. Only rarely, however, has this condition been confirmed, as it has in the eclogites of the Kokchetav massif of Kazakhstan.[123] The potential is there, nevertheless, for the diamond-forming process to become so intense that diamond reaches the abundance of a rock-forming mineral.

Horizons of garnet clinopyroxenites have been found in the Beny-Bouchers massif of North Africa with diamond pseudomorphs of graphite octahedra with flat facets and sharp

---

[161] *Ed. note: The author considers that granite magma distorts the conduit through which it rises.*

edges, having a mass up to one carat and comprising two to five percent and locally up to fifteen percent of the rock.

The interpretation for these primary diamonds was that volatile components had accumulated in a residual magma.[161] More likely, the diamond formed from jets of hydrocarbon gas that permeated the porous horizons from below. Passage of such comparatively cold gas would have lowered the temperature sufficiently for the crystallization of diamonds as the gas was destroyed. Then, after exhaustion of the gas, high temperature was restored, and the diamonds turned to graphite.[162,163]

Even today we cannot produce artificial diamonds of the quality and sizes achieved by nature. According to the theory presented here it should be possible to simulate and imitate the natural diamond-forming process, to improve on it, and to grow crystals larger and of better quality. According to the publications available,[154,155,162, 164,165,166,167,168] there have been no such attempts.

Only J. B.. Hannay, who published the positive results of his experiments in 1880, re-created conditions remotely similar to the natural circumstances of diamond growth. He put a mixture of 90 % hydrocarbons (kerosene-paraffin), 10 % bone oil, and 4 g of lithium into steel tubes, the edges of which were welded. The tubes were then put in a furnace, brought to red heat, and kept in that state for fourteen hours.

In most cases the tubes could not stand the stress and exploded. Only three tubes out of 80 [and by another report, 4 of 34] survived. After the experiment, there were found to be eleven diamonds with average sizes of 0.4 x 0.2 x 0.1 mm,

---

[162] *Burmin 1984*

[163] *Ed. note: Moissanite, SiC, occurs in Yakutia sometimes in association with diamonds.*

[164] *Vasiljev & Belykh 1983*

[165] *Rich & Chernenko 1976*

[166] *Vilke 1977*

[167] *Nemilova 1956*

[168] *Elwell 1986*

carbon content of 97.85 %, and specific gravity of 3.54.  In 1943, K. Lonsdale confirmed by X-ray spectrography that these crystals are diamonds.

Hannay's diamonds are in the British Museum of Natural History in London under the label "Hannay's artificial diamonds."  It is to be noted that the pressures in Hannay's tubes were not more than 100 to 200 atmospheres.  As a consequence, some "experts" have suggested that Hannay's diamonds "may not be of artificial origin" [164] ... however, almost in the next sentence "this riddle has not been solved yet!"

My theory of diamond formation in nature may not yet be taken as faultless in all respects.  But the reader should recognize that it is internally cohesive and offers a resolution to the general problems of diamond geology.  Derivation of new meanings from old ideas and implications of a new theory in contrast to older ones may encourage the reader's indulgence.  If there are aspects to the theory that need verification, correction, or emphasis, let the reader take a charitable view to possible mistakes.

# LABORATORY WORK ON DIAMONDS

### C. WARREN HUNT

The foregoing analysis of diamond growth is substantially confirmed by laboratory diamond-growth experiments. A review in Engineering and Science[169] provides some insights that bear directly on the Skobelin theory:

1. A mixture of hydrogen and hydrocarbon gas heated by a high energy source, such as an electric arc or a hot tungsten filament, results in dissociation of hydrocarbons into elemental hydrogen and solid hydrocarbon "radicals" [partial molecules].

2. The radicals in the high-energy environment initially assume configurations of tetrahedra joined at their corners.

3. The high-energy environment causes hydrogen atoms at the surfaces to evaporate and be captured by nascent hydrogen of the atmosphere.

4. Exposed carbon is, thus, left in the isometric crystal form of diamond. Crystals continue to grow as hydrogen is stripped away.

5. The process can operate at 1 600°C and deposit diamond on a substrate with a temperature as low as 700°C.

These results demonstrate the crucial importance of nascent hydrogen in diamond formation, where it preserves isometric diamond crystals and prevents their recrystallization as graphite with hexagonal symmetry.

Thus, where Skobelin anticipated hydrocarbon dissociation by heat in highly-reducing conditions, and correctly interpreted that diamond will form at temperatures below presumed Mantle

---

[169] *Smith 1991*

temperatures, he did not anticipate the imperative for nascent hydrogen in the surrounding atmosphere. Goodwin shows that **nascent hydrogen is essential** for the preservation of isometric carbon.

In an unrelated study by Oded Navon (1991) on the gas in micro-inclusions in diamonds from southern Africa, it is shown that pressures of entrapped gas fall in the 1.5-2.1 GPa range. Taking into account the elasticity of the diamond crystals leads to interpretations of entrapment pressures in the 4-6 GPa range. That implies depths of 170-240 km and temperatures of 900-1 300°C. Diamond is stable in that PT range.

As this work is based on sound calibration of infrared absorption pressure shifts, **it is necessary to recognize Mantle provenance for the diamonds that were analyzed.** Shallow diamond origin [at depths of petroliferous sedimentary basins] is incongruent with the data from inclusions in these diamonds.

However, Skobelin did not specify a shallow depth constraint for his theory of diamond formation; and **the fact that hydrocarbons rise through the Mantle, in some instances, allows interpretation of diamond formation as a hydrocarbon degradation process of magmatic heat at Mantle depths.**

# EVOLUTION OF THE YELLOWSTONE HOTSPOT:

## ITS SILICA, HEAT, AND WATER

The prevalent understanding of explosive volcanism is that it represents the effect of sudden decompression of magma. The deposits that result, whether airborne or subsurface, are functions of the energy released and the volatile component of the magma. In an analysis of the geodetics of the Yellowstone caldera, Vasco, Smith and Taylor (1990) interpret that new rock is being added at the Yellowstone "hotspot" at the rate of 0.028 $km^3$/yr. "Groundswelling" is their term for the net effect.

The caldera area, measured as the area of the last collapse, is 3 375 $km^2$. Seismicity and gravity data imply that most intrusive activity occurs above nine-km depth; but earthquakes to depths of 15 km occur north and south of the caldera center, being situated to 30-km distance from the center. Activity has spanned two million years.

Let us calculate what has entered the Crust by this endogeny. First, we are justified by assuming that surface dilation is a direct function of silica being added in the subsurface. Our theory is that this is due to silane influx. The silane-generated rock component then comprises newly added mass in a 9-km-deep cylinder beneath the caldera. This volume is 30 375 $km^3$.

To the cylinder mass we can add 3 000 $km^3$ for ejecta, as estimated by Vasco et al.(1990) Then we must take into account the volume of the zone of recorded deeper tremors, which is broader than the surface caldera. This will encompass a truncated cone between depths of nine and fifteen km. It would seem reasonable to assume that 50 % of this volume is new siliceous rock, a further 23 675 $km^3$ thusly to be added. We now have a total introduced rock volume of 57 000 $km^3$. **Taking its silica component as 73 %, we have a net crustal addition of 41 610 $km^3$ of silica.**

Geoid elevations in the western USA (color plate, reverse of Frontispiece) illustrate with elegance that a feature much larger than just the immediate Yellowstone caldera terrane is emergent. A 1 600-km (1,000-mile) diameter welt, in fact, surrounds the Yellowstone volcanic apex. The welt results from diapiric endogeny and includes within its topology many regions of alpine prominence, which result from isostatic compensation for local mass deletion.

In addition to mountains of the immediate Yellowstone area, the ranges of the Idaho batholith, the Canadian Rockies, the Colorado Rockies, and the British Columbia Coast Ranges all express positive features superimposed on the great welt and its extensions to the northwest and south. Crustal volume is increasing by phreatic endogeny, which dilates the welt directly and by volume additions that result from oxidation of silanes from the Mantle.

Returning to the immediate Yellowstone environs, if the replaced rock in the cylinder below the caldera has a density of 2.6, the mass that has been added is $1.08 \times 10^{20}$g. Heat released in the process of creating silica from silane is 326.2 kCal/g. This leads to a figure for **total heat generated of $3.5 \times 10^{25}$ calories,** an exothermic heat "event" spread over two million years.

Applying the approximate largest published figure for regional crustal heat evolution, $10^{-5}$ cal/cm$^2$/sec., to the area of the Yellowstone calderas, we obtain a **total heat-loss over the 2-my period of $2.1 \times 10^{25}$** calories, 60 % of the heat generation figure, but quite comparable in view of the quality of assumed data as to rock masses involved.

**Water yield from the oxidation of silanes to produce $1.08 \times 10^{20}$g (41 610 km$^3$) of silica is $0.845 \times 10^{20}$g (84 667 km$^3$).**

## IMPLICATIONS FOR THE GENERATION OF EARTH'S CRUST AND HYDROSPHERE

Let us examine what is generated by silane reactivity. Earth's Crust is estimated as $24 \times 10^{24}$ g, $2.4 \times 10^{22}$ kg:

$$SiH_4 + 2O_2 = SiO_2 + 2H_2O$$

| | | | | | |
|---|---|---|---|---|---|
| 32 g | | 64 g | 60 g | | 36 g |

| | | | |
|---|---|---|---|
| .125 H | | .4667 Si | .1111 H |
| .875 Si | | .5333 O | .8889 O |

.3333 $SiH_4$   .6667 $O_2$   .625 $SiO_2$   .375 $H_2O$

If the Crust has on average 55 % silica ($SiO_2$), the water generated to create it from silane is: $.55 \times 2.4 \times 10^{22}$ kg $= 1.32 \times 10^{22}$ kg $SiO_2$ This amount of silica could be produced by the following reaction:

$$SiH_4 + 2O_2 = SiO_2 + 2H_2O$$

$7.0393 \times 10^{21}$ kg $+ 1.4081 \times 10^{22}$ kg $= 1.32 \times 10^{22}$ kg $+ 7.92 \times 10^{21}$ kg

$2.112 \times 10^{22}$ kg total          $2.112 \times 10^{22}$ kg total

That is to say,

$7.0393 \times 10^{21}$ kg of $SiH_4$ will make $6.1604 \times 10^{21}$ kg Si in continental rock, and $8.7991 \times 10^{20}$ kg of $H_2$ released would make $7.92 \times 10^{21}$ kg of water

Earth's oceans are estimated to have a total volume of $1.35 \times 10^{21}$ kg. The water that would have resulted from silane replacement activity to produce the entire continental Crust would have amounted to nearly six times that in the present oceans. Generation of all this rock by silane reactions is, however, unlikely, because most continental rock must have been present as primordial silicates. The huge "water equivalent" of silica brings up an important aspect of carbide-hydride systematics that was mentioned in Chapter I: **Oxidation of each molecule of either methane or silane produces two water molecules. Furthermore, it is noteworthy that oxide molecules are bulky and contribute substantially to the expansion of the upper geospheres.**

Thus, whereas it might be convenient to imagine that continental growth and ocean water both result from silane replacement of mafic rock, there are constraints to consider. Furthermore, cometary water and meteoritic rock have also contributed to planet growth over geological time at rates that are not well understood and may be quite large (e.g. Frank 1990).

Thus we are left with a quandry. Silane oxidation is quite able to account for the hydrogen in ocean water in "events" that would easily deplete the atmosphere of its oxygen. Let us consider what minimum volume of silane would eliminate oxygen from the atmosphere as it was oxidized:

Oxygen in the atmosphere weighs $.001 \times 10^{21}$ kg

Matching exhaled hydrogen weight $.000125 \times 10^{21}$ kg could convert it to $H_2O$

Matching silicon in silane is $.0008749 \times 10^{21}$ kg, which is added as crustal rock

Mass of added rock $.001875 \times 10^{21}$ kg

Crust mass given above $2.4 \times 10^{22}$ kg

Increment $= 7.8 \times 10^{-5}, .000078, .0078$ %

**These figures would indicate that the creation of 800 km³ of quartz sand by silane oxidation could remove the entire**

oxygen component of the atmosphere. Since hydrocarbons flowing with the silanes would have a similar effect on atmospheric oxygen, a smaller exhalation of sand could very well represent an atmospheric toxification event with consequences for iron deposition or biotic extinction.

Gottfried gives a figure for $CO_2$ bound in carbonates of $2 \times 10^{20}$ kg. This contains about 140 times the amount of oxygen present in the aforesaid 800 km$^3$ of silicon dioxide.

These calculations suggest that endogenic events that are quite small in terms of geological feature generation could have biologically catastrophic side effects through elimination of oxygen from the atmosphere, total toxification, that is to say. However, atmospheric toxification for most larger life forms would require only a fractional $O_2$ removal, perhaps as little as 20 %.

Hydrocarbon combustion that yielded initially ground-hugging $CO_2$ would kill most air-breathing life forms long before the oxygen was totally eliminated from the atmosphere. Local atmospheric toxification of this type is exemplified by the $CO_2$ belch from the water-filled volcanic crater known as Lake Nyos in Cameroon. Some 1,600 people and numbers of livestock were asphyxiated in a few minutes.

Many "events" cast in stone suggest this scenario to the geologist, among them the following:

(1) Frequently, thin sedimentary layers are strewn with killed organisms. Such fish kills are prominent in Silurian and Triassic strata where I have worked; and the Belemnites "battlegrounds" of the Jurassic is another such scene that comes immediately to mind. Slaughtered Belemnites resemble toy soldiers slain by a gas attack, their bodies whole and resting all together in a single horizon, evidently struck down suddenly, as in an "event" of toxification.

(2) Major volcanic eruptions are in evidence abundantly throughout the world: thick blankets of tephra, lahars, pyroclastic debris flows covering 10 000 $km^2$ and huge sand exhalations [as described in Chapter VIII]. The all-too-obvious relicts are as plain to see as if they were stone tablets listing "events" of the Gods' Olympics in a bygone era. Toxification is a probable corollary process when dilation and incrementation of the Crust occur, because the generation of silica from silanes unavoidably consumes twice as much oxygen as silicon on a mol-for-mol basis.

**I suggest that toxification of the atmosphere in Earth "belches" of silane + methane along with their derivative products [$CO_2$, CO] have been significant occurrences through geological time, perhaps on $10^5$ year frequencies.**

## Chapter VII

# RESIDUAL GAS IN CRUSTAL ROCKS

**Kimberlite pipe emplacement** was dealt with by Gold (1987) as a mechanism of explosive pulses in which volatiles entraining solid rock fragments are propelled upward through the solid lithosphere, one chamber at a time. He used the apt comparison of the decompression chambers that are occupied by divers.

The dynamics for Gold's mechanism are provided by pressure differentials: the lithostatic pressure in the wall rock beside a chamber versus the internal hydrostatic pressure within the chamber. The former rises with depth according to the mass of the overlying rock; the latter only rises within the chamber from top to bottom by the mass of the overlying volatiles. As the rock weighs between 2.5 and 3.3 times as much as the volatiles there comes a point under the particular conditions of rock and volatile mix at which plastic flow of the walls pinches off the base of the chamber.

Should new seepage into the chamber occur from continued volatile emanation, a new explosive surge upward can follow. **The energy released by this process is, in my judgment, the source of all earthquakes in the Mantle and many in the Crust.**

Skobelin's collapsing steam-filled chambers [Part 4] and my explosively-opening volatile-filled chambers would have similar seismic effects, both being sudden volume changes resulting from gas behavior. The steam-filled chamber is possible only at vadose crustal levels; the volatile-filled chamber is possible in solid [even plastic] Mantle as well as in water-saturated Crust.

Skobelin made an unresolved point as to first-arriving "p-waves" from earthquakes. In various provinces the first-arriving wave is compressional, in others it is tensional. The difference may be the contrasting first energy released in chamber creation (compressional p-wave) and chamber collapse (tensional p-wave).

Mantle-type minerals with chaotic assemblages of rock fragments of disparate origins have burst explosively into shallow crustal situations where they can be examined [and mined]. Kimberlites are distinctive members of this class that more generally can be called explosion breccias. Entrainment of the rock fragments in a chaotic, gas-driven slurry is clearly the only way these bodies could have been produced.

It has long been thought by most geologists that diamonds are brought to surface levels by the upward-bursting kimberlites. This creates a preservation problem, which Gold addressed. He thought that the cooling effect of decompressing gas could preserve the diamonds. The Skobelin concept of diamond generation at shallow levels peripheral to hot intrusives [including but not limited to kimberlites] brilliantly sidesteps the preservation problem and negates the Mantle- and exclusive-kimberlite provenance for diamonds.

Steam is the most abundantly available gas at crustal levels. Other gases are found in ambient form dissolved in pore fluids, and small amounts of residual gas are entrapped in bubbles as fluid inclusions, which show the gas-mix at the time the rock congealed. These gases include hydrogen, carbon dioxide, carbon monoxide, methane and other alkanes, nitrogen, helium, and argon. Hydrocarbon gas is a significant, if not major, component of most of the ambient and included gases. I will now analyze three occurrences that shed light on possible gas evolution.

# THE FUXIAN KIMBERLITE

The Fuxian kimberlite pipe is in China on the Korean border. A report by Leung et al. (1990) indicates a rock matrix with a silicon carbide component as great as 0.15 %. It occurs in two crystal forms, alpha, the "normal" high temperature form, and beta, a lower-temperature secondary form that is believed to have its origin with secondary recrystallization.

The alpha form of silicon carbide occurs in the Fuxian pipe as large [up to 5 mm] dark-blue crystals. It is thought to imply a Mantle provenance because of its abundance in the rock and because the minimal temperature required for its formation is >1 000°C. In my judgment this mineral is some of the deep-Mantle silicon carbide that results in silane and hydrocarbon emanations. It is quite stable once having reached the lower-temperature and pressure regimes of surface rock. Only by a violent mode of emplacement could the mineral and its chaotic associated rock clasts have reached the surface.

The beta form of silicon carbide is clearly a secondary mineral. It occurs in cavities [vesicles] overgrown and intergrown with quartz and often plated with an iridescent coating of native silicon. Diamond occurs in this kimberlite; and in one sample was found to encase a cluster of crystalline beta silicon carbide. The diamond, thus, must be late-formed [as Skobelin theorizes] under low pressure and temperature conditions.

Carbon isotopes have been analyzed to see how the carbon of diamond and of SiC compare. The finding is curious: $^{12}C$ in the SiC is enriched over $^{13}C$ [$\partial^{13}C$ $^o/_{oo}$ = -24 $^o/_{oo}$]. By contrast, in the diamond the comparable figure varies from -2.9 $^o/_{oo}$ to -4.8 $^o/_{oo}$. The authors theorize that fractionation is the cause of the difference: they do not say how. The -24 $^o/_{oo}$ C in the SiC is in the range thought to represent primordial C, "light" C, that is to say. If primordial SiC had been hydridized, and the hydrocarbon vapors subsequently exposed to heat sufficient to dissocaite them, I venture to interpret that *diamond was generated from* **this** *fractionated hydrocarbon:* $^{13}C$ **was retained preferentially by undissociated hydrocarbon, and** $^{12}C$ *was deposited preferentially in diamond.*

A transitory hydridization of carbon as well as silicon in transition from SiC to diamond must have occurred. Hydrogen, acting as a catalyst to break the SiC bond, was quickly driven off by heat [in excess of the heat toleration of hydrocarbon]. That left the dissociated carbon and silicon to crystallize as diamond, and it left elemental silicon deposits on the remaining undissociated beta SiC. Native silicon can, thus, be considered a normal mineral association for diamond.

**The Fuxian Pipe demonstrates the presence in the Mantle of silicon carbide at the level of a rock-forming mineral. Recrystallization of some of it has yielded the lower-temperature beta form. Hydridic dissociation of beta SiC is implied, and appears to have resulted in deposits of silicon metal and of carbon in the form of diamonds.**

Analyses of entrapped gas are not available on the Fuxian pipe, unfortunately. Let us look at a related occurrence.

## THE UDACHNAYA-ZAPADNAYA KIMBERLITE

Gas recovered from vesicles of the Udachnaya-Zapadnaya kimberlite pipe located in the Yakutia region of the USSR was analyzed by Bratus et al. (1988). The gas analyses and isotopic distribution from this region are dismissed by Skobelin as simply adding to the confusion. I do not share his pessimism about the analytical data, and in fact, find it most informative.

Bratus gives us the following analysis for this included gas:

$$9.4 \% \ CH_4 + 39.8 \% \ H_2 + 5.7 \% \ CO_2 + 45.1 \% \ N_2$$

One might conclude on first inspection that hydrogen and nitrogen dominate lithosphere gas in this terrain. The rock making up this pipe is reported by Kravtsov et al. (1976) to be host to:

**"bitumens and ... gases [that] contain carbon with a substantially different isotopic distribution than bitumen and oil and gas in rocks of the sedimentary cover [around the pipe]. Moreover, there are specific features of zoning in the distribution of bitumen and free gases, which are confined to fracture zones."**

If the kimberlite is acknowledged to be Mantle rock, my theory would suggest that Mantle gases, silane and hydrocarbon especially, should be emerging in the lower reaches of the conduits. There, the silicon of the silane should be oxidizing and hydrogen evolving. Higher up, hydrocarbon dissociation should be taking place. Where oxygen is encountered, oxidation of carbon, hydrogen, or both may occur.

If we assume first of all that there is no native hydrogen flowing from the deep Mantle, second that all the carbon and four times its molecular equivalent of the hydrogen derives from methane, and third, that the remaining hydrogen is all derived from silane, the composition of a "Mother Vapor" is determinable as:

$$3.0 \% \ CH_4 + 85.9 \% \ SiH_4 + 11.9 \% \ N_2$$

The profound redistribution represented by these figures is due to the silicon component. As it is oxidized to quartz or dissolved silica, the percentage figures of the surviving volatile components rise. ***Ultimately, the "Mother Vapor" necessarily must have lost 100 % of its silane component, 32 % of its hydrocarbons, and nearly 75 % of its mass to oxidation in transit.***

Where the lost silica goes is an open question and very different in the many vicissitudes of the planetary lithosphere and surface. Among the opportunities for its redistribution are the mineral transformations as described by Collins in Chapters IV and V, siliceous ash, as described in Chapters III and VI, granophyres (not discussed), and quartz sands, as taken up in Chapter VIII. Silicon from Earth's interior, as the provenance for those deposited lithologies is more understandable than the segregation of silica from mafic Mantle silicate rock as existing theories require.

## THE KTB PILOT DRILLHOLE

A 4 000-m drillhole was put down as a precursor to a deep hole to be drilled in the western margin of the Bohemian massif for the German continental drilling program (Emmermann et al. 1990). A saline gaseous aquifer was encountered at 3 447-m depth with compositions in the following weight-percents:

$$3.1 \% \ CO_2 + 69.6 \% \ N_2 + 27.0 \% \ CH_4 + 0.1 \% \ H_2 + 0.2 \% \ He$$

Making the same assumptions as on the Udachnaya pipe, we can interpret a "Mother Vapor" with the following composition:

**60.9 % $N_2$ + 24.6 % $CH_4$ + 14.3 % $SiH_4$ + 0.2 % He**

Clearly, this was a leaner source in silanes than Udachnaya. In their other components, however, the gases are similar. Excluding silanes, we can calculate the following compositions:

Udachnaya  79.9 % $N_2$ + 20.1 % $CH_4$
KTB        71.1 % $N_2$ + 28.7 % $CH_4$ + 0.2 % He

Inasmuch as the Udachnaya terrain is petroliferous, the absolute volumes of exhalative gases emerging through the Crust are likely far greater there than at the KTB site. **The "Mother Vapor" at the KTB site is interpreted to have lost only 12 % of its mass to devolatilization of silane against nearly 75 % in the Udachnaya terrain.**

## MID-CONTINENT HYDROGEN WELLS

Hydrogen is a component in wells of various kinds. From wells drilled for petroleum and natural gas to water wells, seismic shot holes, and grout holes drilled for engineering works, wherever gas is analyzed, hydrogen is a component, often a major one. Most occurrences are too poorly reported to be analyzed further. The following is one of these:

In central California near the city of Stockton, drilling at a water-pumping station by the U.S. Bureau of Reclamation is reported by B. J. Moore (1985) to have drilled a 745-foot borehole for determination of slope stability. A flow of gas at 130 feet was encountered. The gas is reported as 57.7 % hydrogen. Exact location, gas-flow rate, and geology have eluded precise determination, although the serpentinous

Franciscan formation is implied as the host bedrock, slope stability being a concern there.

Better studied are the hydrogen-rich gases of the mid-continental United States, where they emerge above the buried graben known as the mid-continent rift system. Coveney et al. (1987) provide these analyses:

|  |  | Iowa Webster County | Kansas Scott Well | Kansas Hein Well | Michigan Washtenaw County |
|---|---|---|---|---|---|
| $H_2$ | mole % | 96.3 | 34.7 | 29.6 | 26.0 |
| $CH_4$ | mole % | 0.1 | 0.03 | 0.6 | 69.0 |
| $H_2O$ | mole % | - | 2.1 | 0.0 | - |
| $N_2$ | mole % | 3.5 | 61.0 | 67.7 | 3.6 |
| $O_2$ | mole % | - | 4.6 | 4.6 | 0.8 |
| Ar | mole % | - | 1.1 | 0.7 | - |
| $CO_2$ | mole % | - | 0.2 | 0.01 | 0.6 |

The low hydrocarbon and carbon dioxide contents of the Iowa and Kansas occurrences can be interpreted as the result of relatively pure silane reacting with water and liberating hydrogen. One might speculate that the water in the Scott well gas is the reaction product after a small amount of oxygen was available at depth. The hydrocarbon gas in the Michigan well suggests either the mixing with biogenic hydrocarbons or a hydrocarbon-rich endogenic gas source.

In Chapter VIII the St. Peter sandstone, covering six states of the mid-Continent region, is interpreted as a product of silane emission through this same rift system in Ordovician time. The Hein well location is shown with respect to the ultramafics of the mid-continent rift on **Fig. VIII-1**.

Coveney et al. (1987) also presents gas analyses of some foreign hydrogen-rich gases, as follows:

| | Philip-pines | USSR Nish-na-T. | USSR Stav-ropol | Oman Haw-qayn | Poland Lubina | Germany Mul-hausen | East Pacific Rise |
|---|---|---|---|---|---|---|---|
| $H_2$ mole % | 41.4 | 66.5 | 27.3 | 45.0 | 73.1 | 61.5 | 52.46 |
| $CH_4$ mole % | 52.6 | 9.5 | 51.0 | 3.2 | 0.43 | 22.5 | 0.86 |
| $H_2O$ mole % | - | - | - | - | - | - | - |
| $N_2$ mole % | 5.5 | 20.7 | 20.3 | 43.0 | 26.5 | 14.3 | 41.45 |
| $O_2$ mole % | 0.5 | 3.3 | 0.8 | 9.0 | - | - | - |
| Ar mole % | - | 0.2 | - | - | - | - | 0.52 |
| $CO_2$ mole % | - | - | 0.0 | 15.5 | 0.0 | 1.6 | 4.71 |

These data show that significant regions of the planet are permeated by hydrogen and nitrogen, occasionally with major hydrocarbon association. Where the Coveney theory attributes the generation of hydrogen to alteration of ultramafic rock, I would link it to the rifting, which allows rise of ultramafic magma, diapirs, and mantle-sourced silane gas.

## THE LARIN THEORY

Vladimir N. Larin, of the Geological Institute, USSR Academy of Sciences, has develped new theory that draws attention to the association between ambient hydrogen and electrical conductivity in rift zones.[170] He reports on high conductivity at six- to eight-km depths under the Tunka depression of the Baikal rift USSR).[171]

Dr. Larin makes the interesting interpretation that the silicate shell of Earth cannot be demonstrated to extend deeper than 300 to 350 km, and that native metals and metal hydrides likely prevail in the Mantle below. He thinks that rifts allow "tongues of oxygen-free alloys, native silicon and silicides of magnesium and iron primarily," to have penetrated the Crust in some rifts, where they interact with water. Water is hydrolyzed by these metals and hydrides with release of hydrogen and heat. He proposes that copious hydrogen could be produced to order by drilling into the unoxidized metals and hydrides and introducing water to react with them.

[170] *Larin, 1980, 1991, 1992.*

[171] *The Tunka rift depression spans 150 km north-south length and 35 km width in the southern part of the Baikal rift.*

## ZEOLITES AND FULLERENES

### ZEOLITES

Zeolites are known and manufactured by industry while being little understood by most geologists. Partly this is due to complex chemistry. The major deposits have appeared to be bedded tuffs, quite unlike the mineral specimens geologists have been taught to recognize as zeolites. These latter comprise fillings of voids ("amygdules") in lavas, and therefore, comprise only minor mineral contributions from volcanoes. This trivialization of zeolite occurrence has obscured reality. Zeolites comprise the major components of many young, unaltered, bedded tephras. Their thicknesses reach into tens of metres in parts of the Great Basin and Columbia Basin of northwestern North America. Handbooks on industrial minerals fail in some cases even to mention zeolites.[172]

Whereas all zeolites are high in $SiO_2$ 58 to 61 % is usual, clinoptilolite is especially interesting in our consideration of endogeny. Clinoptilolite, with $SiO_2$ content up to 71 %, is the most abundant zeolite in many bedded tephras. Where other zeolites break down with temperature elevation at 250 to 350°C, clinoptilolite is stable to about 700°C.

Regarded as a high-temperature variety of heulandite (another zeolite), clinoptilolite contains more sodium than calcium or potassium. Its composition is approximately the following:

$$(Na_2O)_{0.70} \cdot (CaO)_{0.10} \cdot (K_2O)_{0.15} \cdot (MgO)_{0.05} \cdot Al_2O_3 \cdot (SiO_2)_{8.5\text{-}10.5} \cdot 6\text{-}7H_2O$$

[172] *Johnstone 1954; Gillson 1960*

Clinoptilolite has an interesting porous crystal framework that gives it the unique ability to act as a molecular and cation sieve (**Fig. VII-1**). Among metallic cations, the ammonium ion, $NH^{4+}$, is readily absorbed by clinoptilolite. This adsorptive capacity gives clinoptilolite the industrial property of effectively deodorizing organic materials such as organic wastes.

VII-1
**CRYSTALLINE HABIT OF ZEOLITES AS SEEN
WITH ELECTRON MICROPRIBE MICROSCOPY.
Cl, CLINOPTILOLITE; E, ERIONITE; Ch, CHABAZITE.**

Commercial ammonia is made in a reaction between nitrogen and hydrogen catalyzed by nickel. If excess hydrogen is available to give reducing conditions, ammonium ion prevails over ammonia gas. This is to be expected in crustal volcanogenic situations, as we saw in the previous sections on entrapped gases. Nitrogen and hydrogen are always present in deep crustal residual gas. It would seem appropriate to suggest that these gases could be involved in a natural catalytic reaction to generate ammonium ion in that environment. Its uptake by naturally-occurring clinoptilolite then is foregone.

Clinoptilolite generated under deep-crustal, high-temperature conditions would then be loaded with ammonium ion during volcanic ejection. With decompression the volcanic heat would drive off the ammonium cation as ammonia. This would allow deposition of ammonium-free clinoptilolite.

Deposits of clinoptilolite are widespread among pyroclastic rocks of northwestern North America, as summarized by D. A. Holmes.[173] The clinoptilolite found in these stratified volcanic rocks of the Great Basin is left in a mode that can be considered chemically receptive for recovering the lost ammonia or ammonium ion as well as other compatible cations.

This origin for zeolites in general is contrary to current thought on the subject. Many researchers attribute bedded zeolites primarily to chemical alteration of volcanic glass in alkaline lakes, documenting their deduction with convincing evidence.[174] However, the broad picture of bedded zeolites in the Great Basin shows their great continuity over long distances, often in hilly, even mountainous terrain, where Cenozoic lacustrine conditions are often not readily envisioned.

In short, the geomorphology of the bedded zeolites makes it appear unlikely that *all* of them owe their existence to mineral transformation in alkaline lakes. This is especially the case for clinoptilolite. Its stability up to high temperatures suggests formation under hot subterranean conditions at depth in the Crust *before* volcanic ejection. Origin at depth has all along been recognized as the provenance of the first-recognized zeolites, those found sealed in the amygdules of lava flows. Thus, it seems fair to say that most zeolite mineral originates within crustal volcanic chambers.

A more difficult question, one for which no answer is apparent, is why monoclinic clinoptilolite with its open lattice

---

[173] *Holmes 1990*

[174] *Sheppard 1986; Fisher & Schmincke 1984*

rather than hexagonal quartz with a closed lattice would have been generated in the first place. Perhaps the reducing conditions conducive to ammonium preservation promote it over crystallization of acidic quartz. The recent discoveries of fullerenes suggest possible analogous origins.

## FULLERENES

---

"Amazingly, $C_{60}$ appears to result inevitably when carbon condenses slowly enough and at a high enough temperature."

**CURL, R.F. & SMALLEY, R.E. 1991**

---

In addition to the well-known allotropic forms of carbon, graphite, and diamond, the recent discovery of a new form, known as "fullerene," holds major portent for geology. Fullerene structures are geodesic structures (hence the name commemorating Buckminster Fuller). Shaped variously like hollow balls or cages, fullerenes have multi-faceted exteriors with carbon atoms occupying the corners of pentagons and hexagons. They occur with carbon atoms numbering from 32 to more than 600. Sixty is the number giving the greatest stability, hence the chemical name, $C_{60}$.

Among the remarkable chemical- and physical-properties and characteristics of this interesting form of carbon are the following:

Fullerene production under laboratory conditions is inhibited when $H_2$ is present, and ceases altogether in an atmosphere of 10 % $H_2$.[175] This is because hydrogen bonds to open carbon edges and corners, thus neutralizing their ability to form carbon-to-carbon junctions in the closed-cage form.

---

[175] *De Vries 1991*

Fullerene is chemically remarkably inert, as it cannot be oxidized by perchloric acid and is only partially oxidized by chromic acid.[176]

Even hydrofluoric and hydrochloric acids together (Hf-HCl) do not destroy fullerene.[177]

Where diamond can be combusted at about 500°C, $C_{60}$ is stable up to 570°C, and about 10 % of its mass is found to survive after application of 1 000°C .[176]

Fullerenes are the most volatile modifications of elemental carbon. $C_{60}$ is about as volatile as elemental sulfur.[177]

With chemically contributed electrons, cooling and subsequent warming create an organic magnet.[178]

Buckyball [fullerene] films doped with potassium or rubidium are superconducting up to 45°K. These are among the best superconductors known.[178]

Soot from natural processes such as candle smoke contains up to 9 % fullerenes.[178]

The molecular cage structures can contain metals [such as Pt and Os] and entrap inert gases.[178,179]

The volatility reported by Heyman[177] is also reported by de Vries et al.[175] and seems to conflict with the inertness reported by Gilmour et al.[176] The reason for this is not known. Heyman suggests that fullerene soots form whenever elemental carbon condenses, and that they are carriers of inert gases. Adsorbed

[176] *Gilmour et al. 1991*

[177] *Heyman 1991*

[178] *Pennisi 1991*

[179] *Curl Smalley 1991*

inert gas may afford some protection from chemical attack. Whether this is so or not, the role of fullerenes is remarkably similar to the behavior of the zeolite, clinoptilolite, that is to say, as gas entrapment micro-structures.

The zeolite and fullerene cage structures are comparable for their high-temperature stabilities and their affinities for gas species. A difference between them is that the zeolite silicon is oxidized, fullerene carbon is not. Fullerene magnetism [by acceptance of electrons], a reduction process according to the standard definition of the term, is opposite to zeolite behavior.

Carbon in the several reduced states of charged, partially-dissociated hydrocarbons will be discussed as an agency of coalification of peat in Chapter IX. Reduced carbon, fully hydrogenated, as the fullerene, $C_{60}H_{60}$, while theoretically possible,[179] has yet to be found in nature. Geodesic encapsulation within the geodesic fullerene structures of metal atoms could explain some forms of metal ore generation and the not-infrequent association of ore metals with carbon and hydrocarbon. Fullerenes could well be instrumental in the effusion of endogenic carbon from deep levels of the planet.

It is beyond the scope of this book to try to analyze what physical/chemical conditions yield zeolites and fullerenes instead of the less-elaborate crystalline forms of silica and carbon. The opening quotation of Curl and Smalley, **"Amazingly, $C_{60}$ appears to result inevitably when carbon condenses slowly enough and at a high enough temperature,"** appears to describe the carbon fact succinctly. We should anticipate that the very abundance of clinoptilolite and fullerenes in nature purport future discoveries of new and unsuspected properties related to their structure and chemical/electrical behavior.

In closing, we might suggest that silicon should, like carbon, occur in geodesic fullerene forms. Geodesic "silerenes" would have to be constructed of silica rather than elemental silicon because of the reactivity of the native metal. In fact, we need look no farther than the natural world to find the intriguing form of $C_{60}$ elegantly duplicated in silica. The shells of living radiolarians (**Plate VII-2**) are built in the familiar pentagon-hexagon composite shells. They comprise millions of silica molecules, whereas $C_{60}$ is created from just sixty or variously from 32 to more than 600 carbon atoms. Nevertheless, the propensity for growth of organic as well as inorganic structures in this form must give us pause to ponder the reason.

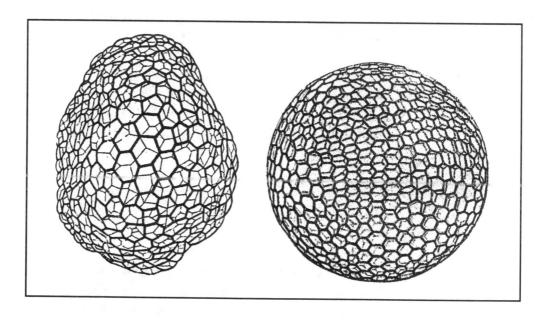

VII-2
**GEODESIC SHELL STRUCTURE IN SILICA AS PRODUCED BY MODERN RADIOLARIAN PROTOZOANS (from Thompson, D'Arcy, 1917 "ON GROWTH AND FORM")**

CHAPTER VIII

# PULSES OF ENDOGENY

## ENDOGENIC QUARTZ SANDS AND QUARTZITES

**C. WARREN HUNT**

Writing on the Cargo Muchacho Mountains, L. G. Collins described mafic igneous rock transformation into both intermediate igneous rock and high-grade metamorphic lithologies. "Metasedimentary" layered rock, appearing to be phyllites and quartzites, aluminum- and silicon-rich, are not what they seem, but the evolutionary results of silane pervasion of pre-existing mafic rock. Thus, the Cargo Muchacho terrane was originally not sedimentary, but rather "pseudo-metasedimentary."

An entirely different geological development follows in the case of the direct oxidation of silanes. This produces quartz sand and water, a slurry of low-density relative to surrounding rock. Such a sand slurry, responding to hydrodynamic entrainment, would rise in conduits and, with sufficient hydrodynamic impulse, flow out upon the surface. Wherever the sand might be when hydrodynamic propulsion ceased, "sedimentation" would occur, whether in fracture and joint conduits of the host terrane or upon its surface. Thus, it would create sandstone and quartzite dikes and flood deposits.

Some characteristics may be expected of the sands deposited by this process:

Firstly, uniform clast size [due to crystallization in residual water, where crystals are suspended in the water].

Secondly, rounding and "frosting" of crystal facets as a function of flow distance.

Thirdly, interstitial cementation restricted to minerals of silane combustion and cation replacement [quartz, dolomite, magnesite, hematite, chert, chalcedony, opal, zeolites, and others].

Fourth, absence of mafic sand grains.

Fifth, lack of large conglomerate [+.5 cm] clasts.

Sixth, mineralogy indicative of reducing conditions [because of accompanying methane, and free hydrogen].

Under conditions of abundance of water, if the hydrogen is not entrapped, the escape of hydrous fluids, including slurries, should follow hydrodynamic flow paths according to endogenic pressures and available channels, entraining and sweeping along crystalline quartz grains. The retention in place or escape of this sort of slurry, as mentioned, produces sandstone and quartzite dikes or extrusive sedimentary deposits.

Water has been found deep in the Crust, an occurrence that evoked great surprise during the drilling of the 11-km deep well on the Kola Peninsula, Russia. Such water should intercept rising silane down to Mantle depth, perhaps deeper.

In the special case where very little water is available to oxidize pervading silane gas, the country rock becomes permeated and subject to mineral transformations. Incremental substitution of higher-silicon petrologies for mafic predecessors then proceed apace. **This transformation is able to go right on**

**up to pure, white orthoquartzite, a petrofabric wholly lacking in relict grain textures.** Examples of this kind of quartzite include the Killarney quartzite [description on p316 ahead] and the quartzites of the Cargo Muchacho Mountains [described in Chapter IV].

**The silane reaction where water is available, can produce quartz sand slurried in compressed steam and rich in hydrogen gas. This may debouch on the surface quiescently, convulsively, or explosively, depending on volume and its charge of dissolved gas. The term "sands of endogeny" appropriately depicts particulate quartz that has first crystallized after silane combustion as grains, and then has been expelled.**

First moving in conduits, these slurried sands may not flow out on the surface. The conduits, if left filled with sand, become quartzite or sandstone dikes or irregular, pseudo-sedimentary apophyses. They may occur in unlikely sites, such as surrounded by serpentinous, ultramafic rock, for example. The resemblance of their fabric to sedimentary sand deposits is misleading. Better they be regarded as "ficto-sedimentary," superficially sedimentary in appearance. Dikes of this type, while important as clues to endogeny, are not major features of the Earth.

The "Sands of Endogeny", on the other hand, comprise voluminous sand expulsions that have surged in slurry form and burst through surficial rock, debouching on the erosion surface in great sand heaps. They comprise huge manifestations of silicon endogeny, true sedimentary rock, but with distinctive character.

In varying degrees the debouched sand deposits have been redistributed by meteoric waters of local erosional systems. The characteristics to be expected for primary endogenous sand deposits include all six foregoing clast and matrix characteristics but, in addition, two more features of sedimentary structure.

These, the seventh and eighth characteristics are:

**Seventh, a "pulse" of emission should produce a massive deposit, internally lacking stratification and other fluvial features [e.g. thin bedding; gradational change of clast size]. These constraints do not preclude changes in lithology in successive pulses. Neither do they restrict the development of sedimentary features by surface processes away from the vent. In fact, with increasing distance from a vent, one should find that redistributive erosion has introduced fluvial features as well as materials from foreign sources.**

**Eighth, thickness should be greatest near the vent, a heaping up, in fact, and distal redistributive thinning by meteoric waters away from the vent.**

Let us consider some examples:

The **Hayfork Creek apophyses** of late Jurassic age, Trinity County, California;

The **Mount Wilson quartzite** of late Ordovician age in British Columbia and Alberta;

The **St. Peter sandstone** of middle Ordovician age in the mid-continental USA, occurring over most of six States, centering on Illinois and Missouri;

The **Oriskany sandstone** of Devonian age, occurring in New York State and the adjacent Appalachians, USA;

The **Gog Group** of the preCambrian-Cambrian transition in Alberta and British Columbia;

The **Belt Supergroup** of late preCambrian age in Montana, Idaho, and British Columbia;

The **Minnes/Ellerslie/Nikinassin/Kootenay group** of the Jurassic-Cretaceous transition, western Alberta and northeastern British Columbia;

**"Earthquake Sand" Expulsions**

    **The Charleston Earthquake Sand**, a Holocene expulsion;

    **Submarine "sand dunes,"** Pleistocene sand expulsions;

    **The Winterburn Silt**, a late Devonian sand expulsion;

    **The Ellerslie Sandstone**, a late-Jurassic to early-Cretaceous sand expulsion;

    **The Athabasca sandstone,** a mid-Proterozoic sand expulsion northern Saskatchewan and Alberta;

    **The Killarney Quartzite**, a late preCambrian (Grenville-age) replacement, Ontario.

The geologist will note that all entries in this list fall close in time to major upheavals in geological time, the "Revolutions," as they used to be called:

The Athabasca, Belt, and Gog are associated with the turbulent preCambrian-Cambrian transition, the Athabasca and Belt being deposited long before the Gog.

The Mt. Wilson and St. Peter are contemporaries of the early "Caledonian" orogeny.

The Oriskany has an age close to the "Frasnian" revolution; and the Winterburn expulsion was soon after.

The Minnes/Ellerslie/Nikinassin/Kootenay group and the Rattlesnake dikes were deposited in the time of the "Nevadan" orogeny.

The Cargo Muchacho dikes (Chapter IV) may coincide with the "Laramide" revolution of the late Eocene.

The submarine dunes and earthquake sands are contemporary, part of our own Quaternary times of diastrophic violence.

It is the point of this chapter that the coincidence of siliceous sedimentary deposition and crustal upheaval are not randomly coincident features. Rather, they are the concomitant effects of endogeny and new insights for interpretation of prehistory.

I take up the Hayfork apophyses first because they follow the not dissimilar features described in the Collins paper on the Cargo Muchacho Mountains.

Then I take up the St. Peter and Mt. Wilson occurrences, because they provide the most comprehensive records. Following are the Gog and its relative, the Belt supergroup. I compare the Oriskany with the other Paleozoic sands.

Probably the most interesting formation of the group, as the reader will see, is the Athabasca formation, first because it is the largest exhalation of sand, and second because economic implications of major importance are connected with its interpretation.

Next I take up the Minnes/Ellerslie/Nikinassin/Kootenay group ("MENK").

I take up the earthquake- and fault-related occurrences last, showing the similarities between the Quaternary earthquake-related sands and prehistoric sand exhalations of the western Canada basin, the Winterburn and Ellerslie formations.

## THE HAYFORK CREEK APOPHYSES

At the headwaters of Stringbean Creek, a tributary of Hayfork Creek in the "Western Jurassic (Rattlesnake Creek) terrane" of the Klamath Mountains a northwest-trending fault intersects a body of harzburgite ultramafic rock. The harzburgite on opposite sides of the fault has slightly different trace-metal contents, nickel-iron inclusions, and other siderophile minerals, thus implying different magmas in fault contact.

Scattered along the fault trend are small apophyses of diorite, dark chert or quartzite on scales of a few metres diameter. Where the diorites are replacement bodies, the cherts are chemical precipitates, and the quartzites are dikes of particulate quartz. The quartzites are even-textured, fine- to medium-grained, sugary rock fully reminiscent of sedimentary habitats.

Unlike quartzites from sedimentary habitats, this quartzite as well as the chert and diorite apophyses are wholly foreign to their host. They occur in isolation amid ultramafic rocks, far from any recognizable source of granular or dissolved silica. Whereas the chert and diorite can be interpreted as replacements, the quartzite clearly must have flowed into place as a sand slurry, which then was cemented by precipitation of dissolved silica. The presumed Mantle origin of the siderophile metals and the ultramafic rock implies a related source for the silica itself. **Quartz sand of high purity and probable Mantle origin has flowed in slurry form into fractured ultramafic rock to create quartzite apophyses on upper Hayfork Creek, California.**

## THE MT. WILSON QUARTZITE

This late Ordovician formation is found outcropping on the east flank of the Rocky Mountain trench in the vicinity of Golden, British Columbia. Two quarries produce commercial silica from it, the Mountain Minerals Moberley Mine east of Golden, and the Hunt Mine [named for the author] south of Golden. At the Hunt site the sand is lithified to quartzite [**Fig. VIII-1**] and crops out as three distinct pods >100 m in thickness resting unconformably on sericitic Cambrian shales.

For 100 km or so to the northeast and southeast in the main ranges of the mountains, the Mount Wilson formation is repeated in fault blocks overthrust westward toward the Rocky

VIII-1
PHOTO SCENES OF MASSIVE QUARTZITE, MOUNT WILSON
FORMATION, HUNT AND MOBERLEY MINES, GOLDEN,
BRITISH COLUMBIA

Mountain trench or eastward toward the Alberta basin. There are several remarkable features about these occurrences. The first is purity: 99.5+ % $SiO_2$, 0.05 % dolomitic intergranular cement. The second is the massive character: the main body of the deposit lacks detectable bedding planes. The uppermost and basal contacts are bedded, showing fluvial reworking of the original sands. The mass lacks other sedimentary features as well as bedding, that is to say, foresets, current beds, shale partings, or horizons of oversize clasts that imply original attitude. Pale streaks in the white quarry walls are seen under the microscope to be no more than tints to the intergranular cement. There are no dark sand grains at all; all sand grains are white quartz.

The size distribution of sand is a third surprise. Crushing returns this rock to its original grain sizes; and these show uncommon uniformity. The following size distributions illustrate the uniformity. The first is taken from the Hunt minesite, the second from an outcrop on the White River, 100 km to the southeast:

| U.S. Standard Screen Size | Hunt Mine % Retained | White River % Retained |
|---|---|---|
| 30 | 15.7 | 8.5 |
| 40 | 45.8 | 30.5 |
| 50* | 23.3 | 30.5 |
| 70 | 8.5 | 25.0 |
| 100 | 4.2 | 1.0 |
| 140 | 1.1 | 1.0 |
| 200 | 1.1 | 1.5 |
| -200 | 0.3 | 2.0 |
| *=0.5 mm | 100.0 | 100.0 |

Middle- and late-Ordovician sedimentary strata of this part of the Canadian Rockies have strong volcanic affinities. Flecks, patches, and marbling with bright red hematite are common; and the carbonate deposits are indurated with chemically precipitated silica. Small igneous dikes and mafic flows sporadically appear along the crest of the main ranges.

There is no obvious source for this Ordovician sand, certainly not for pure quartz sand such as the Mount Wilson comprises. Its absence of bedding implies sudden deposition, as though the entire mass arrived in a single pulse of deposition without dilution from volcanic or other foreign matter.

The Hunt and Moberly quarries are on the edge of the Rocky Mountain trench, an elongate graben bisecting the mountain uplift into Rocky Mountains to the east and Selkirk and Purcell Mountains to the west, a structural depression some 700 km in length. Although the trench exhibits some right-lateral shear tendencies, it is basically a tension suture, a dropped keystone at the crest of the mountain arch, a zone of crustal weakness, where deep-seated [asthenospheric] intrusion has dilated the cover in Cenozoic times.

The deduction was made in the EV volume that the first such major dilation of the Crust happened about 38 my ago and that additional uplift has followed intermittently to the present. This was the Eocene "Laramide revolution," when Rocky Mountain gravity-glide overthrusting occurred. The Rocky Mountain trench and lesser sub-parallel trenches near the crest of the dilationary arching are products of subsequent relaxation of the dilatancy, dropped keystone blocks on the collapsed arch. The Rocky Mountain trench is the largest of these keystone blocks.

The much-earlier Ordovician performance along this suture system is less clear because of the intervening 350 my of overprinting. Besides the afore-mentioned volcanic contributions along the crest of the Rocky Mountain main ranges, a calc-alkalic igneous complex 50 km or so east of the Hunt Minesite, known as the Ice River complex, received intrusive mineralization in Ordovician time, but was likely a well-established basement horst before late Cambrian time

(Gussow & Hunt 1958). Thus, the long-standing crustal weakness and lithospheric suturing along the trend of the Rocky Mountain trench is apparent. It was a suture of this type in Ordovician time through which the Mount Wilson sands were expelled in one or more pulses upon the clays of the late-Cambrian seafloor.

**The Mt. Wilson quartz sands and water were most likely generated by silane combustion deep in the Crust. Ordovician volcanism opened a lithospheric suture of the ancestral Rocky Mountain trench system, whence the previously-trapped sand slurry, buoyant, due to its lesser density than rock, surged out upon the erosion surface. Accumulating in unstructured mounds close to the expulsion vents, the sand was lithified by accompanying dissolved silica and minor dolomite. The Hunt Mine orebody is one of these mounds.**

Beds of redistributed sand spread from the mounded accumulations by meteoric water in the drainage system of the day. Such bedded, broadcast sands comprise the sand distribution now known, reaching 75 km or so to the southeast and north. The redistributed sands show normal fluvial bedding; the sands at the Hunt and Moberly Mines do not, excepting at the very top and base of the mounds.

## THE ST. PETER SANDSTONE

Staying with Ordovician deposits for the moment, we can leave the Canadian Rockies. The middle-Ordovician St. Peter Sandstone is another pure quartz sand, which is remarkable for the great extent of its occurrence [**Fig. VIII-2**].

Like the Mt. Wilson, the St. Peter has a pure quartz character. It maintains this character over the entire extent of its occurrence but for a small area in the southeastern extension of the St. Peter basin, the Ozarks, where carbonates are interbedded. Deposition took place under shallow marine conditions between periods of carbonate rock deposition.

The beds of sandstone are laid upon one another without interleaves of disparate lithology [except in the Ozarks]. The pile resembles successive basalt flows, one after the other with distinct, parallel contacts.

The changeover from underlying carbonates to sand deposition and then back again after deposition of the St. Peter is itself notable, because it shows that nearby lands were not contributing significant amounts of quartz sand to the marine embayment, either before or after St. Peter deposition. One must wonder what caused such complete switchovers. How could a complete change suddenly come about? From no sand the system is suddenly flooded copiously with absolutely pure quartz sand.

Ordinary erosion processes would hardly suffice. Thickness averages 17 m, although locally in excess of 100 m. From that figure one can estimate a total volume of 25 000 km$^3$ of sand. Northern areas are thought to have been deposited somewhat earlier than southern areas.

Is it even rational to believe that such a huge volume of sand could be winnowed by wave action to 99+ % purity, much less that it could be deposited without introduction of contaminating debris from nearby erosional sources? Even without the introduction of foreign matter through surface runoff, there would be wind contributions from volcanic and other sources during the long period that must be assumed for normal erosion and redeposition of such a large mass of rock. These wind-blown contributions to sedimentation should be easily visible as layers in the otherwise-pure quartz milieu.

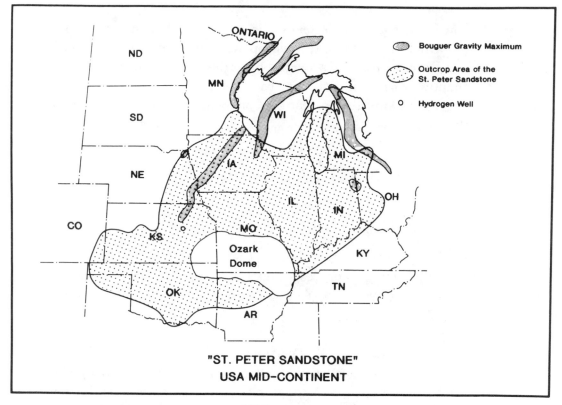

**VIII-2**
**TOPOLOGY OF THE ST. PETER SANDSTONE**

St. Peter sandstone is monotonous in coloration and almost entirely without fossils. Only rare shell casts attest to the existence of contemporary life. Like the Mt. Wilson, there are no signs of volcanic contributions in the sands. Where mid-Ordovician volcanism in the Canadian Rocky Mountains has already been mentioned, there are also bentonite clays [volcanic in origin] and at least one basalt flow in the St. Peter environs in strata of mid- and late-Ordovician age. These would be expected in view of the worldwide volcanism of the Period.

**Consideration of the facts surrounding the 25 000 km³ St. Peter sandstone clearly support an endogenic provenance. Perhaps the most convincing evidence is its position astraddle the seismically-active basement trench known as the "mid-**

continent rift." This feature is surely a lithospheric suture, on which a vent could have opened and expelled entrapped sand and water in a series of pulses of endogeny. The shallow-marine depositional basin of the St. Peter deposition suggests a subsidence feature consequent upon expulsion of the sand mass.

## THE ORISKANY SANDSTONE

The Oriskany formation is another quite pure quartz sandstone, but is only a metre or so in thickness. It differs from the St. Peter and Mt. Wilson in that it has shale partings. It is Devonian in age and rests unconformably on early-Devonian shales in eastern New York, and on Silurian strata in western New York.

A feature that is highly pertinent to the question of endogeny of high-purity quartz sands is examined by Raymond Moore (1933) in connection with this thin and persistent sandstone. Moore reports **"... the occurrence of fissures in the pre-Oriskany rocks (possibly formed by earthquake shocks) that are filled with Oriskany sand."** These fissures appear to be the submarine exhalation fractures through which the Oriskany sands were debouched in pulses on the seafloor. Between pulses, clays were deposited.

New York was an arm of a much deeper and more extensive lower-Devonian basin occupying the maritime provinces of Canada. Extrusive and intrusive volcanic rocks and granite batholiths were emplaced in that extended basin contemporaneously with Oriskany deposition. A common origin of silane systematics is shared between the igneous rocks and the Oriskany.

# THE GOG GROUP

The Gog Group is a sedimentary pile of lower-Cambrian age found along the crest of the Canadian Rocky Mountains. It is up to three km in thickness,(Young 1979) and exhibits lithologies of quartzite, shale, and quartzite-clast conglomerate. According to Hein & Arnott (1983) and Hein (1987), deposition occurred after an abrupt "change in sedimentary patterns ... from deep water, marine [character exhibited by the] preCambrian Miette Group to shallow marine clastics [of the Gog Group]."

Evidence for "abrupt change" between preCambrian and Cambrian time is elegantly expressed by the transition from preCambrian Miette conglomerates to lower Cambrian Gog quartzites. From 5 to 40 % (averaging 20 to 25 %), feldspar content in the Miette decreases to less than 1 % in the Gog.(Charlesworth et al. 1967) To these observations I would add that, whereas Gog clasts are white and off-white in the main, Miette clasts have more varied and stronger coloration.

A provenance for the pure quartz sand is as problematical for the Gog as it is for the later, superjacent Mt. Wilson. To winnow a clean quartz sand with less than 1 % feldspars from igneous rock that is 20 % feldspathic, while at the same time destroying all mafic grains is not a realistic possibility. Contaminating material from surrounding terrain would be constantly added to the winnowing system by water and wind, and constantly would set back the completion of winnowing. Some dark, mafic minerals from the Miette and from surrounding terrain are as abrasion-resistant as quartz, and could not be winnowed out to leave only quartz.

Hein describes the "massive sand beds" of the Gog as containing "virtual absence of sedimentary structures, [which makes them] difficult to interpret." **Clearly, this character represents sudden removal from source of the mass in each** such bed, its expulsion to the surface in a slurry, its spreading, and rapid deposition, all in a single pulse until complete. This is the only appropriate interpretation for these sedimentary beds.

Needless to say, no source can be identified, nor should one be expected to occur amid the preserved geology. The clasts, however, pose a problem that was not met in analyzing the Mt. Wilson, Oriskany, and St. Peter strata. The Gog conglomerate clasts as well as the matrix appear to be very highly quartzose, perhaps 99 %. **The appropriate inference that can be drawn from this evidence is that early-deposited sands were cemented by colloidal silica and compacted into orthoquartzites. These were later disrupted, corraded into their present rounded state, and the conglomerate slurry flushed, along with *a later generation of sand,* out upon the land surface where they now comprise the thick conglomerates of the Gog scenery.**

## THE BELT SUPERGROUP

The Belt Supergroup comprises a huge thickness of siliceous sedimentary rocks deposited in a 200 000 km$^2$, subsident, intra-continental depression centered approximately on the Rocky Mountain trench, primarily in western Montana, northern Idaho, and southeastern British Columbia. (**Fig. VIII-3**) Maximum thickness of the Supergroup reaches at least 18 km, and volume, therefore, may originally have approached 3 600 000 km$^3$. The age is not precisely settled, but falls within the period 1.9 to 0.8 by.(Winston 1989)

The orientation of this basin of the Belt Supergroup astride the Rocky Mountain trench suggests possible rift behavior. Repeatedly in this chapter, as well as in the EV volume, I describe regional, post-Belt dilation, quartz sand effusions, and volcanic activities in the preCambrian-Cambrian transition, the

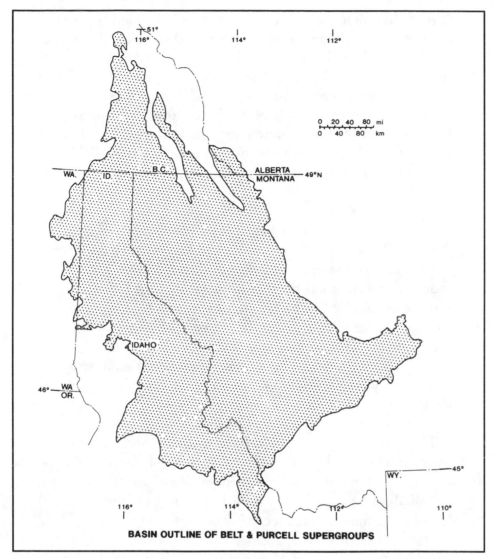

BASIN OUTLINE OF BELT & PURCELL SUPERGROUPS

VIII-3
TOPOLOGY OF THE BELT SUPERGROUP

late Ordovician, the Jurassic-Cretaceous transition, the late
Eocene Laramide orogeny, and later Cenozoic times. Many of
these are oriented to an axial line coincident approximately
with the Rocky Mountain trench. The Belt is the largest-
volume member of the group.

Belt lithologies are monotonous, thinly-bedded quartzites, arenites, and argillites. Representing shallow water deposits of mud, silt, and sand, the Belt is but rarely conglomeratic, very rarely sulphidic, and deficient in carbon and carbonates. Bright colors, including green, red, purple, and rust are common, and result from elevated levels of oxides of iron and manganese. These silica-rich, Proterozoic sediments contrast starkly with the dominantly-gray, carbonate- and organics-rich, Paleozoic strata that make up most of the main ranges of the Rocky Mountains.

Referring to the Colorado Rocky Mountains, Collins & Davis (1992) say that "the great amount of introduced silica ... suggests that large volume changes occurred deep in the Mantle ... during preCambrian time. Although siliceous sedimentation is the rule for the Rocky Mountains in Proterozoic time, the volume of the Belt in a relatively small basin is especially noteworthy. What agency, one must question, would debouch 3.6 million km$^3$ of silica and alumina on a continental shield in such thickness as to depress it 18 km? I find this a striking apparition.

The western part of the Belt basin is arched into an anticlinorium today.(Yoos et al. 1991) "Intense thermal arching" that produced a "continent-scale rift system" in which "peralkaline basalts flowed from hot spots" (Whipple 1989) describe the most comprehensive, cautious, and correct view of the Belt basin by modern geology. But no one has ventured to interpret a source for the sediment, which is only tenuously linked to peralkaline basalt sources.

If carbide-hydride systematics is the responsible agency – and it surely can produce the silica and alumina – why is there not more carbonate and carbon in evidence? The answer could be that silicides in the deep Earth were primarily interactive with Core hydrogen at the time. Perhaps in Proterozoic time the hydrogen of the Core had no exposure to the trove of carbides, which have been reacting throughout Phanerozoic time.

The abrupt, profound, and seemingly permanent change in chemical makeup of oceans and atmosphere happened in early Cambrian time, long after Belt sedimentation. The changeover from silica to carbonate prominence, long a puzzle to geologists, coincided with an astonishing proliferation of new life forms. Nineteen new phyla occur in the "Burgess shale" deposit of Cambrian age within the northern limits of the Belt basin. (Gould 1990)

The environmental change reflected by the carbonate rocks likely depicts the first ample supply of carbon in oceans and atmosphere, the first chance life forms had to build calcic skeletons and, generally, to proliferate. Gould shows how the concept of the tree of life, speciation from a single ancestor, is an upside-down concept. All but four of the nineteen phyla that inhabited Earth in Cambrian Burgess time had gone extinct again before mid-Paleozoic. The demise of fifteen phyla suggest ecological disaster. In fact, however, the decimation was spread over two hundred million years.

Popular astonishment over the meteoric biotic flowering and subsequent demise, ignores the causes of both the procreation of nineteen new phyla and the extermination of fifteen of them. The most impressive fact of the endogeny of this scenario is the inescapable association of vast quantities of toxifying gas to the huge outpourings of endogenic silica that geology records.

With respect to the Belt supergroup the factors that imply an endogenic source for the content of its sediments are the following:

**1. Siliceous and aluminous lithologies are collaterals of hydridic effusion.**

**2. The enormous volume of the Belt, which is certainly 1.6 million km³, and may approach**

3.6 million km³, implies a local source. There is no evidence for removal of that volume of material from anywhere else in the neighborhood.

3. The orientation of the basin of the Belt supergroup in an alignment over the Rocky Mountain trench suggests that Beltian rift behavior was later mimicked by dilation and effusions of siliceous materials in the preCambrian-Cambrian transition, the later Ordovician, the Jurassic-Cretaceous transition, the Laramide orogeny, and in the late-Pleistocene and Pleistocene-Holocene transition periods.

## THE MINNES/ELLERSLIE/NIKINASSIN/KOOTENAY ("MENK") GROUP

The names, Minnes, Ellerslie, Nikinassin, and Kootenay,[180] [the "MENK" group] are local names for a clastic series of sediments, mainly thinly-bedded sandstones and shales, but also coals. All but the Ellerslie were distributed as a thick apron on the eastern slopes of the Canadian Rocky Mountains. They extend 300 km east across the western Canada basin, in the middle of which they correlate with the Ellerslie. Southward they extend far into northwestern Montana.

The mountain belt had yet to rise in late Jurassic (Portlandian) to early-Cretaceous (Barremian) time, when the thick MENK apron was deposited. It locally reaches two km in thickness, apparently deriving sediment from the west, as the strata thin toward the basin in Alberta and disappear on its east flank.

---

[180] *The name, Minnes, was intended to include two phases of sedimentation, the continental strata of the Kootenay formation and marine-to brackish-water deposits of the Nikinassin and Ellerslie. In reality, these phases interfinger laterally and partially succeed each other vertically. Thus, the four terms are combined for reference purposes in this book, or Minnes-Ellerslie-Nikinassin-Kootenay, = MENK.*

The strata comprise a distinctive pile of continental and marine sandstones and shales. Identifiable fossils are rare, although evidence of benthonic activity is often abundant in the thinly bedded, dark gray to black, carbonaceous shales and interbeds of coal. The sandstones tend to be light gray, quartzitic, fine-grained and even-textured, their grains mainly quartz with well-preserved crystal facets. Local floods of chert grains and conglomerates fill out the lithology menu. The distinctive features are the sparkling crystalline quartz of the sandstones and the dark coloration of the shales, which phases to black where coal content increases.

The underlying Jurassic Fernie formation is very-calcareous, marine shale, quite unlike the MENK strata. The overlying Cretaceous Blairmore (Mannville) formation is non-marine, brackish-water deposition with calcareous members, coal members, shales with volcanic-sedimentary affinities, and at least one basalt lava flow [the Crowsnest volcanics].

Whether we look west or east from the axis of the Alberta basin, there is no apparent quartz source for the sands of the MENK. Whereas pebbles found in the overlying Blairmore can be ascribed a derivation from the gneissic terrane immediately west of the Rocky Mountain trench, the infrequent conglomerates of the MENK cannot be so identified. A recent Geological Survey of Canada interpretation makes the following observations:

> **"It is interesting ... that conglomerates of the [MENK] clastic wedge, which have commonly been interpreted as having their source in the region of origin of the upper Blairmore pebbles, [the crystalline terrane west of the Rocky Mountain trench, that is to say], do not contain pebbles of this type".**[181]

---

[181]  *Price & Mountjoy 1970*

Their observation is backed by dating of the pebbles found in the Blairmore to K-Ar ages, ranging from 113 to 174 my. Thus, although the presumed source terrane was creating rock debris at the time of MENK deposition, none of it reached the MENK site until much later.

The features about the MENK that suggest endogenic input are:

**(1) angularity of quartz clasts and mono-mineralogy sandstone: a "non-detrital" character, and**

**(2) high carbon-content shales: their pervasive dark color.**

Angularity of clasts along with mono-mineralogy is inconsistent with winnowing by fluvial action. In the winnowing process, tumbling corrades the softer lithologies to micron sizes, while harder clasts, such as quartz, may be preserved. Inevitably, however, the harder clasts are rounded in the process. Angularity cannot be preserved at the same time as winnowing ("cleaning up") of the sand proceeds.

The argument may be raised that the angular quartz is a product of recrystallization. The argument against this is that sand beds of similar permeabilities in the Blairmore immediately above and others subjacent to the MENK strata have well-rounded clasts. Recrystallization of certain strata only, while others nearby are left unaffected, does not seem likely.

**A plausible explanation for much of the MENK sand is endogenic expulsion during the repetitious dilation and subsidence of the proto-Rocky Mountain belt.**

The origin of the carbon content in the MENK can also be attributed to the carbon fraction of the silane system of endogeny. This subject will be deferred until the later section on COAL.

## "EARTHQUAKE SAND EXPULSIONS"

In the previous sections dealing with the Mount Wilson, St. Peter, Gog, Belt, and MENK sandstones there was an inference of their spatial relationships to major faults or rifts; but precise positioning for the break through which the sand emerged could not be achieved.  In this section the reader is apprised of some sand emanations that can be specifically attributed to known faults.

## The Charleston Earthquake Sand Expulsion

**VIII-4**
**OLD PHOTOGRAPH OF THE**
**"EARTHQUAKE SAND" EXPULSION,**
**CHARLESTON, SOUTH CAROLINA, 1885**

The record of sand expelled by an earthquake at Charleston, South Carolina, in 1886 is told by **(Fig. VIII-4)**.[182] The mechanism by which an aquifer is squeezed, expelling its water and entrained sand, is illustrated. As the caption implies, the sand appears to be mainly quartz. Although it is likely to be merely reactivated local sand from a buried Tertiary coastal source, the possibility cannot be discounted that it could be endogenic sand instead.

## Submarine "Sand Dunes"

A large field of "sand dunes" is reported in 3,000 feet of water on the floor of the Gulf of Mexico.[183] The term used here, dunes, can only be the product of wind activity. Since the floor of the Gulf has not been emergent recently, these features cannot properly be called dunes. Furthermore, they are too far from land to have come into the Gulf with river water. More likely, this is sand expelled by endogeny, perhaps the expulsion product of earthquake energy. Unlike the Charleston area, where buried Tertiary beach sands might have been mobilized by hydrological activity connected with an earthquake, there are no buried beach sands this far from shore in the Gulf of Mexico. Hence, further study of these sands is likely to establish them as truly endogenic.

## THE WINTERBURN SILT

In late-Devonian times a 200-km series of reefs, known as the Meadowbrook-Rimbey chain, grew along a north-trending fault scarp near the middle of the western Canada sedimentary basin. Their morphology and history have become well known, because of their oil and gas productivity. The existence of the fault is inferential, on the basis of the long, straight trend of reef growth.

[182] *Photo from USGS Annual Report, 1887-1888; caption by Geotimes, 1990.*

[183] *A headline in Science News, 1985 (v128:p191) entitled "A Systematic Sounding of the Sea Floor"*

Deepening of the basin in time drowned the reefs and buried them under clays. Subsequent basin shallowing then allowed biostromal reef, the Nisku formation, to blanket the area, a stage of carbonate sedimentation that was terminated abruptly, when quartz silt was laid over everything to a depth of 30 to 40 m. The Winterburn silt is clear quartz. Its equant crystals are extremely uniform in size and exhibit well-preserved crystal faces and sharp corners. Total volume of quartz silt is on the order of 1 000 km³.

This terrain was entirely at sea. Besides emergent carbonate banks far to the north, only a few low atolls stood within hundreds of km. There was absolutely no source that might have shed any erosional debris, much less quartz silt with the purity of Winterburn lithology.

**The only obvious source for 1 000 km³ of Winterburn quartz silt is effusion through the fault that defines the west flank of the Meadowbrook-Rimbey reef chain.**

## THE ELLERSLIE SANDSTONES

These sandstones of the Jurassic/Cretaceous transition were briefly discussed with their correlatives of the Rocky Mountains, the Minnes, Nikinassin, and Kootenay groups. The sandstones are near-100 % quartz lithology, uniform, fine- to medium-grained, lacking in fossils, and interbedded with dark shales, which are also largely barren of megafossils.

Unlike its correlatives in the mountains, the Ellerslie is situated near the center of the western Canada basin, the same area as the Winterburn. Its thickness is about the same as that of the Winterburn, and its volume is likely similar, about 1 000

km³. Also, like the Winterburn, the best development of the Ellerslie is aligned along the Meadowbrook-Rimbey reef chain, and the inferred underlying fault.

The Ellerslie rests on detrital and regolithic material of the bevelled Paleozoic carbonate erosion surface. The detrital materials are calcareous, heterogeneous, and wholly incompatible for generation of quartz sand. No possible quartz-sand source is apparent, a fact which has puzzled me from 1950, when I named and described the formation (Hunt 1950) at its type locality beside the Leduc reef oilfield.

Endogeny offers, for the first time, a solution to the enigma. **The obvious source for 1 000 km³ of Ellerslie quartz sand is effusion through the fault that defines the west flank of the reef Meadowbrook-Rimbey reef chain.**

## THE KILLARNEY QUARTZITE

### L. G. COLLINS

On the northwestern shore of Lake Ontario, Canada, the granitic, preCambrian shield terrain is interrupted by a prominent northeast-trending dislocation known as the Grenville fault. On the west side of the fault is a wedge of fine-grained pink granite (aplite) that is essentially devoid of ferromagnesian content and exhibits cataclastic texture (**Fig. VIII-5**). The pink color is caused by hematite.

Two km east of the town of Killarney the aplite hosts a white orthoquartzite of nearly-pure quartz composition striking parallel with the Grenville fault. Dikes and veins of quartzite intersect the fine-grained granite. The quartzite, lacking the hematite found in the aplite, has an off-white coloration.

Relict sedimentary granularity, a feature that is usually well-preserved in sedimentary quartzites, is entirely lacking, as are

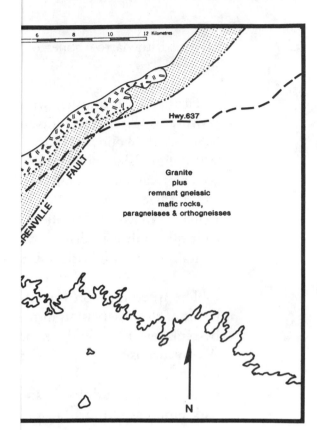

Granite
plus
remnant gneissic
mafic rocks,
paragneisses & orthogneisses

Hwy.637

GRENVILLE FAULT

-5
TTING OF THE
ZITE, ONTARIO

t would normally be preserved
ea of exposed quartzite is 100 x
bstantial offshore, beneath the

l adjacent aplite to the west is
e controlled by cataclasis.
less cataclastic; and broken
r and quartz crystals appear in
spar grains. This continues for
n abruptly changes in a three-
ssive, coarse-grained Killarney

granite with narrow zones of cataclastic granulation and mafic enclaves.

Farther west, another wide zone of aplite veins and dikes, representing replacement and preserving mafic enclaves, is found at the boundary of the Killarney granite on the east side of the Pine Island Channel. These dikes and veins are narrow zones of cataclastic granulation, the ferromagnesian silicates having been replaced by quartz, magnetite, muscovite, and hematite films. The mafic enclaves are biotite- and enstatite-rich quartz diorite dissected by aplite and granite veins that contain myrmekite with tiny quartz vermicules.

The interpretation I would advance for this terrane is the result of the observation in thin sections that quartz has formed by replacement of both ferromagnesian silicates and feldspars. A continuous series of transformations is present from Killarney granite to quartzite. Where the aplite grades to quartzite [either small stringers and dikelets or massive quartzite] the aplite is dark pink to red near the granite and progressively-lighter pink toward the quartzite as the mafic silicates are replaced. Hematite and ferromagnesian silicates are removed together, and rock color changes from red through pink, and pale pink to white. The dikelets and stringers appear to represent cataclastically permeable avenues for silane entry and transformation.

Thus, deformation older than the Grenville fault (880 to 1,000 my) must have created cataclasis, which then allowed silane permeation and conversion of former mafic rocks, leaving partially transformed lithologies and angular enclaves of host rocks (**Fig. VIII-6**). This process has recurred in many episodes and produced the Killarney, Bell, and Chief Lake plutons. Its most thoroughgoing conversion has been the aforesaid quartzite east of Killarney.

## VIII-6
### PHOTO SCENES SHOWING THE TRANSITION FROM
### DIORITE TO KILLARNEY QUARTZITE

A

B

*Scene 1: A.* Unaltered diorite pierced by granite vein, proving the granite to be younger. *B.* Cataclastic zone mainly a sea of fine quartz and feldspar in center, gradationally changing to granite outward, with coarser texture and mafic mineral.

*Scene 2: C.* Mafic granite transformed from diorite on left in sharp contact with pink, less-mafic granite containing myrmekite. *D.* Gradational contact from mafic granite with pink K-feldspar megacrysts bordered with myrmekite to pink myrmekite-bearing granite in which quartz has replaced mafic minerals.

*Scene 3: E.* Grey dike of quartzite intruded into pink aplite. *F.* Grey, branching quartzite intruding pink quartzitic aplite.

G

H

*Scene 4: G.* Patchwork of white quartzite injected into pink quartzitic aplite. *H.* scenery of massive white quartzite introduced into pink aplite but leaving island remnants of partially converted pink quartzitic aplite. The pink color of the aplite in frames F, G, and H is caused by iron and manganese in vestigial feldspars, the rock being largely replaced by quartz.

Needless to say, the literature on the region presents a very different picture. T. T. Quirke[184] thought the quartzite a relict of former sedimentary rock enveloped in magmatic granite. Debicki[185] describes the sedimentary features of nearby terranes that should be found in a quartzite remnant, such as Quirke interprets. The Killarney quartzite lacks such features.

Clues to a different origin are readily found. The secondary hematite coatings do not fit a magmatic origin, but are readily explained as redeposited iron after pervasive silane transformation. Instead of the mechanism of sand accumulation within crustal crypts, and its subsequent expulsion in a pulse, as exhibited by other terranes, the Killarney silane oxidation was contemporaneous with its injection into cataclastic breccias. This resulted in minerals being transformed *in situ* to quartz.

The petrological data demonstrate with elegance the process of pervasion of water-saturated, shattered rock fabrics by volatile silane in the absence of oxygen. The progressive cation stripping-out and flushing-upward with excess water and liberated hydrogen has left transformed silicate rock fabrics and, ultimately, pure quartz.

**In summary, all field and microscopic evidence in the Killarney terrane supports an interpretation of replacement *and quartz deposition in situ* by pervasive silane infusion in tectonic shear zones. The granite and aplite are earlier stages, massive quartzite the final stage of this replacement. Structural lineation of the terrane components with the regional strike of the Grenville shear pattern implies a temporal association between silane infusion and tectonism.**

[184]  *T. T. Quirke 1927, 1940*

[185]  *Delbiki 1982*

# THE ATHABASCA SANDSTONE

This massive sand deposit of northern Saskatchewan and Alberta was deposited to a thickness of 1.6 km or so over an area of about 75 000 km² in middle Proterozoic time, 1.3 billion years ago. A basal debris flow commenced the deposition, and copious sand flows followed. The sand is unmetamorphosed, 95 % or more pure quartz, but off-white due to minor clay content, and in places prominently-mottled to hematite-pink by iron oxide.(**Fig. VIII-7**)

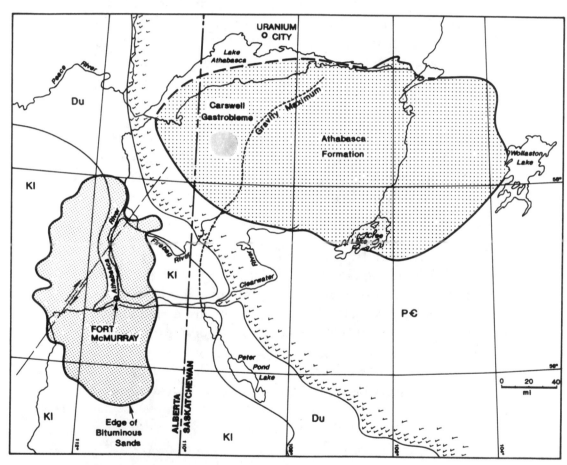

VIII-7
MAP OF THE ATHABASCA REGION,
ALBERTA AND SASKATCHEWAN

Like the St. Peter, the Athabasca sands rest unconformably in a saucer-like depression floored with preCambrain granitic rocks.[186]  At a west-central position in the sandstone terrane, a 20-km-diameter crater-form feature, known as the Carswell structure, exposes a diapiric core of gneissic metamorphic rock. The peripheries of the gneissic core are brecciated and mylonitized.  Beyond them, an outer, 36-km-diameter rim of mid-Proterozoic dolostones encircles the crater-form.[186]  Beyond the dolostones the Athabasca sandstones encirle the Carswell feature.

The Carswell "event," the diapiric rise of the gneissic plug through the Athabasca sandstone series, was a time of intense hydrothermal activity, mylonitization, and the tectonism that accompanies a mechanically injected plug.  It is dated to Ordovician time, 480 million years ago.

The durability of the 1.3-by cradle of sandstone amid erosion-resistant granites is remarkable.  Its preservation through the "Carswell event" of 480 my age is even more remarkable.

Sand volume within the depression around the Carswell crater-form amounts to approximately 50 000 km$^3$.  R. W. Johns[187] described the sand as having a "quartz content approach[ing] 100 % and consist[ing] of pure creamy quartz sandstone over thicknesses of thousands of feet.  The lack of variation is quite astonishing ... [so much so, in fact, that] rare beds with grain size as large as 1/10th inch are called 'micro-conglomerates'."

Johns' focus on conglomerates arose from the belief that conglomerates indicate uranium mineralization;[187] the particular interest that he shared with later authors,[186] who found enough clay in the sandstone and between its flows to differentiate it into members.

---

[186]  *Lainé et al., Eds. 1985; readers should note in particular the Hendry & Wheatley and Pagel & Wheatley papers*

[187]  *Johns 1970*

**The enigma Johns recognized [and I emphasize] is the unprecedented size of the sand body and its complete disjunction from any evident surface source.** Quoting an unnamed old prospector, Johns poses the question, **"Where the Hell did all the quartz come from?"**

Clearly, the mass of quartz is too large and too pure to be the result of weathering and winnowing of local shield granite. And clearly, its deep bowl habitat is anomalous. If the Crust had foundered, creating a great depression, the bowl would have filled with a much more diverse assemblage of erosional debris from the vicinity, not primarily relatively pure quartz sand. Silane activity could produce such a quantity of quartz sand. It is difficult to conceive any other origin; and the basin itself could only have been created by the weight of sand on the Crust over an evacuated system of crypts and chambers.

Salient features of the Carswell structure as reported by eleven author groups[186] permit some generalizations:

An astrobleme origin for the structure was suspected because of two particular features: **(1)** the dolostones are overturned outward, away from the center of the crater-form feature, and **(2)** the breccias and mylonites contain pseudotachylite (false contact melt rock). However, a diapiric rise of the gneissic core would also lead to outwardly overturned dolostone beds; and pseudotachylite is known as a product of endogeny in metasedimentary terranes, such as the Archaean Witwatersrand gold mining terrane (discussed in the EV volume in connection with the Vredefort structure).

There are, on the other hand, two good and sufficient reasons why the Carswell crater is not an astrobleme. First, there is no associated ejecta blanket; and second, there is no glass lake or permeation of underlying and surrounding bedrock with glass from impact melt (note Ries crater in the EV volume). Both of these features should be prominent in an astrobleme crater of Carswell dimensions. And, even if one temporizes that erosion could have removed the ejecta blanket and the glass lake, there is no possibility it could have removed glass that must have permeated the fractured crater floor beneath a relict astrobleme.

If it is not an astrobleme, can the Carswell crater-form be explained as a product of endogeny? The answer, as suggested already, is indisputably affirmative. **First** of all, the 20-km diameter center of the "crater" is a gneissic complex that has risen diapirically, breaching the cover of sedimentary dolostone, which now encircles it as a collar of 36-km diameter.

**Second**, the periphery of the gneissic core is laced with breccias and pseudotachylite-permeated mylonites. This combination of features is typical of mechanical intrusion originating with endogeny.

The literature[186] provides seven further points of support for the endogenic origin of the Athabasca sandstone and the Carswell structure:

**Third**, debris flows underlie the sandstone facies and comprise the basal member of the Athabasca group. These are characteristically features of volcanism.

**Fourth**, the sandstones have been divided into a series of members based on the rather subtle mineralogy of their minor clay content. This implies a series of pulses to expel each diverse lithology.

**Fifth**, the Ordovician "Carswell event" subjected the gneissic core and adjacent rock to intense hydrothermal activity. It must have been density reduction consequent upon the hydrothermal alteration that caused the diapirism of the gneissic plug in the first place.

**Sixth,** the crystalline diapiric core can be divided into a clutch of associated granitic "domes," which carried with them mafic enclaves as they rose isostatically through their dolostone cover. The petrology of the transformed core gneisses and their isostatic rise is precisely in accordance with behavioral predictions of silane systematics.

**Seventh,** melt-derived cryptocrystalline breccias were developed and moved *upward* (with the diapir), not downward.

**Eighth,** no coesite or stishovite (very high temperature forms of $SiO_2$) is associated with the structure.

**Ninth,** altered volcanic shards indicative of contemporaneous silicic volcanism are found in mudstone and siltstone facies of the Athabasca group.

**The Carswell crater-form defines a vent through which, in mid-Proterozoic time, an enormous mass of sand was convulsively exhaled. The granitic Crust bowed beneath the sand mass, creating a sand-filled crustal depression that survives intact today. In Ordovician time, 820 million years after it exhaled sand, the vent was reactivated as a conduit for hydrothermal fluids. These fluids altered minerals of the Archaean rocks below the old crater, lightening them. The altered core rose diapirically through its cover rocks, which included dolostones that had been deposited in the old caldera before Cambrian time. These characteristics define the volcanic behavior of a long-standing gastrobleme.**

That covers the crater origin and the nature of the diapiric core that now occupies it. The generation of sand, its storage underground, and its expulsion 1.3 by ago now merit our attention. Since the reactivity of silanes with free oxygen is so great that silanes will never be detected directly on the surface, their existence must be posed as a "given." Certainly, if silanes exist, they could have produced the sands by simple oxidation.

If we assume that silanes rise from the Mantle, the question remains, "How could they penetrate brittle Crust?" The answer is clearly given by R. C. Bailey[188], who considers the hydrodynamic behavior of fluid that accumulates under pressure in the lower Crust. At the crustal transition zone, where rock turns from ductile to brittle deformation behavior (at depths that are usually in the range of 10 to 15 km), upward hydrodynamic fluid movement is deflected. Bailey finds the "brittle-ductile transition" to be an impenetrable barrier to continued upward migration of fluids.

However, the barrier does not stop lateral migration. After fluid accumulation and pressure buildup in excess of lithostatic pressure, the trapped migrating fluids hydraulically fracture a lateral escape route. Such hydraulic fractures are horizontal; none is vertical. The integrity of brittle caprock at mid-crustal levels is preserved. "Fracked" reservoirs with sill configuration are created with intergranular sand porosity and permeability beneath the brittle caprock.

If silanes are involved in upward-migrating gas mixtures, quartz sand is created where water is first encountered. If such sands are entrained by hydrodynamic forcing of horizontal fractures, sand should prop the fractures open after pressure is relaxed. This must have happened on a massive scale concomitantly with the production of 50 000 km³ of sand that was later to be expelled convulsively to form the Athabasca formation.

Horizontal hydraulic fractures of the kind envisioned here conform well to Skobelin's Siberian sill emplacement mechanism (Chapter VI). The difference is that in Siberia the fractures were filled with magma; here they became packed with sand. The fracking and injection concept is the same in both cases.

---

[188]  *Bailey 1990*

We may be confident, then, that before the 1.3 by eruption, there was in place a well-developed set of sand-filled horizontal crypts beneath the brittle upper-Crust around the Carswell gastrobleme. Up to the time of eruption, continued inflow of sand forced the crypts wider and farther laterally, the oversill granite (to use Skobelin's term) must have dilated upward in a cymatogen (to use King's term). **Silane-generated quartz sand and water in crustal chambers then existed in a gravitational inversion, a state of astatic stability. A breach at the crest of the crustal oversill complex would be followed by convulsive emission of sand from the crypts.**

**The release, when it actually occurred in mid-Proterozoic time,** *flowed in all directions from the vent. Surface runoff, by gravity, gave this flow a bias toward the east, where it flowed 100 km from the vent.* **Had the vent been central on the top of the welt, the sand would be deposited symmetrically around it. Instead, its preferred eastward flood direction shows that** *the Carswell vent must have been situated on the east flank of the welt.*

The volcanic release of sand must have been accompanied by heat from hydrogen ignited by sparks between quartz in the erupting slurry and quartz in the conduit walls. This could explain the anatectic melting of peripheral rock that generated pseudotachylite. Without the energy contributed by burning hydrogen, the emission should have been cooled by the expansion of contained inert gas. Concomitantly, as the subsident granite surface exchanged places with outflowing sand, the light under-mass flowed from its abyssal crypts and settled on the former oversill terrain in a bowl of its own making.

*Certainly, silane endogeny is the best – perhaps the only – rational explanation for the known features of the Athabasca formation.*

## CHAPTER IX

# ENDOGENY AND ECONOMIC MINERALS

## PETROLEUM IN A SUPRA-MANTLE SAND CHAMBER

### C. WARREN HUNT

*"Alkanes occur in rocks of all ages,*
*even older than three billion years."*

**GORDON W. HODGSON, 1972**

In the EV volume I implied that beneath the terrain dilation of the Gros Brukkaros gastrobleme, endogenic hydrocarbons may have accumulated since the last explosive event in late-Cretaceous or early-Tertiary time. That prospectiveness can be expanded now to other regions. The Skobelin description of the Siberian Irelyan oilfield (**Fig. VI-10**) is instructive. Oil is produced from a reservoir at the contact between Phanerozoic sedimentary rocks and crystalline basement. Diamonds are mined from a kimberlite emplaced in young sedimentary cover rock directly above the oilfield. This suggests hydrocarbon emanation from crystalline sources and diamond generation from hydrocarbons.

Another hydrocarbon accumulation that is hard to explain, unless a source in crystalline rock can be invoked, is that of the Siljan Ring. The Siljan structure is a meteorite impact crater in the crystalline shield of northern Sweden. Bitumens are found at the surface in the brecciated crater rocks. On October 7, 1991 a drillhole that was being drilled to test the subsurface potential for petroleum reported that, while drilling at 2 833 m, free oil was recovered with drill cuttings. Inasmuch as a previous test well had produced only questionable results, [although it is claimed that earlier in 1991 it actually produced 80 barrels of oil from the open hole], and further information has not been released since the original announcement, the recovery of free oil must be treated as signifying conclusively that oil does, indeed, rise from depth in granitic shield terranes of the world.

The Alberta bituminous sands terrane is another interesting site where an association is suggested between petroelum seepage and crystalline rocks. Perhaps the largest single petroleum accumulation known on Earth, 209-billion m$^3$, or 1.315-trillion barrels, makes up an accumulation that occupies an area of 28 thousand km$^2$. As mentioned already, these deposits of bitumens are situated beside the Athabasca sandstone of northern Saskatchewan and Alberta (**Fig. VIII-6**). The proximity of this petroliferous terrane to the great Athabasca sand debouchment need not be regarded as coincidental.

The scenario described in Chapter VIII for the origin of the Athabasca sandstones and the Carswell crater-form included sand eruption from a vent on the east side of a crustal welt, and subsidence of the Crust under the erupted sandstone. **The eastward tilt of the Crust from the Alberta bituminous sands to and beyond the Carswell gastrobleme is of critical significance for petroleum accumulations because the formerly-horizontal chambers from which the expelled sand flowed must now dip gently east if the dip is to act as a caprock, to effect containment of hydrocarbons.**

**If sand bodies were left in the western and central parts of the original dilation, as one would expect after exhalation from the eastern limb, there should be hydrocarbon accumulations within them. And, if the granite caprock were to leak generally, the pattern of the seepage should outline the trap and any remaining subsurface reservoir.**

Thus, a reasonable interpretation of known geology in terms of endogenic generation of hydrocarbons and reservoir sand-matrices is that there is a compelling spatial association between the effused endogenic sand to the northeast and the bituminous sands on the southwest. If the bitumens are the product of seepage, the opportunity is open to test the existence of a trap below them by drilling for it.

**<u>In summary:</u> Slurries of quartz sand and water that originated with silane oxidation in mid-Proterozoic time "fracked" open sill-like crypts at the ductile-brittle transition in the middle Crust [at depths of 10 to 15 km] creating planar injectites precisely analogous to Siberian trapp sills (Figs. VI-7, VI-8, VI-9). Continued injection into the crypts caused dilation of the Crust in a 300-km-diameter swell over the combined area of the Athabasca sandstone *and Alberta bituminous sands*. The swell finally burst at the Carswell crater.**

**Hydrogen and hydrocarbons in the volcano made the explosion a fiery one. The point of eruption being east of the center of the swell, the debris and sand flowed mainly eastward, down the flank of the crustal welt in successive waves. The weight of effused material depressed the former "oversill" granite and choked off continued flow.**

**The area around the Carswell vent should now be essentially depleted of sand resources. The sand-filled crypts to the southwest, below the Alberta bituminous sands, should be intact with full charges of injectite sands in place behind the east-tilted granite cap.**

Geologists have speculated on the subject of where the bituminous-sands oil came from for over a century without any plausible explanation being found. As bitumens are too viscous to migrate in fluid form, they are believed necessarily to be residues of more fluidic hydrocarbons. Biodegradation as a factor in the development of the Alberta bituminous sands is, thus, generally accepted.

There is no consensus, however, on where the petroleum came from. Reaching the biosphere in the Mannville formation, migrated gas and oil must have been attacked by bacteria and dehydrogenated *in situ* to bitumens.

Besides its elusive generative source and its present lack of fluidity for migration, there are other factors to be satisfied. They include the following: (1) The low permeability of the lower-Cretaceous Mannville host sandstones. (2) The supersaturation of richer streaks, where the sand grains float in bitumens. (3) The marginally petroliferous quality of the host rock over wide areas of Alberta and Saskatchewan provides no clue to why this area should be so much more petroliferous.

A satisfactory answer to all these enigmas is provided by silane systematics. **As the Athabasca sandstones flowed from the northeastern limb of the crustal welt, and the bituminous sands sit above the western limb, it is reasonable to anticipate that subsurface sand volumes below the Alberta bituminous sands will compare in volume with the extruded sand. That implies a huge reservoir volume and a thick net-oil column, possibly stacked to several kilometres in gross thickness.**

**From mid-Proterozoic time to the present, hydrocarbons emerging from the Mantle should have been trapped in the injectite sands and against the tilted, subsident, crystalline caprock. Reservoir leakage during the last 1.3 billion years has, manifestly, been enormous. As a minimum, the capping oversill can be interpreted to have leaked sufficient feedstock for hydrogen-seeking bacteria to degrade 1.315 trillion barrels of migrated petroleum in the 120 million years since Mannville times.**

However, leakage is not evident today, perhaps because silica-saturated waters have re-sealed the escape fractures used in earlier leakage, or perhaps because no one has taken the trouble to determine hydrocarbon levels in fractures of the granite beneath the bituminous sands.

It is noteworthy in connection with preCambrian sources for hydrocarbons that bitumens occur in the deformed crystalline rocks of the Carswell crater and that they are reported to have "strong negative $^{13}$C values: -48.5 $^o/_{oo}$ to -48.8 $^o/_{oo}$ ( $\pm$ per mil relative to PDB). Even for preCambrian kerogen, these values are too negative to correspond to values from sedimentary organic matter." [189]

Caprock effectiveness is a question critical to analyzing the subsurface prospect. There is no way of telling how competent it may be for petroleum retention. Two factors can be cited as bearing on this: (1) One is favorable, one unfavorable. The unfavorable one is the brittle nature of granite, which mitigates against its acting as a reservoir seal for a highly-pressured reservoir. (2) The favorable factor is that abundant dissolved silica precipitated as interstitial cement in the Athabasca sandstone; and this suggests that water supersaturated with silica would have accompanied any leakage of a supra-Mantle hydrocarbon reservoir situated beneath the Alberta bituminous sands. Silica, thus, would have sealed off the escape conduits, and caprock would have been indurated over time by secondary silicification, a phenomenon frequently seen in oil and gas fields.

On balance, it is not unreasonable to expect significant silicification and improvement of the granitic cap rock through time, thus preserving the reservoirs. The existence today of multiple, stacked, sand-filled, sill-like crypts and a very large accumulation of petroleum under the area defined by the present Alberta bituminous sands is a reasonable

---

[189]   *Lainé et al. Eds. 1985, Note the Landuis and Dereppi paper*

expectation. Silane systematics is the first adequate and rational explanation for the huge bituminous-sands oil accumulation; and it bodes well for pointing the way to remaining accumulations below.

**What sort of structure is implied, then?** The regional gravity scene shows a Bouguer gravity maximum trending northeast [as shown on **Fig. IX-1**] into the center of the Athabasca exposures. This is likely representative of a deep trend of preserved mafic rocks. The Alberta bituminous-sands area is weakly Bouguer-negative. Massive, low-density sand sills and associated silane-transformed rock could explain the regional low gravity beneath the Alberta bituminous sands.

## The Supra-Mantle Reservoir

The seismic response of the ductile/brittle transition within the Crust normally at about 10-km depth is a reflective horizon. If multiple injectite horizons are present, they also should produce good reflections.

The depression of the Crust under the Athabasca basin is shown on Fig. IX-1 as a simple dip. The profile is drawn from the center of the basin southwest and passing through Fort McMurray. The dip is taken as a straight line from 1.6 km below the surface at the center of the basin to the outcrop contact between basement and the Athabasca formation. This slope is simplistic; the basin margins are likely steeper and its floor more nearly flat.[190] However, as a crude approximation of crustal distortion at the 10-km level, it may still be a reasonable approximation of actual conditions.

Using the aforesaid dip, one can draw parallel dip slopes from a 5- and a 10-km depth below the 1.6-km floor at the center of the Athabasca basin. The 10-km point is presumed to be the ductile-brittle transition. Projecting on the

---

[190] **Ramaekers 1990**

Foldout here

**IX-1**
**Athabaska Region: Profile from center of Athabasca sandstone basin**
**southwestward across the Alberta bituminous sands near**
**the town of Fort McMurray, Alberta.**

presumed dip to the center of the bituminous sands area brings the 10-km level up to about 6.2 km (from 11.6 km), and the 5-km depth up to 1.2 km (from 6.6 km). This implies 5.4 km of northeasterly closure for any porosity beneath the bituminous sands area at its midpoint.

Then, if the oilfield precisely underlies the area of bituminous sands, a horizontal line may be drawn west from the intersection of a projected dip-slope (the 5-km slope is used on the profile) with the vertical below the east edge of the bituminous sands area. The graphic solution, as shown on Fig. IX-1, implies vertical closure of 1.3 km above such an hypothetical "water line" in a sand-sill reservoir.

Whereas these prognostications will doubtless be revised by seismic work, they furnish a first approximation of drilling expectations for a "Supra-Mantle" oilfield that could eclipse the Alberta bituminous sands as the largest hydrocarbon accumulation known on Earth.

# COAL AND COALIFICATION

The subject of this section is the enigma posed to science by coal. Seemingly this simple substance is merely the reduction of plant material to nearly pure carbon by the chemical removal of the other elements contained in vegetation, primarily oxygen, hydrogen, and nitrogen. Close scrutiny shows this idea to be simplistic and impossible on many counts.

Indisputably, identifiable plant parts are encased in coals; "macerals", as they are called. Their biological origin as terrigenous plant debris is undeniable. However, **the nature of the coalification process continues to elude science. Let us consider the enigmas. They include the following:**

*Mineral content:* Coals normally have ash content of less than 4 %, often as low as 0.5 %. Plant material usually contains 10 % mineral that reports as ash. If the plants provided the carbon, there would be no escape for this mineral. Trace metal components, in particular germanium, cannot be linked to plants, since their presence in coal is often far higher than the usual clarkes for those metals in vegetation. Gold[191] discussed the exotic, non-crustal aspects of germanium in coals; we will take the discussion further.

Soil is needed to grow the plants, whose spores and macerals are found in coal. They must be rooted where mineral is available to nurture them, not on top of one another or in purely vegetal debris from former plants. This means that soil and regoliths derived from soils should be found in and below coal measures. Especially where coal seams are many metres thick (sometimes over 50 m) the problem is apparent. Coal-ash leaves absolutely nothing to represent the soil!

*Quantity of carbon:* Any given coal thickness would have required tenfold greater thickness of accumulated plant debris

---

[191] *Gold 1987*

just to yield the carbon fraction. Swamps filled with vegetal material to 500-m depth are unknown today and likely never existed. Production of the amazing quantity of carbon in some deposits would be impossible for the most fertile of rainforests.

*Fusain:* Charcoal layers occur in most coal terranes; their porosities belie burial and compaction for their modes of origin. Sometimes they are secondarily lithified with mineral, mainly silica. Like coal measures, fusians can be traced long distances. Their freedom from clastic mineral particles belies deposition from erosional sources, since streams always carry some clastic debris. Petrographically, fusain is the "partially-charred vestige of former woody structures. [It is] invariably accompanied by large quantities of flattened spores...[and has been] laid down under water. It follows that [fusain] is not derived from the transformation of peat bogs into coal."[192]

*Hydrocarbon content:* The variable and often abundant hydrocarbon content in coal measures cannot be considered a coal or terrigenous-plant degradation product. Coal maturation is incapable of producing hydrocarbons. Input of energy is prerequisite to making hydrocarbon from carbon. It is this energy of the hydrogen-carbon chemical bond and that of oxidation of hydrogen and carbon that is liberated in the combustion of hydrocarbon. Certainly heating carbon and water under pressure cannot produce a hydrocarbon; and water is the only possible hydrogen source unless silane is available for making the hydrocarbons that are found in coal.

The paradox of coal formation has been addressed by a great diversity of solutions. There are too many answers rather than too few. For every coalfield, different explanations seem preferable. No single concept appears satisfactory as a generalization.

---

[192]  *Jeffrey 1913*

Let us try to analyze a variety of aspects of coal to see what is possible. Let us consider the:

**origins of hydrocarbons and trace metals in coal measures;**
**coal dikes and pseudo-intrusive bodies;**
**coal in non-marine habitats;**
**selectively coalified strata;**
**time, pressure, and heat as factors in coalification;**
**detrital and marine coals;**
**trace germanium in coal;**
**absence of coal in preCambrian sedimentary rocks; and**
**synthetic coal from lignin.**

Since the reader will anticipate an endogenic connection, let us first consider the possibility of the general process of oxidation that accompanies the emergence of silanes and hydrocarbons from the Mantle. Silicon, hydrogen, and carbon are attracted in decreasing order of strength to oxygen. Silicon is first, hydrogen second, carbon last.

Carbon becomes oxidized where there is sufficient oxygen first to combine with all available silicon and hydrogen. Where carbon is not oxidized, it may remain in one of five forms, diamond, graphite, soot, colloidal, or as radicals in hydrous fluids. The occurrence of diamond and graphite as products of hydrocarbon degeneration under especially constrained conditions has been explained by E. A. Skobelin in Chapter VI.

Graphite residues, prominent members of preCambrian and other terranes, heretofore have been regarded as products of the high PT conditions of the deep Crust or Mantle. Extensive graphitic horizons are thus usually attributed to metamorphism that affected sedimentary sequences that included coal deposits. This contrasts with E. A. Skobelin's explanation of graphite through hydrocarbon dissociation by proximal magmatic heat (high PT conditions). At low PT conditions, on the other hand, the dissociated hydrocarbon should yield particulate carbon, soot, colloidal carbon, or charged radicals dispersed in hydrous fluids.

The industrial production and physical chemistry of carbon black thus become germane to this interpretation. Three processes well known to modern industry are recognized, the "furnace process" using heavy oil as a feedstock, "thermal cracking" of methane, and the "acetylene cracking" process.

Acetylene, $C_2H_2$, is "cracked" by heat and pressure into its components, the hydrogen then used to heat on-coming acetylene to keep the process going. Thermal cracking of methane is essentially the same process acting on methane. The furnace process entails the larger molecules of heavy, carbon-rich oil with measured quantities of air, which result in "cracking" at high temperatures. Carbon black is produced in these processes at 1 250 to 1 350°C, temperature conditions prevalent at upper Mantle depths.

Many possible differences may contrast the industrial and Earth processes. Higher pressures are to be expected in the Earth; and temperatures other than optimum, either higher or lower, can be expected.

Dissociation of hydrocarbons by heat was explained by Skobelin in Chapter VI in connection with the generation of diamond. This is also to be anticipated as a precursor to coalification. Partial dissociation of $CH_4$ would leave particulate carbon or radicals, C, $CH^{3-}$, $CH^{2-}$, or $CH^{1-}$; and the heavier hydrocarbons would dissociate partially to solid hydrocarbon radicals. Negatively-charged solid radicals and colloid-sized particles of carbon suspended in acidic water would simulate "smoke" in the hydrous environment. These radicals and entrained colloidal carbon particles would likely adhere to undissociated residual hydrocarbons, while the hydrogen ion that was dissociated could be neutralized by any available metallic material and then lost as hydrogen gas.

**This scenario is advanced to explain coalification. Charged radicals and colloidal carbon carried in water and expelled on**

the surface, would be attracted to and adhere to decaying carbonaceous matter, especially terrigenous plant material, "peat." Peat is partially-oxidized and positively charged to start with. The positively-charged peat would collect ambient negatively-charged radicals and colloidal carbon.

Decaying plant fragments of the peat would be encased and preserved as "macerals" within amorphous "pure coal" [coal types known variously as "vitrain," and "clarain"].[192,193] Unoxidized methane dissolved in the water with this endogenic carbon could also be entrapped during this collection process, later to be recognized as coal-bed methane.

This concept of the coalification process should work in the groundwaters of swamps, muskegs, bays, and estuaries or in buried biomass accumulations, wherever, in fact, the hydrological conditions would allow penetrative flow.

The evolution of coal by deposition between macerals in once-permeable beds of decayed vegetation is a scenario involving a duality of carbon sources, terrigenous plant debris, and mineral "carbon smoke." It is unique in theories of coalification in explaining the major coal component, "vitrinite" as endogenic carbon black.

## Origins of Hydrocarbons and Trace Metals in Coal

Large methane resources occur in coal measures and much attention is being given to their commercial exploitation today. Not much is being said about their origins, although quite apparently the reducing environment reflected by ambient hydrocarbons is not a product of peatification or coalification by peat "maturation."

[193]   *Smith, R. I. L. & Clymo 1984*

Coalbed methane is understandable as methane that accompanied carbon black during deposition. Such methane occurs in erratic and unpredictable quantities. This can be interpreted as the consequence of local fortuities, mainly the level of original hydrocarbons carried with locally emanating Mantle gases and the degree of oxidation those hydrocarbons experience as heat dissociates them into carbon black and hydrogen.

Trace elements in coals are also distributed erratically. After an exhaustive attempt to find a rationale for such extreme variations among west German coals, the research team of Litke and Haven[194] drew the intriguing conclusion that "differences in elemental compositions of coals of the same organic facies can be explained by trapping of hydrocarbons in the coals." In other words, trace elements are introduced along with *hydrocarbons* into coal and not with other components, such as the macerals.

In another study on the same subject the coals of the US and England are analyzed.[195] These analysts deduce that there is "no relationship between rank, age, and floral composition of the coals and the minor- and trace-element chemistry of the vitrinite concentrations." Lacking a relationship, the minor- and trace-elements of the vitrinite concentrations must have a different origin from age, rank, and floral composition.

If the Littke and Haven theory[182] is correct, as I think it must be, the divergent origins of trace metals from other components of coal are well-demonstrated by the latter study.

## Coal Dikes and Pseudo-Intrusive Bodies

The concept of mineral carbon black as provenance for much of the carbon in coal beds explains many enigmas. For example,

[194] *Litke & Haven 1989*

[195] *Lyons et al. 1989, in Lyons & Alpern, Eds.*

at Corbin, British Columbia, Roddick [196] reports:

> "a large mass of coal [that] forms an 'intrusive body' in strata of the Kootenay formation. Emanating from the 'intrusive' ... is a narrow coal dike ... [extending] to the surface".

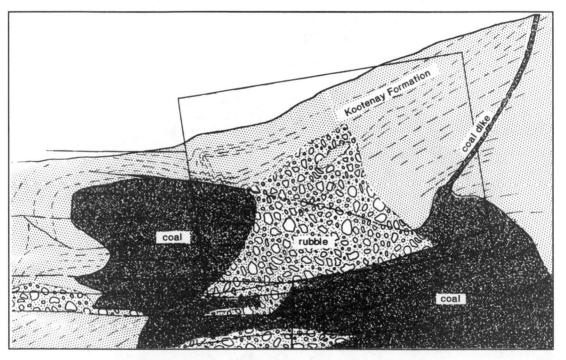

### IX-2
### COAL AS AN INTRUSIVE BODY IN THE KOOTENAY FORMATION, CORBIN, BRITISH COLUMBIA.

The left side shows coal structure similar to the overlying strata, suggesting the coal is squeezed toward the left there. On the right and lower left, however, the coal crosses stratigraphic borders without apparently deforming them; and the extraordinary coal dike to the surface clearly implies a fluid depositional mode. The coal exposure is about 50-m wide.

*Made from a photograph by, and printed courtesy of J. A. Roddick.*

[196] **Roddick 1982**

This mass of coal, is obviously discordant with the sedimentary strata and, hence, could not have been deposited with them by normal sedimentary processes. Neither could it have "intruded" by plastic flow in the manner of a salt plug because, although coal does flow that way, it could not have flowed into the narrow dike that way. The dike resembles an escape orifice, not a welling plug of deforming coal. Hydrous or gaseous fluid escape is essential to explain the dike, a feature that most likely was synchronous with and part of emplacement of the entire mass.

## Coal in Non-Marine Habitats

Thomas Gold[179], in his survey of associations of coal with hydrocarbons, alludes to Geological Survey of Greenland references to:

> "Coal, that is interbedded with volcanic lava and without any sediments ... occurs in the southwest [and] the west coast of Greenland in close proximity to large lava-encrusted lumps of metallic iron, not far from mud volcanoes that issue methane and a rock face which frequently has flames burning in its cracks."

"Mineral hydrocarbon" and "pyrobitumen" are generic terms used by geologists to describe occurrences of asphaltic coal or coaly asphalt. Specific terms include asphaltite, uintaite, wurtzilite, grahamite, libolite, stellarite, impsonite, albertite, and byerite. The names were devised to describe dikes, veins, pods, or dispersed flecks rather than an origin by sedimentation under water. The chemical overlap between petroleum and coal and an absence of evidence for biotic origin is inherent in all of these.

The name libolite originated in Liberia; albertite, byerite, and stellarite in New Brunswick; uintaite and wurtzilite in Utah. Their essential similarities and worldwide occurrences imply an origin that works on a worldwide basis. It is not clear that any fundamental difference exists between these apophyses and Roddick's intrusive body of coal.

## Selective Coalification

In another diagnostic situation Gold describes coal seams of the Donetz basin, USSR, where fossilized tree trunks are clearly rooted in calcareous sedimentary rock below coal seams, pass through the coal, where they are coalified, and then are fossilized but, again, not coalified in the overlying limestone. The enigma is that only the trunk segments within the coal seams are coalified. If heat and the pressure of burial had been able to make coal out of biomass, the conventional interpretation of the origin of coal, the entire trunks should be coalified.

Since trunks were only affected where they passed through coal seams, the cause of coalification must have involved flow across the trunk axes of fluids with the chemical ability to convert wood to coal. In the calcareous rock beneath and above the coal seams, the trunks are "petrified" [that is to say, permeated and replaced chemically] with sediment characteristic of those sedimentary host rocks. "Woody biomass" is implicitly a primary precursor of this coal. Its macerals, their identifiable structure surviving, are now surrounded with vitrinite. Mineral carbon black and solid hydrocarbon radicals emplaced by hydrous flow are, then, implicitly a second precursor of this Donetz coal because of the positioning of coal only in those parts of the tree trunks which intersect the "horizon" in which coal was deposited.

If coal can evolve from the addition of mineral carbon-black and solid hydrocarbon radicals to woody biomass, the process should be amenable to laboratory testing using partially-oxidized woody biomass and hydrous carbon "smoke" with a weak negative valence. I am not aware that anyone has tried to do this.

## Time, Pressure, and Heat in Coalification

Youthful coal is a phenomenon that strikes at the heart of the coalification issue. If depth of burial, time, and moderately high temperatures are required for "maturation" from peat to coal, short life and surface environment are anathematic.

In **Yunnan Province, China,** an elongate basin of 35-million km$^2$ has been filling with Pleistocene sediment for only 600,000 years. Peat beds of 75,000 to 10,000 years old and up to 20 m in thickness and under 170 m of sediments are found throughout the basin. They are partly transformed to soft brown coal.

The authors, Kuili & Yong[197], report basalt flows, andesites, diatomites, and clastic sediments as the basin stratigraphy. A very high geothermal gradient, 99°C/km of depth, is found there. Seventy-one violent earthquakes have rocked the terrane in the 472 years up to 1974. Kuili & Yong vaguely attribute the rapid coalification to these factors. In other words, time and pressure as factors are wholly discounted, whereas heat and volcanic contributions of some sort may have contributed.

The evidence is definitive that time and pressure were not involved in bringing about coalification. And, surely, the association between volcanism and coalification appears strong.

In another part of the world R. S. Clymo[198] reports on an occurrence of peat on Beauchene Island in the South Atlantic

---

[197]   *Kuili & Yong 1989*

[198]   *Clymo 1987; in Lyons & Alpern, Eds., 1989*

Ocean. There, a peat deposit in excess of eleven metres in depth has a basal layer 8,000 to 12,000 years old. It has a "hard cheesy texture, turns black on exposure to air, and breaks with a conchoidal fracture."

"What, [the geologist (Clymo) asks] is the source of structureless peat?" [199]   To answer his own question, this author decides the only feasible explanation is microbial action in which "as much as 99 % of original plant materials may have disappeared during peat formation."

How 99 % of plant matter could be removed in 8,000 to 12,000 years to leave 11 m of "structureless peat" requires an explanation far different than any existent theory can provide. Total biomass compacted to present density and 100 times present thickness would imply an original 1100-m thickness. No coal or peat deposit known today has such thickness. Furthermore, peat swamps are characteristically poorly drained. Even if a microbial leaching could have occurred in such a pile, where would all the effluent degradation products have gone?

*An alternative explanation, introduction of carbon as radicals or colloids, makes a far less labored explanation for the Beauchene Island coalified peat.* There is no need for great peat accumulation thickness, superior microbes, or a comprehensive waste disposal system that would have boggled the imagination of a modern engineer.

In a similar vein, P. D. Moore[200] says that "only the estuarine **Sarawak** peat that is being deposited today resembles the bog conditions of Paleozoic coals." Even so, he continues, "it is difficult to envisage hydrological conditions which permitted the very extensive development of bog forests such as those indicated by the vast spatial areas ... and thicknesses of some coal seams."

[199]   *Williams 1886*

[200]   *Moore, P. D. 1987*

## Detrital and Marine Coals

**Indian coal** deposits have long puzzled geologists. Francis [201] says **Northumberland and Durham coals of England** have marine interbeds and that Pakistan coals and the **Geiseltal coals of Germany** encase marine fossils. Hacquebard & Cameron [202] describe detrital coal in turbidite strata offshore of **Nova Scotia**.

Francis' description is most telling where he describes **Indian coal measures** with "upright tree trunks of dry-land types" rooted in the coal measure, which evidently accumulated as "drift" around them. Silicified wood fragments are common. "No Indian coals have plant stems rooted in underlying formations." **These arguments are powerful advances toward the conception of hydrous flow bringing carbon into a surficial or near-surface porous stratum and depositing carbon as vitrain.**

## Germanium and Coal

The leading source of germanium for industry is flue deposition after coal combustion. The quantities carried in various coals are extremely erratic, however. **The range of germanium in common rocks is approximately the following:**[203]

| | | | |
|---|---|---|---|
| Ultramafic rock | .00015 % Ge | Clays | .00020 % Ge |
| Mafic  " | .00013 | Shales | .00016 |
| Granodiorite | .00013 | Sandstones | .00008 |
| Granite | .00013 | Carbonate rock | .00002 |
| Syenite | .00010 | | |

---

[201]  *Francis 1954*

[202]  *Hacquebard & Cameron 1989*

[203]  *Turekian & Wedepohl 1961*

By contrast we can tabulate the germanium content of some coals:[204]

**Tokoku Mining, District, Japan:** 600,000,000 tons of lignite: .000036 % Ge. 20 % of deposit is silicified tree trunks.

From this it may be deduced that germanium is not concentrated by terrigenous plants, since the Ge values reported in the lignite are less than the leanest of rock averages excepting in carbonates.

**English coals** may run much richer: .05-.10 % Ge

**Russia** (unnamed district) A 130-m coal rests on granite basement, which assays > .3 % Ge. The coal is 90 % fusainized (gellified matrix of clarain-durain, and averages .1 % Ge. The coal macerals appear to "float" in desmite, a transparent high-grade coal. Ge and other metals (Re, Mo, W, U, Be) are concentrated irregularly near the base of the coal.

Clearly, the germanium could not be contributed by terrigenous vegetation that had been nourished on this granite. Conversely, there are no known terranes where germanium is dispersed throughout the country rock at regional levels even close to these values.

The statement is made in Vlasov's volume [204] that "Maximum germanium occurs in clarain and clarain-durain, which are said to contain up to 10 % components [macerals]. It is significant that when germanium accumulates in coal strata which are not of vitrain composition the maximum germanium [is] nevertheless associated with minute vitrain lenses or interlayers." **If only 10 % of the germanium-bearing coal comprises macerals, the germanium must be related more to the vitrain carbon, and both must be endogenic.**

---

[204] *Vlasov 1966*

# Lang Bay, British Columbia.

A small depression in the Coast Batholith is host to 450 m of Eocene, clastic, terrigenous sediments. Tree fragments up to 9-m length x 4-cm diameter have been encased in clastic sediment and coalified to lignite grade. These are all mineralized with germanium up to about .01 % Ge.[205] The coal ash, however, runs nearly 6.5 % Ge.[197]

The last statement above is most significant. Normally growing vegetation is about 10 % ash by weight and made up of minerals soluble in groundwater. The coalified wood comprising the Lang Bay lignite could not possibly have 6.5 % Ge from groundwater of meteoric origins. An external source is requisite.

**Volcanic associations of germanium** are considered in a study of germanium in iron ores worldwide, Grigor'yev & Zelenov[206] found enhanced germanium in iron and manganese deposits that develop around submarine volcanoes, but not around continental volcanos or in iron ores of sedimentary type. They conclude that "analyses of iron ores related to modern and ancient volcanism show quite convincingly that germanium is a component of volcanic gases [emitted] during submarine volcanic reactivity."

**Tsumeb and Kipushi Mines, Namibia** produce oxidized ores that grade up to 50 to 60 % germanite, renierite, and gallite,[207] but typically average nearer 10 to 20 %. The minerals occur in pipes in carbonate host rock. There is no depth zoning, and the ore is thoroughly oxidized to 280-m depth.

---

[205]  *White, G. V. 1985*

[206]  *Grigor'yev & Zelenov reference detail missing*

[207]  *Germanite: $Cu_3(Ge,Fe)S_4$. Renierite: $Cu_3(Fe,Ge)S_4$.*
       *Gallite: $CuGaS_2$.*

Tsumeb germanium is clearly a product of endogeny, "hydrothermal" in conventional terms, as there is no possible source and no concentration mechanism that could produce it from crustal rock. Volcanic germanium associated with submarine iron ores is, likewise, obviously endogenic. Elevated levels of germanium in coals must be endogenic, because they exceed levels achievable by plant uptake from groundwater.[204]

**Germanium in all of the foregoing natural scenarios most likely arrives at surface levels in the form of gaseous germanes.** The chemical behavior of this gaseous material was expressed in Chapter IV. Under oxidizing environment:

$$GeH_4 + 3O_2 \rightarrow 2H_2GeO_3 \qquad (9)$$

We should now rewrite the formula, $2H_2GeO_3$, a weak germanic acid, to reflect hydrated germanic oxide, $GeO_2 \cdot 2H_2O$. The reader will recall that $H_2GeO_3$ is explosive at elevated temperatures (+330°C).

Hydrated germanic oxide would tend to preserve reducing conditions, preempt oxidation of free carbon, and flow as a colloidal solid with charged colloidal carbon. **Germanic oxide and colloidal carbon flowing in hydrous emanations together should end up adsorbed on oxidized, terrigenous-plant macerals in vitrain. Where hydrothermal and volcanic occurrences lack colloidal carbon (the associations at Tsumeb and in the iron formations), other depositional associations of germanium result.**

## ABSENCE OF COAL IN PRECAMBRIAN STRATA

Worldwide sedimentation in the biosphere changed in lower Cambrian times from a predominantly siliceous, carbon-deficient, and argillaceous mode of nearly four billion years (preCambrian time) to biospheric sedimentation with increased contributions of calcareous, carbon-rich, and argillaceous compositions.

The great geological change was accompanied by an equally great biological change. First there was a proliferation of entirely new phylogenies. Second, animal exo-skeletons became calcareous. Exoskeletal shells fixed carbon and calcium, which then were deposited as carbonate rock. Simultaneously, terrigenous flora appear for the first time to have sequestered atmospheric carbon into coal and carbonaceous sedimentary rock.

The popular explanation for the absence of abundant biotic carbonates and coal before the Phanerozoic is that the organisms that deposit them had not evolved. Alternatively, the geological change could have been first, the response in the biota a consequence. Although the question appears to be a chicken-or-egg one, I suggest that the evidence is skewed toward initial geological change. Otherwise, the simultaneous development of coal and of calcareous skeletons becomes inexplicably coincidental. A shift in the makeup of effusing Mantle gas in early Cambrian time from mostly silane to mixed silane and hydrocarbon gas is the best explanation for observed facts.

## SYNTHETIC COAL FROM LIGNIN

Experimental work attempting to synthesize coal from lignins in the presence of montmorillonite clay achieved some limited success.[208] Lignins, which are defined as highly aromatic molecular components of woody tissue, held at 150°C for a "few months" in the presence of montmorillonite clay, apparently are converted to coal. The clay is thought to play an "acid-catalytic role ... causing fatty acids to form 'alignite,' an important lipid constituent." The fatty acids are thought to be a natural derivative of the remains of algae.

---

[208]   *Larsen 1985*

Talk of "acid" contribution implies free hydrogen ion and base exchange. The agglomeration by biomass of endogenic, negatively-charged carbon in radicals or colloidal form is similar. The artificial conditions of the laboratory make it possible for lignin (rather than more ponderous biomass) to be allowed to react in isolation. This is an interesting exercise, but hardly touches the complexities of the many geological questions of the nature of coalification in open systems.

**To explain coalification, we have sought resolution of the enigmas of (1) the impossibly-large biomass requirements without juxtaposed soil horizons; (2) a lack of indigenous sources for the contained hydrocarbons and trace metals; (3) the evident selectivity, temporal and topological of coalification; and (4) the puzzling question of how coaly shale or mineral charcoal [fusain] could ever be deposited. These are but a few of the more prominent enigmas.**

**The best resolution, albeit lacking in appropriate laboratory backup, is the addition of endogenic carbon to the coalification process. It would be useful if the efficacy of the idea were tested in the laboratory. The coalification of biomass exposed selectively to negatively-charged colloidal carbon and to hydrocarbon radicals should be instructive.**

## FORMATION OF METAL ORES

### L. G. COLLINS

Formation of metallic, hydrothermal-vein ore deposits has a logical explanation where granitic rocks are being formed *in situ* by introduction of silanes. Generally, the modern view is that ore metals are derived either

(1) by "mobilization" of trace metals from wall rocks by meteoric water, a process that could be set in motion by heat from a magma body or

(2) from residual fluids in mafic magmas that have differentiated in progressive stages through intermediate compositions to granitic end-products. In this second hypothesis the metals are theorized to have been concentrated in hydrous fluids secreted during late stages of granite differentiation, and then to have escaped from the solidifying magma to favorable sites of deposition.

Field evidence does not support the second theory. Granitic plutons have singularly barren quartz veins and pegmatites associated with them. Metals that commonly occur in hydrothermal deposits are mainly absent. Neither is there pervasive metallization of proximal granitic wall rocks near these pegmatites and quartz veins.

If a solidifying pluton is assumed to crystallize from the rim inward, and if metal-bearing veins originate from residual hydrous fluids that became concentrated in its core; then metallized veins should increase inward toward the core. This is not observed in the field. What is observed is that metals such as zinc, iron, and copper tend to crystallize early, dispersed in sulfides and ferromagnesian silicates within the crystallizing *mafic magmas*.[209] Therefore, one must conclude that the metals never comprised a residual concentration.

That leaves us with the first theory, "mobilization" from the country rocks. In this respect it must be acknowledged that magmatic heat may, indeed, produce some metal concentrations. However, the silane model may work better toward this end.

As plagioclase in a mafic pluton is replaced by K-feldspar and myrmekite, ferromagnesian silicates are replaced by quartz, releasing metals, which are then free to move in hydrous fluids, whence they may precipitate elsewhere in fracture systems as vein deposits.[210] The introduction of silanes into mafic igneous

---

[209] *Boyle 1970; Krauskopf 1971*

[210] *Collins 1988a*

and metamorphic rocks to form granitic rocks *in situ* offers a plausible method for metal concentration in hydrothermal veins.

I have drawn the conclusion that myrmekite in myrmekoid associations is a clue to the existence of former mafic rocks that once contained metals potentially available for ore deposits. Coarse vermicules in myrmekite indicate a parent rock of gabbroic composition; intermediate-sized vermicules, a rock of diorite composition; and narrow vermicules, a rock of still more granitic composition.

Unreplaced remnants of the mafic parent rock may be found with granitic rocks now containing wartlike myrmekite. Identifying the parent rock is important if one wants to recognize which trace-metals were replaced by silanes from ferromagnesian silicates and hence, possible candidates for concentration.

The following ore deposits, some of them famous, lend themselves to this interpretation:

## Sterling Hill Zinc Mine, New Jersey

The Sterling Hill zinc mine is near Ogdensburg, New Jersey. Ores of this mine are generally free of sulphur and lead, common associations in most zinc mineralization.[211] The Median Gneiss that forms the wall rock to ore concentrations contains both mafic diorite and felsic granitic layers. Wartlike myrmekite occurs in the granitic layers with quartz vermicules of intermediate-size as expected where diorite is replaced by granite.

Electron microprobe analyses of hypersthene, hornblende, and biotite in least-sheared, least-replaced mafic layers reveal an

[211]   *Hague et al. 1956; Metsger et al. 1958*

absence of sulphur and lead along with anomalously abundant zinc, 800 to 4,800 ppm. Adjacent myrmekite-bearing granitic layers contain the same minerals but with less zinc than can be detected by the instrument (<50 ppm). These metal relationships suggest that zinc was released during replacement from ferromagnesian silicates and concentrated in ore zones.

## Broken Hill Zinc Mines, Australia

Wartlike myrmekite with coarse quartz vermicules occurs abundantly in the Potosi Gneiss that contains the sphalerite ore zones in the Broken Hill mining district.[212] Distant from ore zones, myrmekite-free layers contain biotite and plagioclase in which the calcium (anorthite) content is ($An_{80}$) or higher. The gneiss contains abundant trace metals to an average of 158 ppm zinc, 113 ppm lead, and 31 ppm copper.[213] Replacement of gneiss by silanes to form myrmekite-bearing rocks readily explains the release of metallic elements and their subsequent concentration in orebodies.

## Dover Iron Mine, New Jersey

In Chapter IV the magnetite concentrations near Dover, New Jersey were shown to be accumulations of metal derived from iron-rich biotite and hypersthene minerals replaced by quartz as mafic rock was converted to granitic rock *in situ*.

## Pine Creek Tungsten Mine, California

West of Bishop, California, is the Pine Creek tungsten mine, where scheelite is concentrated in a contact "skarn" between alaskite granite and metasedimentary rocks.[214] Wartlike

[212] *Phillips & Ransom 1970; Phillips & Vernon 1972; Vernon 1969*

[213] *Drake 1974*

[214] *Bateman 1965*

myrmekite in the alaskite occurs in coarse vermicules. The presence of myrmekite suggests that diorite replacement occurred. Its texture reflects high calcium content in the plagioclase of adjacent diorites, the logical source of tungsten in the skarn.

## American Girl Gold Mine, California

Gold is sufficiently abundant in low grades in the granitic rocks of the Cargo Muchacho Mountains of southeastern California that open-pit mining of the granite surrounding the former American Girl underground mine is in progress. The dispersion of the gold implies pervasive movement of hydrous fluids through the host rock.

The geology of this region was treated in Chapter IV; and gradational conversions of diorite and mafic quartz monzonite to granite *in situ* were demonstrated. All granitic rocks of the area contain abundant wartlike myrmekite with thick quartz vermicules that suggest replacement from rock at least as mafic as diorite.

## Boulder Batholith Sulphide Ores, Montana

Sulfide ores of the Boulder batholith near Butte, Montana, are surrounded by quartz monzonite that contains wartlike myrmekite with coarse quartz vermicules.[215] Portions of the batholith contain plagioclase, ranging from $An_{55}$ to $An_{75}$. "An" values at this level are indicative of vermicule coarseness. Biotite in the quartz monzonite contains abundant copper, averaging 700 ppm.[216] Therefore, during replacement much of this copper was released from the former mafic rocks, whence it was captured and deposited in sulfide concentrations.

[215] *Smedes 1965; Becraft et al. 1963; Robertson 1962*

[216] *Al-Hashimi & Brownlow 1970*

## Cooma Granodiorite, Australia

The Cooma Granodiorite in southeastern Australia is a pluton without known metal concentrations, that will serve adequately as a contrasting illustration.

Wartlike myrmekite in the pluton contains narrow quartz vermicules that indicate a parent rock with relatively-sodic plagioclase. This characterizes the plagioclase of the metasedimentary country rock ($An_{10}$ to $An_{25}$), on which chemical analyses report low trace metal: averaging 21 ppm copper, 96 ppm zinc, and 23 ppm lead. (Munksgaard 1988) Clearly, abundant trace metal was not available to be concentrated.

## Carlin Gold Mines, Nevada

### C. WARREN HUNT

In the gold deposits in the Carlin area, Nevada, the association of carbon with metal ore has been recognized in the literature since 1968 and in the field, much earlier. Disseminated gold mineralization occurs with carbon in Silurian and Devonian sedimentary rock in which liquid petroleum is also occasionally present.

These associations led to the interpretation of "syn-deposition," gold precipitated by active natural carbon, an idea that prevailed for about fifteen years. Detailed petrography and statistical geochemistry by the US Geological Survey[217] finally discredited the idea. However, no very convincing alternative has been advanced to explain the dispersed gold deposition.

The Mine is flanked to the north by Rodeo Creek valley, where my own work in opening the Bullion Monarch and Goldstrike Mines gave me detailed insight into local geological

---

[217] *Harris, M. & Radtke 1976*

structure. Although not recorded anywhere in the extensive Carlin literature, so far as I know, the depression occupied by Rodeo Creek adjacent to these two mines is a caldera of late Cenozoic age that has been infilled with unmineralized, late-arrived tuff and agglomeratic debris. The petroleum and gold in the wallrock, then, are most likely earlier emanations of the volcanic system, not products of ancient syngenetic Paleozoic marine seafloor deposition, as they were regarded. Most geologists now recognize the volcanic association for the metal, if not the endogenic provenance of the petroleum and carbon.

## Nanisivik Lead-Zinc Mine, Baffin Island, NWT, Canada

The mineralization in the Nanisivik Mine is understood to be a "Mississippi Valley-type" [i.e. stratiform] deposit in the Proterozoic metasedimentary rocks of the area. Sulphur isotope studies show clearly that *in situ* thermochemical reduction of sea-water sulfate by petroleum could have produced $H_2S$ and the chemical conditions capable of precipitating metal ions[218] **But where could the metals have originated?**

It is interesting that this occurrence is just across Davis Strait from the Greenland sites mentioned earlier in this Chapter for their association of petroleum, coal, basalt, mud volcanos, and native iron. Local sources for metal or hydrocarbons are not apparent. The next best option for the provenance of Nanisivik ore is emanation along with hydrocarbons from depth.

## The Mount Brussilof Magnesite Deposits, British Columbia

In the Canadian Rocky Mountains immediately west of the continental divide, south of Banff townsite, huge injections of massive, 97+ % pure magnesite, $MgCO_3$, have been deposited

---

[218] *Gazban et al., 1990*

in bedded middle-Cambrian carbonates, which characterize that scenery[219].(Frontispiece) Rare, erratic veinlets and void fillings of quartz and metal sulphides, which are quite foreign to the terrane, demonstrate that mineralization came from depth rather than from local sources.

The magnesites are aligned parallel to the mountain range in Cambrian carbonates, which are part of a depositional bank known as the Cathedral escarpment. It is unclear just when they were injected. Their orientation corresponds to the definable edge of a fault block that was already upthrown at the time of sedimentation. The block presently comprises the height of land of the Canadian Rocky Mountains.

This escarpment is the same one that farther north was the scene of talus accumulations that buried the notorious "Burgess fauna." [220] Its first movement, therefore, must have occurred in earliest Cambrian time; and a crustal fracture that has persisted for over 500 million years must underlie the alignment.

From this lithosphere fracture, magnesium-bearing hydrous fluids rose after silane replacement of mafic and ultramafic Mantle rocks deep in the Crust. The fluids must have been accommodated by cataclasis in the fault zone as they replaced calcium in existing Cambrian limestones saturated with $CO_2$-rich seawater. This might have occurred within biohermal or biostromal reef structures that had previously been established on the Cambrian scarp.

No local source could have provided the large volume of magnesium that is present in these deposits. The accompanying minor sulphides, which include tetrahedrite [$(Cu,Fe)_{12}Sb_4S_{13}$], fersmite [$(Ca,Ce,Na)(Nb,Ti,Fe,Al)_2(O,OH,F)_6$], and boulangerite [$Pb_4Sb_4S_{11}$] are likewise, as already mentioned, minerals that can only have come from deep sources.[207]

[219] *Simandl et al., 1991*

[220] *Gould 1989*

It should not be taken as coincidence that the magnesite bodies are aligned near the crest of lately-uplifted mountains. L. C. King's term for a broad crustal uplift, "cymatogen" is appropriate for describing such a crest. It is precisely on such a crest where crustal arching causes maximal tensional rifting (King, 1983), where silane penetrates most effectively, where mafic rock is transformed to reduced-density rock, and where a mountain range may be expected to rise buoyantly.

The circle of logic closes, then, with this example of coincidence among otherwise diverse phenomena:

**First we have emplacement of deep-seated mafic or ultramafic rock (Geoid, back of Frontispiece). Next we have the deep-seated mafic-rock replacement of mafic cations, their removal to surficial levels and their deposition in metal-ore. Lastly, isostatic differentials bring about emergent mountains.**

These processes have operated over a span of more than 550 million years in the Mt. Brussilof terrane. Only silane systematics coupled into a major lithosphere fracture can explain all these features in a single scenario.

# EPILOGUE

### _The coverage in this book virtually defies being summarized._

We set forth by describing disarray in geological thinking as a consequence of the inadequacy of existing Earth theories. We proposed **an improved Earth theory** that avoids the problems inherent in presently held theories.

One of us (Skobelin) then presented an entirely **new theory on the nature of the Moho.**

Another (Hunt) explained **the fundamentals of carbide-hydride theory, rise of mountain ranges, continental growth, and depression of the oceanic Crust** peripheral to continents due to off-loading of material from the continents.

Collins then explained _in-situ_ **transformation of mafic rock to granite, the origins of granites, aplites, pegmatites, migmatites, aluminum enrichment, and polonium halos in biotite and fluorite as processes of silane-generated endogeny**.

Two of us (Collins and Hunt) attribute to **silane systematics** a whole panoply of observed features, including especially **lithosphere discontinuities, flood basalts, anatexis, zoned plutons, phenocryst growth in porphyries, and age-dating enigmas that arise from the fractionation of isotopes.**

Skobelin explores **the tectonic nature of earthquakes** and presents a fundamentally **new theory** of their cause.

Hunt considers the **relationship of earthquakes to faulting and aseismic creep** in the light of known fault behavior, and advances aseismic creep as a major factor in terrain distortion and fault dislocations. He develops theories to show how the enigmatic deep-focus earthquakes, Benioff zones, and EM effects that are associated with seismicity are accommodated better by an explanation of faulting that originates with silane systematics than by currently-popular alternative explanations.

Skobelin next explores the mode of **formation of volcanic pipes, diapirs, and sills**, along with their emplacement by crustal dilation and either diapiric piercement or hydraulic fracturing of the brittle Crust.

He analyzes **diamond genesis** through destruction of petroleum by magmatic heat.

Hunt takes up the nature of the **Yellowstone hotspot** and shows how it can be interpreted as a manifestation of silane systematics.

He follows by analyzing **gas recovered from deep drillholes, gas trapped in kimberlites, and hydrogen in near-surface habitats**. He attributes their nature to derivation from a silane-bearing "mother gas," and calculates their original silane fractions.

He considers the origin of the **high-temperature zeolite, and clinoptilolite in volcanic tuffs**, noting its molecular seive structure, and drawing attention to the porosity of related molecular seives, the **geodesic orbs of fullerene carbon** and the **geodesic silica shells of radiolaria**.

Hunt describes **expulsions of quartz sand as "pulses of endogeny,"** manifestations of silane systematics.

Collins describes **quartzites generated *in situ***, an alternative outcome of silane pervasion.

Economic minerals from silane systematics are considered, first by Hunt, who concludes that **petroleum should be found in silane-generated sands that have been hydraulically-"fracked" into sill-like crypts of the middle Crust.**

He moves on to **vitrinite coal and the coalification process, which he attributes to carbon radicals** released after partial oxidation of hydrocarbons introduced by carbide-hydride systematics.

Collins provides a lucid explanation for the **deposition of metal ores from cation-rich hydrous fluids** that are flushed upward after rock transformations by silanes at depth.

Concomitantly, **mountain scenery rises isostatically atop cymatogens** that swell above rifts through which pervasive silane infusion has occurred. The two scenes of the frontispiece illustrate the grandeur of this story.

Thus, we are brought full circle to our starting point. The answer to a question that was not asked is now apparent, "How are magnificent scenery and provident mineral wealth created in nature?" The answer, "By silane systematics," is quite different from what we had thought!

Should we be surprised by all these unexpected linkages in the systems of "natural philosophy" we call silane systematics? H. L. Mencken is quoted as having said that, "Every philosopher spends his career proving that every other philosopher is a jackass, and remarkably they all succeed!" That we did not set out to prove these things in the first place may only show that we are not proper philosophers.

In any case, this volume has been a voyage of discovery for its authors, and, we hope, for the readers. Hypothetical carbide-hydride endogeny opens new horizons in geology. There are still substantial mysteries to silane behavior that forestall certainty in our conclusions. Especially problematical is the fact that silanes are waylaid deep underground by water and oxygen long before they come into range of our scientific instruments so that they can never be seen or directly detected in nature. The evidence for the existence of silane effusion and the processes that follow, although abundant, is necessarily entirely circumstantial.

Contemplating the staggering weight of paper and ink that has been devoted to these very problems in the past, one should hardly hope to solve them. However, most previous research has been directed into selected, isolated aspects of our science, into *closed systems*, in Collins' terms, into **reductionist approaches** in mine. These approaches have failed because the system of analysis is inadequate for the task. Hardly any attempt has been made in recent years to find alternative, comprehensive, and internally consistent resolutions for the observable behavior of the ***open systems*** of our planet.

Carbide-hydride systematics is such a comprehensive resolution, a fortuitous cerebration that gave birth to discovery. Starting with a search for a deep hydrocarbon source, adjunctive discovery of the mechanism for mobilization and transfer of silicon from deep Mantle levels to the surface of the Earth resulted. This came about in isolation from funded geoscience, mathematical modelling, reductionist imperatives, or interference from sharp-elbowed editors or pretentious "peer reviewers." The "accepted" pattern of scientific innovation was by-passed.

Perhaps the theory will achieve studious neglect from the influential members of the science community. That can only happen, however, to the time a new generation decides to think for itself and realizes that the dim light of present theories offers nothing comparably cogent for resolving the geological enigmas that surround us.

Currently-popular reductionist procedures offer only partial solutions and no organized insight into the grand scheme of things. Carbide-hydride systematics does provide such a simple and comprehensive solution and the enormous source of power the geological community has needed and sought for many decades. The breadth of the concept must be immediately apparent to anyone familiar with the natural world.

Now, our voyage of exploration and discovery is over. It is for others to make their own voyages to see whether we deserve to fall victim to Mencken's dictum, whether our assembled wisdom, our "philosophy" in Mencken's terms, is an advance or just another aberrative detour from the modellers' guide to the galaxy.

It has been my great pleasure to receive and to have the privilege of editing the major contributions from E. A. Skobelin and L. G. Collins and to have received necessary technical assistance from the latter. Philosophy means love of wisdom. And all three of us have followed our favorite pursuit in this book. The results of our voyage of discovery, our interpretations, hold momentous purport for all of geology. The experience of their unfolding has quite astonished and delighted us.

# Acknowledgments

Thanks are due to the following persons variously, for helpful commentary, slides, drawings, and separates of their own papers, which have been used by or assisted the editor in preparation of this book.

**Dr. John Badding**, Geophysical Laboratory for High Pressure Research, Carnegie Institution of Washington, 5251 Broad Branch Road, N.W., Washington D.C.

**Dr. Irene S. Leung**, Department of Geology, Lehman College, City University of New York, Bronx, NY 10468

**Dr. Paul D. Lowman, Jr.**, Geophysics Branch, Code 622, NASA/Goddard Space Flight Center, Greenbelt, MD 20771

**Bruce D. Martin, Ph.D,** Bruce Martin Associates, Inc., P.O. Box 234, Leonardtown, Maryland, 20650-0234

**Professor Anne Meltzer**, Department of Geological Sciences, Lehigh University, Bethlehem, PA 18015-3188

**Dr. Dennis G. Milbert**, Advanced Geodetic Science Branch, National Geodetic Survey, Rockville, Maryland 20852

**Dr. J. A. Roddick**, Geological Survey of Canada, 400-100 W. Pender St., Vancouver, B.C. V6B 1R8

Appendix

# MYRMEKITE:
# ITS NATURE AND ORIGIN

*Myrmekite is not a mineral but an association of two minerals, quartz and plagioclase.* The two-mineral association is named from two separate Greek roots. One describes its wartlike outward shapes (of the plagioclase portion that enclose the quartz). The other describes the tunnel-like quartz replacements themselves, which resemble underground ant tunnels. "Myrmecology" is ant science. The quartz-filled tunnels are vermicular, "wormtube-like," that is to say (Fig. IV-2).

Wartlike myrmekite is generally less than one millimetre across and constitutes less than 0.5 percent of a rock. The *plagioclase phase of myrmekite* always occurs in or attached to the outside of plagioclase, which is devoid of quartz. This quartz-free plagioclase *may be "zoned," that is to say, of different chemical makeup in layers from inside outward.* Whether zoned or unzoned, the plagioclase is always altered in complex ways, and parts of it converted to myrmekite. Myrmekite is usually adjacent to K-feldspar and often encasing it or appearing to replace it.

*Thus we have plagioclase, along with secondary quartz growths, comprising the myrmekite mineral-doublet and commonly associated with K-feldspar. This combination is generated in situ by permeation of cataclastically prepared solid rock by silanes. This occurs at sub-solidus temperatures and results in selective mobilization, relocation, and recrystallization of cations.*

*Myrmekite takes three forms: (1) Wart-like or cauliflower-like plagioclase-crystal growths on larger, quartz-free*

*plagioclase crystals, the warts and cauliflowers themselves enclosing tiny, quartz vermicules [forms resembling "worm tubes"] and projecting into adjacent K-feldspar grains. (2) Rims on zoned plagioclase grains. These also border on K-feldspar and often project into it. (3) Isolated plagioclase/quartz intergrowths in rocks devoid of K-feldspar.*

The hypothesis of C. Warren Hunt that silanes rise from the Mantle opens new vistas on the origin of myrmekite in all three of its phases. The fluids that accompany the silanes enable replacement to proceed. Wartlike and rim myrmekite are normally the two types that are discussed in hypotheses describing origins. The wartlike type gets most attention.

The tiny size and protrusion of "warts" into large K-feldspar crystals result in the generally-accepted hypothesis that myrmekite must have replaced the K-feldspar from its exterior progressively inward. Any suggestion to the contrary evokes disbelief. Nevertheless, **my studies show that myrmekite is co-generated with K-feldspar.**[221]

## History of Myrmekite Hypotheses

Myrmekite has been studied for more than 117 years.[222] In the 1930s and 1940s intense interest resulted in at least six major hypotheses being advanced to explain its origin. These include: (1) simultaneous or direct crystallization from a melt, (2) replacement of K-feldspar by plagioclase, (3) replacement of plagioclase by K-feldspar, (4) replacement of plagioclase by quartz, (5) growth of blastic plagioclase around residual, recrystallized quartz, and (6) exsolution of myrmekite from high-temperature K-feldspar.[223] Various combinations of these six hypotheses have added several more explanations,[224] including a polygenetic origin.[223]

---

[221] *Collins 1988a*

[222] *Michel-Levy 1974*

[223] *Phillips 1974; 1980*

[224] *Hibbard 1979*

Nearly all early hypotheses, however, favor an origin (in plutonic igneous and metamorphic rocks) under subsolidus conditions during hydrous alteration of K-feldspar. A contrary hypothesis which proposed simultaneous crystallization from magma of quartz and plagioclase in myrmekite is discredited by the fact that myrmekite is present in non-magmatic metamorphic rocks.

Since 1979, two contrasting new proposals have been made. One of these is my explanation: myrmekite is the product of replacement of primary plagioclase by secondary K-feldspar.[225] The other is a variation of an older idea: myrmekite forms when primary K-feldspar is replaced by secondary plagioclase in igneous bodies or mylonite.[226] Possible transformations will make these options clearer.

## Equations of the Old Hypotheses

Two kinds of balanced mass-for-mass chemical equations are the usual ones given to explain myrmekite.

### [1]: Replacement of K-feldspar by calcium and sodium to form myrmekite

Plagioclase ($Ca^{2+}$, $Na1+$) replaces K-feldspar to form myrmekite;

$$KAlSi_3O_8 + Na^{1+} = NaAlSi_3O_8 + K1+$$
$$2KAlSi_3O_8 + Ca^{2+} = CaAl_2Si_2O_8 + 4SiO_2 + 2K^{1+}$$

K-feldspar                [ Myrmekite ]

[225] **Collins 1988a,b**

[226] **Hopson & Ramseyer 1990a,b; La Tour 1987; La Tour & Barnett 1987; La Tour & Thomsom 1988; Simpson 1984, 1985; Simpson & Wintsch 1989; Moore 1987; Stel & Breedveld 1990**

## [2]: Exsolution of myrmekite from high-T K-feldspar

High temperature K-feldspar is exsolved of its extra $Ca^{2+}$, $Na^{1+}$, $Al^{3+}$, and $Si^{4+}$ (in solid solution), which then may be redeposited on the rims of K-feldspar grains as myrmekite.

$$KAlSi_3O_8 \quad = \quad KAlSi_3O_8 + NaAlSi_3O_8 + 4SiO_2$$
$$NaAlSi_3O_8 \qquad\qquad CaAl_2Si_2O_8$$
$$Ca(AlSi_3O_8)_2$$

High-T K-feldspar     K-feldspar   [ Myrmekite ]

In both equations the *volume* of quartz in vermicules is a function of $Ca^{2+}$ in the host plagioclase. There are, however, inconsistencies inherent in the equations, which are avoided by my alternative hypothesis. Following is evidence of the fallacies inherent in the equations.

In my study of the Temecula terrane, California,[225] I had found that a diorite facies of the Bonsall tonalite, which is devoid of K-feldspar, is progressively replaced by a myrmekite-bearing granite (a facies of Woodson Mountain granodiorite). Beginning one to four metres from granite in deformed and slightly foliated diorite I found progressive breakdown of the ferromagnesian silicates (biotite and hornblende). The replacement consists of quartz replacement (either entirely to quartz or to a quartz sieve texture (**Fig. IV-3**). Other hypotheses for the origin of myrmekite ignore the ferromagnesian silicates despite the fact that they also are replaced by quartz as plagioclase is replaced by K-feldspar.

The quartz with sieve texture in partial replacements can be considered a result of simultaneous crystallization in biotite or hornblende within magma. However, two relationships discredit this idea. Firstly, quartz sieve texture appears only in deformed rocks transitional to myrmekite-bearing granite. It is absent from undeformed mafic rocks nearby.

Secondly, in transition zones where diorite is being converted to tonalite, *ferromagnesian silicates decrease and quartz increases.* **Fig. IV-3** illustrates this: all hornblende and some biotite is replaced by quartz; and quartz volume has increased from <1 % to >15 %.

Diminished ferromagnesian minerals in the Temecula myrmekite-bearing granite adjacent to diorite implies that *in-situ* transformation rather than magmatic emplacement has occurred. In magmatic intrusion of granite into the diorite, assimilated ferromagnesian components should have resulted in increased ferromagnesian minerals in the granite, not diminishment. If the narrow wedge-shaped granite dike had been injected into the diorite from the main granite body, it should at the very least be as mafic as the granite which fed it. More likely, it should be more mafic due to assimilation of diorite. In fact, neither of these occurs.

The dike lacks K-feldspar and myrmekite, both of which the main granite body contains. *Its composition exhibits depletion of biotite and hornblende, which precludes magmatic origin. The dike is, thus, in an early stage of transformation, having not acquired the pervasive myrmekite of the main body of granite.*

This initial stage of transformation leading to myrmekite may be represented by *equations* **[3] and [4]** as follows:

### [3]: Replacement of Hornblende by Quartz

$$(Ca,Na)_{2-3}(Mg,Fe,Al)_5Si_6(Si,Al)_2O_{22}(OH)_2 \quad + \quad SiO_2 \rightarrow$$
$$\text{hornblende} \qquad\qquad\qquad\qquad\qquad\qquad \text{silica}$$

$$SiO_2 \quad + Al^{3+} + Ca^{2+} + Na^{1+} + Fe^{2+} + Mg^{2+}$$
$$\text{quartz (in sieve textures or individual grains)}$$

### [4]: Replacement of biotite by quartz

$K(Mg,Fe)_3(AlSi_3O_{10})(OH)_2 + SiO_2 \rightarrow$
biotite                              silica

$$SiO_2 + Al^{3+} + K^{1+} + Fe^{2+} + Mg^{2+}$$
quartz (in sieve textures or individual grains)

In other terranes pyroxenes or olivine could also be replaced by quartz in transition zones (reactions are not included here).

The above reactions are not written with equal signs as in equations but with arrows to indicate trend. This format is to emphasize that **replacements are volume-for-volume rather than equated mass-for-mass exchanges.** No attempt is made to balance reactants with products. This procedure is generally justified by the absence of any field evidence for volume change where granitic dikes and granodiorite penetrate diorite.[225]

## Alteration of Primary Plagioclase

The second stage of myrmekite formation occurs simultaneously with late stages of quartz replacement of ferromagnesian silicates in diorite. It begins with sequential alteration of stressed, deformed, and/or cataclastically broken plagioclase grains.

Alterations of these types are shown at Temecula in continuous thin sections across a narrow, 10-cm transition zone between diorite and granite. Electron-microprobe studies of 80 different grains of plagioclase across this transition zone show a progression through the following changes:

Plagioclase grains are first altered by $Ca^{2+}$ and $Al^{3+}$ subtraction from ***interiors*** of grains. This subtraction

causes zoned and albite-twinned plagioclase grains [their cores about $An_{40}$, their rims $An_{20}$] to take on a mottled appearance under crossed nicols. As the mottling appears, zoning and twinning progressively disappear.

In later stages, mottled grains become optically clear, with nearly uniform composition of about $An_{20}$ in both cores and rims. In final stages, prior to formation of myrmekite, many optically-clear plagioclase grains develop a speckled, pitted appearance, cores about $An_{2.5}$, and rims about $An_{17-20}$ (**Figs. IV-2 and IV-4**).

Scanning electron microscopy of these pitted grains at 1,600x and 8,000x magnification shows that the pitted appearance results from tiny holes, 1-8 microns wide and up to 10 microns long (**Fig. IV-4**). These holes have an irregular but linear arrangement that is not seen at the 100x magnification of the petrographic microscope. Their linearity suggest that removal of $Na^{1+}$, $Ca^{2+}$, and $Al^{3+}$ ions creates voids that are controlled by lattice orientations parallel to cleavage or twin planes. Thusly altered, plagioclase grains are rendered virtual sieves to fluid and ion migration.

**Holes of this type would be expected from replacement but not from magmatic processes. In fact, the chemical, mineralogical, and textural changes that produce sieve-like plagioclase cannot result from any other process but replacement *in situ*.**

**Electron microprobe analysis of this pitted feldspar (Fig. IV-4) shows that it contains excess silica, more, that is to say, than what is required for feldspar composition. There is, however, no visible quartz nor any other optical clue to the presence of excess silica!**

The explanation for this apparent paradox is that subtraction of some $Ca^{2+}$ and $Al^{3+}$ from plagioclase yields excess silica, which remains undetectable within the pitted host mineral.

The behavior of potassium is similar. Building up before the first appearance of wartlike myrmekite (**Fig. IV-2**), the presence of $K_2O$ is shown both by cathodoluminescence microscopy and electron-microprobe analysis. Plagioclase grains with $An_{2-5}$ cores and $An_{17-20}$ rims gain increasing amounts of $K_2O$ until as much as fifty percent of the lattice positions for $K^{1+}$, $Na^{1+}$, and $Ca^{2+}$ are filled with $K^{1+}$. The early stages of internal growth of $K^{1+}$ in plagioclase (forming K-feldspar) are depicted in **Fig. IV-4**.

These altered grains are pitted but still optically clear and without the gridiron twinning common in microcline. They are undifferentiable from other altered, pitted plagioclase grains.

***Thus, in the same manner as silica is sequestered in the plagioclase, there is no optical clue to the existence of this abundant $K^{1+}$!*** The $K^{1+}$ introduction occurs in conjunction with the second stage in myrmekite development, when K-feldspar is formed.

Tiny islands of aligned K-feldspar flank aligned holes in **Fig. IV-4**. These islands of K-feldspar in the cores of plagioclase grains are striking examples of potassium ion being introduced into plagioclase. Their presence contrasts with their absence from the calcic cores of zoned plagioclase in the original diorite. It is clear that plagioclase must be replaced by K-feldspar from the core outward.

The replacement process is evinced by the distribution of barium feldspar, "celsian" ($BaAl_2Si_2O_8$). This mineral is a secondary one that occurs on corners of secondary K-feldspar

crystals or in rims of altered plagioclase grains (**Fig. IV-4**). The concentration of $Ba^{2+}$ in altered plagioclase would be anticipated as an adjunct to the stripping of $Al^{3+}$, $Ca^{2+}$, and $Na^{1+}$ from relatively calcic cores of plagioclase. Barium can substitute for potassium because of its nearly identical atomic radius [contrasting with sodium and calcium].

Thus, ambient barium that was absent from primary plagioclase is sequestered in its holes by secondary growth of K-feldspar and celsian. This gives a biased distribution that would be impossible in crystallization from magma, where the barium should be uniformly distributed or zonally patterned from core to rim as the K-feldspar or plagioclase grew.

## Replacement Processes

The final stage of myrmekite formation is shown by electron microprobe analysis to comprise replacement of altered plagioclase from the interior outward by $Na^{1+}$ and $K^{1+}$. Four different products can result: **(A) K-feldspar alone, (B) K-feldspar containing albite lamellae, (C) K-feldspar containing quartz blebs, and (D) recrystallized relatively-sodic plagioclase.**

*(A) K-feldspar alone:* Where $Na^{1+}$, $Ca^{2+}$, and some $Al^{3+}$ have been removed, altered plagioclase grains are replaced from the interior outward to form K-feldspar (*reaction [5]*). K-feldspar crystals can grow larger than original crystals by conjunctive growth. The common K-feldspar which does this is microcline; the less common is orthoclase.[227]

---

[227] *Collins 1988a; Augustithis 1990*

### [5]: Complete replacement of primary plagioclase
### by microcline (or orthoclase)

$CaAlSi_2O_8 + K^{1+} + SiO_2 \rightarrow KAlSi_3O_8 + Ca^{2+} + Na^{1+} + Al^{3+}$
$NaAlSi_3O_8$
plagioclase         silica    microcline

**(B) K-feldspar containing albite lamellae:** This may occur where centers of $K^{1+}$ replacement occlude islands of $Na^{1+}$. Excess sodium ion, which cannot be displaced by potassium ion, recrystallizes as albite lamellae, "perthite" (in a process different from that described by Alling).[228]

### [6]: Production of albite lamellae in K-feldspar
### that replaces primary plagioclase incompletely

$CaAl_2Si_2O_8 + K^{1+} + SiO_2 \rightarrow KAlSi_3O_8 + NaAlSi_3O_8 + Ca^{2+} + Al^{3+}$
$NaAlSi_3O_8$
plagioclase        silica  microcline   albite lamellae

Occasionally plagioclase in perthite consists of irregular islands or spindles distributed irregularly in K-feldspar. Elsewhere it is uniformly distributed. Uniform distribution is especially characteristic of granulites and charnockites, where the plagioclase is coarse-crystalline and abundant, and where coexisting myrmekite contains coarse quartz vermicules.

**(C) K-feldspar containing quartz blebs:** This is produced where $Al^{3+}$ is extracted from the plagioclase lattice to a greater degree than necessary to make K-feldspar and simultaneously to utilize all residual $SiO_2$. The excess silica remains as quartz blebs in K-feldspar (*reaction* **[7]**).

At Temecula the quartz blebs occur in clusters that are barely visible under high (100x) magnification. I have called them "ghost myrmekite" because of their elusive habit. In other

---

[228]  **Alling 1932, 1938**

terranes, where parent plagioclase has higher calcium content ($>An_{24}$ on average), the blebs are larger in proportion to the calcium content of the parent plagioclase.

### [7] Quartz bleb formation (ghost myrmekite) in K-feldspar that completely replaces primary plagioclase

$$CaAl_2Si_2O_8 + K^{1+} + SiO_2 \rightarrow KAlSi_3O_8 + SiO_2 + Ca^{2+} + Na^{1+} + Al^{3+}$$
$$NaAlSi_3O_8$$

plagioclase　　　　silica　microcline　quartz blebs

Combinations of *reactions* **[6], [7], and [8]** can occur in different places in the same plagioclase grain. Which occurs is a function of local alteration and the degree of cation subtraction before the entry of potassium.

**(D) Recrystallized relatively-sodic plagioclase:** In addition to replacement of altered plagioclase grains by $K^{1+}$, electron microprobe analysis shows that other grains are replaced by $Na^{1+}$. In early stages of replacement $Na^{1+}$ [left over in the lattice after $Ca^{2+}$ has been subtracted], compositions are changed, so that the cores are $An_{2.5}$ and the more calcic rims are $An_{17-20}$(**Figs. IV-3 and IV-4**). This change in composition occurs in altered plagioclase grains of all sizes. As $K^{1+}$ replacement occurs, some displaced sodium ions in ambient fluids enter nearby plagioclase, causing its recrystallization as relatively-sodic plagioclase. Recrystallized sodic plagioclase of this type is optically clear, shows albite twinning, but lacks zoning.

### [8]: Conversion of altered primary plagioclase to recrystallized plagioclase of lower An content

$$CaAl_2Si_2O_8 + Na^{1+} + SiO_2 \rightarrow CaAl_2Si_2O_8 + Ca^{2+} + Al^{3+}$$
$$NaAlSi_3O_8 \qquad\qquad\qquad NaAlSi_3O_8$$

altered plagioclase　　　silica　　　more-sodic plagioclase

## *Myrmekite Formations*

Interior introduction of $K^{1+}$ or $Na^{1+}$ displaces most $Ca^{2+}$ in altered plagioclase grains and allows recrystallization either as K-feldspar or relatively sodic plagioclase. In some circumstances, however, some residual $Ca^{2+}$ is left and combines with residual $Al^{3+}$ and $Na^{1+}$ as secondary plagioclase which has lower An value than the parent because of partial removal of $Ca^{2+}$. Calcium plagioclase requires more $Al^{3+}$ and *less* $Si^{4+}$ than sodic plagioclase. Thus, where calcium is entrapped, there will not be a balance among the cations of calcium, sodium, aluminum, and silicon. **Excess silicon of this origin permits formation of the quartz vermicules; and the altered plagioclase from this process becomes the host for quartz vermicules in myrmekite (reaction [9]).** This silicon has an internal, non-silane source. Silanes still may contribute silica to quartz in isolated myrmekite (discussed ahead).

### [9] Formation of wartlike myrmekite projecting into K-feldspar that incompletely replaces primary plagioclase

$$CaAl_2Si_2O_8 + K^{1+} + SiO_2 \rightarrow KAlSi_3O_8 + CaAl_2Si_2O_8 + Ca^{2+} + Al^{3+} + SiO_2$$
$$NaAlSi_3O_8 \qquad\qquad\qquad\qquad NaAlSi_3O_8$$

plagioclase      silica   microcline   more-sodic   plagioclase     quartz
$$\{ \leftarrow \quad m\,y\,r\,m\,e\,k\,i\,t\,e \quad \rightarrow \}$$

Wartlike myrmekite occurs in two forms: (a) tiny chain-like clusters between two K-feldspar crystals, and (b) bulbous, larger grains attached to plagioclase crystals but projecting into K-feldspar. Variation in the style of replacement of the plagioclase determines which form develops.

The tiny chain-like form develops in *enclaves of cataclastic plagioclase* fragments between two or more centers of K-feldspar growth. Growth of K-feldspar from different centers occludes and traps ambient cations ($Na^{1+}$, $Ca^{2+}$, $Al^{3+}$). Unable to escape

and present in the wrong proportions relative to ambient $Si^{4+}$ to form only plagioclase, these elements crystallize as chain-like myrmekite between K-feldspar crystals (*reaction* [9]). The grain size of this kind of myrmekite is always small because it is made up of reformed cataclastic fragments of plagioclase. The K-feldspar that grows around this myrmekite may become much larger.

The second kind of wartlike myrmekite, the bulbous large grains found on borders of plagioclase and projecting into K-feldspar, develops where an altered plagioclase grain is trapped between *two different kinds of centers of replacement*. At one center the $K^{1+}$ replaces altered plagioclase from the interior outward. At the other center $Na^{1+}$ replaces altered plagioclase from the interior outward to form sodic plagioclase (**Fig. IV-3**, right side). Ion imbalance, as described for chain-like myrmekite development, results in bulbous myrmekite. $Ca^{2+}$ trapped between incoming $Na^{1+}$ from one side and $K^{1+}$ from the other side can result in coarse vermicules and An values that are elevated above the values in the newly formed sodic plagioclase that is devoid of quartz vermicules.[225]

In any case, newly created myrmekite is a reformation of original plagioclase. One part of it is free of quartz vermicules, while the other is loaded with them. Both parts are in parallel optical orientation (albite twinning is continuous between them) and the part with the vermicules projects into the other part in a bulbous fashion. This arrangement gives the illusion that plagioclase has replaced K-feldspar, when, in fact, both the non-myrmekitic part and the K-feldspar are replacements of former plagioclase.

## Simultaneous Origin of K-feldspar and Myrmekite

In the 10-cm continuous section at Temecula, the progressive alteration of primary plagioclase is seen to lead into coeval K-

feldspar and myrmekite. Here, tiny plagioclase grains with quartz vermicules of chain-like myrmekite first develop adjacent to tiny altered plagioclase grains with sodic cores ($An_{2-5}$) and calcic rims ($An_{17-20}$). Progressively the tiny plagioclase grains become myrmekitic.

Conversion begins where plagioclase has been strained and altered so that micro-cracks with lattice orientations extend into the grain interiors. Hydrous fluids with potassium ions are able then to penetrate the grain interiors.

In the early stages small isolated blebs or irregular islands of K-feldspar expand to envelope enclaves of plagioclase entirely with K-feldspar. These enclaves recrystallize, as above described, into myrmekite. **In the early stages myrmekite *far exceeds* K-feldspar in volume. Only later does the K-feldspar catch up and exceed myrmekite in volume. Myrmekite cannot have developed from K-feldspar as theorized in *equation* [2].**

Thus, the implication of co-generation from altered mafic rock is inescapable, when we find the first appearance of myrmekite in altered plagioclase enclosed by K-feldspar adjacent to potassium-loaded, pitted plagioclase cores (to 50 % K). **These relationships negate *equation* [1].**

Wherever wartlike myrmekite is formed, whole-rock chemistry and electron-microprobe analysis confirm that $Ca^{2+}$ has been removed, not added, and that plagioclase in myrmekite is *secondary*, not primary. Here, $Ca^{2+}$ has been removed from the interiors of strained plagioclase grains by hydrous fluids because the relatively calcic cores were less resistive to replacement than the sodic rims. This contravenes the assumptions leading to *equations* [1] and [2] in which $Ca^{2+}$ is added and K-feldspar is taken to be primary. Not only is $Ca^{2+}$ removed from plagioclase cores, but it is also removed from the whole rock fabric, which is then transformed into myrmekite-bearing granite.

As wartlike myrmekite is produced, one-half the original plagioclase may be transformed to K-feldspar and the other one-half to relatively-sodic plagioclase. At the same time the mafic cations of the ferromagnesian silicates are replaced by $Si^{4+}$ to form quartz. These transformations yield final compositions of granite or granodiorite. Virtually all trace of original mafic rock is removed (compare top and bottom, Fig. IV-3). *The wartlike myrmekite may thus provide the only clue to former mafic character.*

**Silica in the** *reactions* **[3], [4], [5], [6], [7], [8], and [9]** all show $SiO_2$ being added to the left sides. This is somewhat misleading for *reactions* **[5], [6], [7],** and **[9]**, because added silica is not essential to them unless the parent plagioclase is silica-deficient (high in An value). Such plagioclase has more $Ca^{2+}$ cations than $An_{24}$, and insufficient silica to convert it fully to K-feldspar. An external source for the additional silica is needed. Since full conversion is not uncommon, the existence of an external silica source is implied. Ambient silanes are interpreted as the most likely source.

Only in the case of parent plagioclase with calcium values less than $An_{24}$, is there plenty of silica to produce equal volumes of K-feldspar from plagioclase. In *reaction* **[8]**, where sodic plagioclase is created, $SiO_2$ must be added because relatively-sodic plagioclase requires more silica than any relatively-calcic parent plagioclase can provide.

Where silica from an external source enters into these reactions, it does not contribute to quartz in ghost myrmekite or to quartz vermicules in myrmekite as shown on the right sides of *reactions* **[7]** and **[9]** because, once K-feldspar has formed, newly introduced silica cannot penetrate the newly formed feldspar lattice. **Therefore, silica introduced from silanes is not the source for isolated quartz blebs in K-feldspar or quartz vermicules in wartlike myrmekite. Island quartz blebs inside K-feldspar grains and quartz vermicules in wartlike myrmekite are derived only from residual $SiO_2$. The silica on**

the opposite sides of *reactions* [7] and [9] have separate origins.

## Constant Versus Variable Volume Changes

Although replacements represented by *reactions* [3], [4], [5], [6], [7], [8], and [9] are volume-for-volume, they may not result in constant volume where they affect other simultaneous reactions. Inhomogeneous original lithology or secondary interstitial porosity can receive deposition of quartz introduced by silanes, for example. There is no simple way to equate the masses or volumes of all these reactions; and balancing the equations is not attempted.

## Quartz Vermicule Sizes

In *reaction* [9] a direct correlation is present between $Ca^{2+}$ in the plagioclase in myrmekite and the coarseness of quartz vermicules that result. In all wartlike myrmekite, quartz vermicules exhibit tapered thickness, from a maximum in plagioclase away from K-feldspar. This is true whether or not an albite rim separates the myrmekite from the K-feldspar.

Maximum thickness, however, correlates with plagioclase An content of the precursor mafic rock. For example, small, narrow quartz vermicules in myrmekite ($An_{12-15}$ [**Fig. IV-2**]) occur where Woodson Mountain granite replaces Bonsall tonalite (average $An_{24}$). In contrast, relatively-coarse quartz vermicules in myrmekite ($An_{18-24}$) occur where granitic rock has replaced diorite of the Kernville and Sacatar plutons (original plagioclase contains calcium at levels greater than $An_{40}$ [**Fig. IV-3,D,E**]). In general, for granitic rocks of nearly the same modal and chemical compositions, quartz vermicules in myrmekite can range from fine and barely visible under high power

magnification to intermediate to coarse. Maximum coarseness of vermicules occurs where gabbro has been replaced by granite.

**There is a direct relationship between vermicule coarseness and calcium content of plagioclase in the parent rock. The more calcic the plagioclase, the coarser the vermicules in myrmekite (***reaction* **[9]).** Relatively calcic plagioclase composition in parent rock correlates with elevated levels of retained $Ca^{2+}$ (relative to residual $Al^{3+}$) in the transformed rock. The explanation for this relationship is that in the recrystallization process, retained $Ca^{2+}$ preferentially combines with $Al^{3+}$. This leaves an excess of $Si^{4+}$ from which quartz can be deposited in vermicules.

Since maximum coarseness can be correlated with parent rock character, it can be used as a clue to the original mafic parent. The usefulness of the concept is exemplified by the cordierite-bearing two-mica Clotty granite, which occurs near Perrault Falls, western Ontario.[229] Recognizing coarse vermicules in this rock, I anticipated its source to be a biotite-rich diorite or gabbro. I was able to verify my prediction with 45 thin sections across the transition zone from Clotty granite into adjacent older biotite diorite.

In another study, this one on a deformed part of the South Mountain batholith, Nova Scotia,[230] barely visible vermicules are found in wartlike myrmekite on oriented K-feldspar megacrysts. From this, a relatively-sodic parent rock would be expected. A nearby granodiorite with plagioclase having $An_{25}$ fills the requirement very well.

---

[229]  *Morin & Turnock 1975*

[230]  *Abbott 1989; personal communication 1989*

---

## Rim Myrmekite Versus Wartlike Myrmekite

In many terranes, instead of wartlike myrmekite, rim myrmekite grows on borders of zoned plagioclase in parallel optical orientation while projecting as a bulbous protrusion into K-feldspar. Characteristically, rocks with rim myrmekite exhibit only minimal deformation, and then primarily the breaking of seals along grain boundaries and the stressing of plagioclase grain margins. It is the stressed margins of zoned plagioclase grains which are affected by fluids, losing $Ca^{2+}$, $Na^{1+}$, and $Al^{3+}$, and then being replaced either by silica from silanes [to form untapered vermicules in the rim myrmekite] or by both silica [from silanes] and $K^{1+}$ [to form K-feldspar and tapered vermicules that abut K-feldspar].

Introduced silica from silanes may crystallize as quartz vermicules in rim myrmekite because fluids have access to the rims. These vermicules are untapered because of the uniformity of composition of the enclosing plagioclase. Untapered vermicules occur particularly where bordering K-feldspar is very scarce. Where K-feldspar is abundant, $K^{1+}$ displaces $Na^{1+}$ and $Ca^{2+}$, and *vermicules taper toward the K-feldspar*, adjusting for calcium variations in the adjacent plagioclase of the myrmekite.

Since rim myrmekite prefers less-deformed rocks in contrast to wartlike myrmekite, it is oriented to rocks where less penetration of silanes is possible and, thus, where K-feldspar development should be less. In fact, this is what we find: K-feldspar levels of < 5 % of rock volume and grain size ranging from barely visible films adjacent to rim myrmekite up to relatively large grains that partly enclose bulbous projections of rim myrmekite. This rim is the recrystallized margin of parent plagioclase, an interstitial filling, "trapped," so to speak, between secondary K-feldspar and relatively unstressed zoned plagioclase. It conveys the false impression that myrmekite replaces K-feldspar from the outside inward.

Thus it should be borne in mind that the kind of myrmekite that occurs is a function of the degree of prior deformation. This is why terranes of least deformation display only rim myrmekite, terranes of somewhat greater deformation have both kinds, and strongly deformed terranes have only wartlike myrmekite. However, rim myrmekite does not convert to wartlike with increasing deformation, and, in regions of intense deformation and fluid movement, wartlike myrmekite never forms or may be obliterated.

The disappearance of wartlike myrmekite or its failure to develop occurs where broken and deformed crystals are annealed, and the open system permits escape of $Ca^{2+}$ and introduction of excessive amounts of $Na^{1+}$ and $K^{1+}$. **An excess of $Na^{1+}$ prevents myrmekite formation.** It is suggested that the introduction of hydrous fluids containing silanes plus $Na^{1+}$ ion can explain sodium metasomatism and lack of myrmekite in these rocks.

An example of the disappearance of myrmekite or its lack of formation occurs in a zone of progressive cataclasis in the seismically active Peninsular Ranges of southern California. A mylonite zone[231] has been subject to $Na^{1+}$ metasomatism. My own sampling and 44 thin sections along highway S22, west of Borrego Springs, shows that less-deformed (Anderson zones A and B) rocks contain myrmekite, while more-deformed rocks implying a system "open" to fluid exchange (Anderson zones C and D) do not. The lack of myrmekite on K-feldspar borders in the "open" system rocks is contrary to expectations based on *equations* [1] and [2]. An "open" system in theory favors myrmekite production.

Extreme brittle deformation is another situation in which myrmekite is not found. It is not apparent whether this is due to low temperatures, or to destruction in cataclasis.[232]

---

[231]  *Anderson, J. R. 1983*

[232]  *Higgins 1971*

## Isolated Myrmekite Versus Wartlike Myrmekite

The openness of altered plagioclase is seen at 8,000x magnifications as aligned holes in pitted, altered grains. The opportunity for fluid incursions into these holes suggests that quartz vermicules and blebs in plagioclase are simply cavity fillings of $SiO_2$. This kind of deposition does seem to occur, and to produce vermicules that do not taper. I call this kind of myrmekite "isolated myrmekite" because no K-feldspar coexists with it. Its generation is explained as a result of a lack of $K^{1+}$ or $Na^{1+}$ to replace the $Ca^{2+}$, $Na^{1+}$, and $Al^{3+}$ that have been removed to create the holes. Introduced silanes have entered freely, reacted with excess oxygen, deposited quartz, and released water.

Isolated myrmekite is not common, however. I have found it only twice, in both cases in altered gabbro. One of these is the Elk Creek hornblende-augite gabbro, Colorado.[233] The other is a hypersthene gabbro near Dover, New Jersey. Both of these are described in Chapter III.[227] Augustithus, in addition, provides further descriptions of isolated myrmekite.

---

[233] *Collins & Davis 1992*

# REFERENCES

**Abbott** R.N., Jr. 1989, "INTERNAL STRUCTURES IN PART OF THE SOUTH MOUNTAIN BATHOLITH, NOVA SCOTIA, CANADA" GSA Bull., v101, p1493-1506

**Adams**, M.A. 1990 "SOME OBSERVATIONS OF ELECTROMAGNETIC SIGNALS PRIOR TO SOME CALIFORNIA EARTHQUAKES" Jour. Sci. Exploration, v4, #2, p137-152

**Albuquerque**, C. De, 1971 "PETROCHEMISTRY OF A SERIES OF GRANITIC ROCKS FROM NORTHERN PORTUGAL" GSA Bull., v82, p2783-2798

**Alderman**, A.R. 1936, "ECLOGITES FROM THE NEIGHBORHOOD OF GLENELG, INVERNESS-SHIRE" Quart. Journ. Geol. Soc., v92, p488-530

**Alekseyev**, A.S., Vanjan, L.L., Berdichevsky, M.N., Nikoolayev, A.N., Okulesky, E.A., Rjaboy, V.Z. 1977 "CHEMISTRY OF THE ASTHENOSPHEREIC ZONES OF THE SOVIET UNION" In Rept. of Akad.Ski.USSR, v234,#4,p790-793
-Rjaboy, V.Z., 1978 "EARTH'S ASTHENOSPHERE" In Earth and Universe, #5, p36-42

**Al-Hashimi**, A.R.K., Brownlow, A.H. 1970, "COPPER CONTENT OF BIOTITES FROM THE BOULDER BATHOLITH, MONTANA" Econ. Geol. v65, p985-992

**Alling**, H.L. 1932, "PERTHITES" Amer. Min., v17, p43-65. 1938, "PLUTONIC PERTHITES" J. Geol., v46, p142-165

**Anderson**, H.J. 1990, "THE 1989 MACQUARRIE RIDGE EARTHQUAKE AND ITS CONTRIBUTION TO THE REGIONAL SEISMIC MOMENT BUDGET" Geoph Lett. v17, 7, p1013-1016

**Anderson**, J.L. 1983, "PROTEROZOIC ANOROGENIC GRANITE PLUTONISM IN NORTH AMERICA" GSA Mem. 161, p133-154

**Anderson**, J.R. 1983, "PETROLOGY OF A PORTION OF THE EASTERN PENINSULAR RANGES MYLONITE ZONE, SOUTHERN CALIFORNIA" Contrib.
Mineral. Petrol., v84, p253-271

**Antonov**, V. E., Belash, I.T., Degryareva, E.G., Ponyatovskii, E.G., Shiraev, I. 1980 "OBTAINING IRON HYDRIDE UNDER HIGH HYDROGEN PRESSURE" Sov. Phys. Dokl. v25, #490

**Arp**, H. 1987 "QUASARS, RED SHIFTS AND CONTROVERSIES" Interstellar Media, Berkeley, 195pp

**Artjushkov**, E.V. 1979 "GEODYNAMICS" Moscow, Nauka, 328pp

**Aston**, S.R. 1983, "SILICON GEOCHEMISTRY AND BIOGEOCHEMISTRY" Academic Press

**Augustithis**, S. S. 1990 "ATLAS OF METAMORPHIC-METASOMATIC TEXTURES AND PROCESSES" Elsevier, Amsterdam, 228pp

**Badding**, J,V. 1990, personal commnication  1990, Hemley, R.J., Mao, H.K. 1990  *IN SITU* OBSERVATION OF IRON HYDRIDE" EOS v71, #17, p528 Abs
-1991, Hemley, R.J., Mao, H.K., "HIGH-PRESSURE CHEMISTRY OF HYDROGEN IN METALS *IN SITU* STUDY OF IRON HYDRIDE"

**Bailey**, R.C. 1990 "TRAPPING OF AQUEOUS FLUIDS IN THE DEEP CRUST" Geoph. Res. Lett. v17, #8, p1129-1132

**Bak**, J., Korstgard, J., Sorenson, K. 1975 "A MAJOR SHEAR ZONE WITHIN THE NAGSSUGTOQIDIAN OF WEST GREENLAND" Tectonophys. v27, p191-207

**Barzangi**, M., Dorman, J., 1969 "WORLD SEISMICITY MAPS, 1961-1967" Bull. GSA, v59, #1, p369-380

**Bateman**, P.C. 1965, "GEOLOGY AND TUNGSTEN MINERALIZATION OF THE BISHOP DISTRICT, CALIFORNIA" USGS Prof. Pap. 470, 208p
-Wahrhaftig, C., 1966, "GEOLOGY OF THE SIERRA NEVADA" in Bailey, E.H., ed., "Geology of Northern California" Calif. Div. Mines & Geol. Bull. 190., p107-172

**Becker**, Robert O., M.D. 1990, "CROSS CURRENTS", Jeremy P. Tarcher, Inc.

**Becraft**, G.E., Pinckney, D.M., Rosenbaum, S. 1963, "GEOLOGY AND MINERAL DEPOSITS OF THE JEFFERSON CITY QUADRANGLE, JEFFERSON AND LEWIS AND CLARK COUNTIES, MONT." USGS Prof. Pap. 428, 101p

**Bedell,** R.L. 1985, "MADAWASKA MINES, BANCROFT, ONTARIO DEFORMATION OF THE FARADAY METAGABBRO COMPLEX AND ITS INFLUENCE ON URANIFEROUS PEGMATITE EMPLACEMENT AND ORE DEPOSITION" Unpubl. master's thesis, Univ. of Toronto, 177pp
-1982 "MAP P.2523, BANCROFT AREA, WESTERN PART" Ontario Geological Survey

**Behr**, J., Bilham, R, Bodin, P., Burford, R.O., Burgman, R. 1990, "ASEISMIC SLIP ON THE SAF SOUTH OF LOMA PRIETA" Geoph. Res. Lett. v17, #9, p1445-1448

**Belousov**, V.V. 1967, "ABOUT STRUCTURE AND DEVELOPMENT OF CONTINENTAL TECTONOSPHERES" In Regularities of Mineral Emplacement, v8, Moscow, Nauka, p5-39
-1987, "SEISMIC EXHALATIONS IN THE DJIRGATALSKY EARTHQUAKE, 1984" Problems in Engineering Geology, #28, p30-34

**Berman**, B.I. 1973, "CRYPTOVOLCANIC MOBILIZATION AND ITS ROLE IN THE FORMATION OF ORE COMPLEXES" Soviet Geol. #4, p24-36

**Bezrukov**, G.N., 1974 "GENESIS OF DIAMOND IN THE LIGHT OF EXPERIMENTS IN ITS SYNTHESIS" Soviet Geol., 1974 #11,p31-40
-Butuzov, V.P., Samoilovich, M.I. 1976 "SYNTHETIC DIAMOND" Moscow, Nedra 119p

**Bobrievich**, A.P., Sobolev, V.S. 1957, "ECLOGITIZATION OF CRYSTALLINE PYROXENE-BEARING ARGILLITES OF ARCHEAN AGE" Rept. of VMO, pt86, #1, p3-17

**Bois,** C., Cazes, M., Hirn, A., Mascie, A., Matte, P., Montadert, L., Pinet, B., 1988 "CONTRIBUTION OF DEEP SEISMIC PROFILING TO THE KNOWLEDGE OF THE LOWER CRUST IN FRANCE AND NEIGHBORING AREAS" Tectonohysics, v145, #3-4, p253-275

**Botkunov**, A.I. 1964, "PATTERNS OF DISTRIBUTION OF DIAMONDS IN THE MIR PIPE" Repts. VMO 1964 pt93, #4, p424-435

**Bott**, M.H.P. 1974, "THE INTERIOR OF THE EARTH" Moscow, Mir, 375pp

**Boyle**, R.W. 1970, "THE SOURCE OF METALS AND GANGE ELEMENTS IN HYDROTHERMAL DEPOSITS" In Intern. Union Geol. Sci. Series A, No. 2, "Problems of Hydrothermal Ore Deposition," Stuttgart, Schweizerbartsche, p3-6

**Bratus**, M.D., Tatarintsev, V.I., Sakhro, B.E. 1988 "FLUID-INCLUSION COMPOSITIONS IN CHILLED PARTICLES FROM EXPLOSIVE RING STRUCTURES AND KIMBERLITE PIPES" Geochem. Int., v25, #6, June, 1988 p50-55

**Brocher**, T.M., Taber, J.J. 1987 "CRUSTAL STRUCTURE BENEATH EXPOSED ACCRETED TERRANES OF SOUTHERN ALASKA" Geoph. Jour. Roy. Astr. Soc., v89, #1, p73-78

**Braunmiller**, J., Nabelek, J. 1990, "RUPTURE PROCESS OF THE MACQUARRIE RIDGE EARTHQUAKE OF MAY 23, 1989" Geoph. Res. Lett. v17,#7,p1017-1020

**Burmin**, G.S. 1984, "WONDERFUL STONE" Moscow: Znanie, 168pp

**Byerly**, P. 1987 "DETERMINATION OF FAULT EPICENTERS FROM SEISMIC DATA" In The Earth's Crust, Moscow, IL, p89-100

**Carey**, S.W., 1976, "THE EXPANDING EARTH" New York, Elsevier, 488p

**Carlson**, K.J., Ed. 1990 "THE RESTLESS EARTH" Nobel Conference XXIV, p59-111 Harper & Rowe

**Cater**, F.W. 1982, "INTRUSIVE ROCKS OF THE HOLDEN AND LUCERNE QUADRANGLES, WASHINGTON - THE RELATION OF DEPTH ZONES, COMPOSITION, TEXTURES, AND EMPLACEMENT OF PLUTONS" USGS Prof.Pap. 1220, 108p

**Charlesworth**, H.A.K., Weiner, J.L., Akehurst, A.J., Bielenstein, A.J., Evans, H.U., Griffiths, C.R., Remington, D.B., Stauffer, M.R., Steiner, J. 1967, "PRECAMBRIAN GEOLOGY OF THE JASPER REGION" Alberta Res. Council Bull. 23

**Chaudhuri**, N.K., Iyer, R.H. 1980, "ORIGIN OF UNUSUAL RADIOACTIVE HALOES" Radiation Effects, v53, p1-6

**Clemons**, R.E., Long, L.E. 1971, "PETROLOGIC AND Rb-Sr ISOTOPIC STUDY OF THE CHIQUIMULA PLUTON, SOUTHEASTERN GUATEMALA" GSA Bull., v82, p2729-2740

**Cloud**, P.E., Jr., 1965, "SIGNIFICANCE OF THE GUNFLINT (PRECAMBRIAN) FLORA" Science, v148, p27-35

**Clowes**, R.M., Yorath, C.J., Hyndman, R.D., 1987 "REFLECTION MAPPING ACROSS THE CONVERGENT MARGIN OF WESTERN CANADA" Geophys. Jour. Roy. Astron. Soc., 1987, v89 #1 p79-84

**Clymo**, R.S., 1987 "COAL AND COAL-BEARING STRATA: RECENT ADVANCES AND FUTURE PROSPECTS" Blackwell Scientific Publ. Inc.

**Collins**, L.G. 1959, "GEOLOGY OF THE MAGNETITE DEPOSITS AND ASSOCIATED GNEISSES NEAR AUSABLE FORKS, NEW YORK" Dissert. Abstr. v20, 1739pp
  -1969a, "REGIONAL RECRYSTALLIZATION AND THE FORMATION OF MAGNETITE CONCENTRATIONS, DOVER MAGNETITE DISTRICT, NEW JERSEY" Econ. Geol. v64, p17-33

-1969b, "HOST ROCK ORIGIN OF MAGNETITE IN PYROXENE SKARN AND GNEISS AND ITS RELATION TO ALASKITE AND HORNBLENDE GRANITE" Economic Geology. v64, p191-201

-1988a, HYDROTHERMAL DIFFERENTIATION AND MYRMEKITE A CLUE TO MANY GEOLOGIC PUZZLES" Theophrastus Publications S.A., Athens, Greece, 382pp

-1988b, "MYRMEKITE, A MYSTERY SOLVED NEAR TEMECULA, RIVERSIDE COUNTY, CALIFORNIA," California Geol., v41, p276-281

-1989, "ORIGIN OF THE ISABELLA PLUTON, AND ITS ENCLAVES, KERN COUNTY, CALIFORNIA" Calif. Geol., v42, p53-59

-1990, "CATHODOLUMINESCENCE MICROSCOPY OF MYRMEKITE COMMENT" Geology, v18, p1163

- Davis, T.E., 1992, "ORIGIN OF HIGH-GRADE BIOTITE-SILLIMANITE-GARNET-CORDIERITE GNEISSES BY HYDROTHERMAL DIFFERENTIATION, COLORADO" In Barto-Kyriakidis, A., ed., "High-Grade Metamorphics," Athens, Theophrastus Publ., p297-338

Condie, K.C., 1982, "PLATE TECTONICS & CRUSTAL EVOLUTION" 2nd edit., New York, Pergamon Press, 310p

Coveney, R.M.,Jr., Goebel, E.D., Zeller, E.J., Dreschhoff, G.A.M., Angino, E.E. 1987 "SERPENTINIZATION AND THE ORIGIN OF HYDROGEN GAS IN KANSAS" AAPG Bull. v71, #1, p39-48

Crowder, D.F., Ross, D.C. 1973, "PETROGRAPHY OF SOME GRANITIC BODIES IN THE NORTHERN WHITE MOUNTAINS, CALIFORNIA-NEVADA" USGS Prof. Pap. 775, 28pp

Curl, R.F., Smalley, R.E. 1991, "FULLERENES" Sci. Amer. 10/91, p54-63

Debicki, R.L. 1982, "GEOLOGY AND SCENERY, KILLARNEY PROVINCIAL PARK AREA, ONTARIO" OGS Guidebook #6, Ministry of Natural Res., 152pp

Deer, W.A., Howie, R.A., Zussman, J. 1962, "ROCK-FORMING MINERALS" New York, John Wiley & Sons, 333pp.

De Vries, M.S. et al. 1991, "SEARCH FOR HIGH MOLECULAR WEIGHT POLYCYCLIC AROMATIC HYDROCARBONS AND FULLERENES IN CARBONACEOUS METEORITES" NASA, LPSV XXII, p315-316

Dobretsov, N.L., Zuenko, V.V., Kharkiv, A.D. 1972,"FACTORS IN TYPING THE DIAMOND-BEARING KIMBERLITES OF YAKUTIYA" Geol. & Geoph. 1972, #7, p31-39
- 1981, "GLOBAL PETROLOGIC PROCESSES" Moscow, Nedra 236pp

Drake, G.W. 1974, "GEOLOGY OF AN AREA EAST OF PURNAMOOTA, BROKEN HILL" Unpubl. Bachelor of Science honor's thesis, Canberra, Australian Nat. Univ., 80pp

Durrheim, R.J. 1987, "SEISMIC REFLECTION AND REFRACTION STUDIES OF THE DEEP STRUCTURE OF THE AGULHAS BANK" Geophys. Jour. Roy. Astr. Soc. v89, #1, p395-398

Dutch, S. 1983, "LETTERS: CREATIONISM STILL AGAIN" Physics Today, v36, p11-13

DuToit, A.L. 1957, "GEOLOGY OF SOUTH AFRICA" Moscow: IL 1957, 488pp

Efimov, A.O. 1979, "TRAPPS IN THE CAMBRIAN SREDNEBOTUOBINSK DEPOSIT" In Organization of Oil and Gas Fields on the Siberian Platform, Novosibirsk, SMIIGGIMS, p49-54

**Egorkin**, A.V., Zjuganov, S.K., Chernyshov, N.M. 1984, "THE UPPER MANTLE OF SIBERIA" In 27th Int. Geol. Cong., Moscow, Rept. v8, sec5.08, Geophysics, p27-72

**Ekstrom**, G., Romanowicz, B. 1990 "THE 23 MAY 1989 MACQUARRIE RIDGE EARTHQUAKE: A VERY BROAD BAND ANALYSIS" Geoph. Res. Lett. v17, #7, p993-996

**Elwell**, D. 1986, "MAN-MADE GEMSTONES" Moscow, Mir, 160pp

**Emmermann**, R., Dietrich, H.-G., Lauterjung, J., Whorl, T. 1990, "KTB REPORT 90-8" Geol. Surv. of Lower Saxony

**Faure**, G. 1986, "PRINCIPLES OF ISOTOPE GEOLOGY, 2nd edit., New York, John Wiley & Sons, 589pp

**Feather**, N. 1978 "THE UNSOLVED PROBLEM OF Po HALOS IN PRECAMBRIAN BIOTITE AND OTHER OLD MINERALS" Comm. to Roy. Soc. Edinburgh, v11, p147-158

**Feoktistov**, G.D. 1976, "TRAPP SILLS OF GREAT EXTENSION ON THE SOUTH SIBERIAN PLATFORM" Soviet Geol., #12, p122-127
            -1978 "PETROLOGY AND CONDITIONS FOR FORMATION OF TRAPP SILLS" Novosibirsk: Nauka. 168pp

**Fersman**, A.E., 1960 In Selected Works, v6 Moscow, USSR Acad. Sci. 742pp

**Fisher**, R.V., Schmincke, 1984 "PYROCLASTIC ROCKS" Springer-Verlag

**Fon-der-Flaass**, G.S. 1980 "PECULIARITIES IN THE FORMATION OF NECK STRUCTURES AND MAGNETITE MINERALIZATION IN DEPOSITS OF ANGARO-ILIM TYPE" In Endogenous Ore-forming Processes, Sverdlovsk, Acad. Sci. USSR, p101-114

**Francis**, W. 1954, "COAL" E. Arnold Publ. Ltd.

**Frank**, L.A. 1990, "THE BIG SPLASH" Birch Lane Press, 225pp

**Frolich**, C. 1989, "NEW RUMBLES ON DEEP SOURCES" Nature, v341, p687

**Frolov**, B.M., Efimov, M.I., Belozerova, N.N. 1976, "MAIN FEATURES OF THE SEDIMENTARY COVER OF THE SIBERIN PLATFORM" Leningrad, nedra 112pp

**Frondel**, C., Collette, R.L. 1957, "HYDROTHERMAL SYNTHESIS OF ZIRCON, THORITE, AND HUTTONITE" Amer. Min., v42, p759-65

**Fuis**, G.S., Ambos, E.L., Mooney, W.D., Page, R.A., Fisher, M.A., Brocher, T.M., Taber, J.J. 1987, "CRUSTAL STRUCTURE BENEATH EXPOSED ACCRETED TERRANES OF SOUTHERN ALASKA" Geophys. Jour. Roy. Astr. Soc. v89, #1, p73-78

**Fukai**, Y., Akimoto, S. 1983, "HYDROGEN IN THE EARTH'S CORE" Proc. Jap. Acad., v59,Ser. B, p158-162

**Fullagar**, P.D. 1971, "AGE AND ORIGIN OF PLUTONIC INTRUSIONS IN THE PIEDMONT OF THE SOUTHEASTERN APPALACHIANS" GSA Bull., v82, p2845-2862

**Fyfe**, W.S. 1972, "THE GENERATION OF BATHOLITHS" in Wyllie, P.J., ed., "Experimental petrology and global tectonics," Tectonophysics, v17, p273-283

Gable, D.J. 1980."THE BOULDER CREEK BATHOLITH, FRONT RANGE, COLORADO" USGS Prof. Pap. 1101, 88pp
-& Sims, P.K., 1969 "GEOLOGY AND METMORPHISM OF SOME HIGH-GRADE CORDIERITE GNEISSES, FRONT RANGE, COLORADO" Geol. Soc. America, Spec. Paper 128. 87pp

Galson, D.A., Mueller, S.T., 1986 "THE EUROPEAN GEOTRAVERSE PROJECT: RECENT RESULTS AND UPCOMING EXPERIMENTS" Terra Incognita, v6 #4 p603-610

Garcy, 1923 - reference given in Byerly 1987

Gardner, M. 1989, "NOTES OF A FRINGE-WATCHER: ROBERT GENTRY'S TINY MYSTERY" Skeptical Inquirer, v13, p357-361

Garlick, G.D. 1966, "OXYGEN ISOTOPE FRACTIONATION IN IGNEOUS ROCKS" Earth Planet. Sci. Let., v1, p361-368

Gazban, F., Schwarcz, G., Ford, D.C. 1990, "CARBON AND SULFUR ISOTOPE EVIDENCE FOR IN SITU REDUCTION OF SULFATE, NANISIVIK LEAD-ZINC DEPOSITS, NORTHWEST TERRITORIES, BAFFIN ISLAND, CANADA" Ec. Geol. v85, #2, p360-375

Gebauer, D., Grüenenfelder, M. 1973, "VERGLEICHENDE U-Pb- UND Rb/Sr-ALTERBESTIMMUNGEN IM BAYERISCHEN TEIL DES MOLDANUBIKUMS"
1977, "U-Pb ZIRCONDATIERUNGEN MAFISCHER UND ULTRAMAFISCHER GESTEINE" Arbeitstagung der OeMG gemeinsam mit der SMPG (abs.): Salzburg
1979, "U-Th-Pb DATING OF MINERALS" In Jager, E., Hunziker, J.C., eds., "Lectures in isotope geology," Berlin, Springer-Verlag, p105-131

Gelfand, N.I. 1987, "THE QUESTION OF THE NATURE OS SEISMIC BOUNDARIES IN THE EARTH'S CRUST" News of the Higher Educational Institutions: geology and prospecting, v1, p124-127

Gentry, R.V. 1965, "PLEOCHROIC HALOS AND THE AGE OF THE EARTH" Amer. J. Physics, Proc., v33, p878A
-1970, "COSMOLOGICAL IMPLICATIONS OF EXTINCT RADIOACTIVITY FROM PLEOCHROIC HALOS" In Lammerts, W.E., ed.,"Why Not Creation?" Presbyt. and Reformed Publ. Co., p107-113
-1974, "RADIO HALOS IN A RADIOCHRONOLOGICAL AND COSMOLOGICAL PERSPECTIVE" Science, v184, p62-64
-1983, "LETTERS. CREATIONISM AGAIN, THE AUTHOR COMMENTS" Physics Today, v36, p13-15
-1988, "CREATION'S TINY MYSTERY," sec. edit., Knoxville, Earth Sci. Assoc., 348p

Gillson, J.L., Ed., 1960 "INDUSTRIAL MINERALS AND ROCKS", American Institute of Mining, Metallurgical, and Petroleum Engineers.

Gilmour, I, et al. 1991, "A SEARCH FOR THE PRESENCE OF $C_{60}$ AS AN INTERSTELLAR GRAIN IN METEORITES" NASA, LPSV XXII, p445-446

Glockling, F. 1969, "THE CHEMISTRY OF GERMANIUM" Academic Press

Gold, T. 1987, "POWER FROM THE EARTH" J.M. Dent & Sons, London

Goleby, B.R., Wright, C., Kennett, B.L.N. 1987,"PRELIMINARY DEEP REFLECTION

STUDIES IN THE ARUNTA BLOCK, CENTRAL AUSTRALIA" Geophys. Jour. Roy. Astr. Soc. v89, #1, p437-442

**Gorokhov**, S.S. 1972, "POSSIBLE FORMATION OF ECLOGITES FROM ARGILLITES" USSR Acad Sci. v206, #4, p947-950
-1975 "SMALL DIAMOND SOURCES IN PLACER DEPOSITS" USSR Acad. Sci. v225,#5, p1145-1148

**Gottfried**, R. 1990, "ORIGIN AND EVOLUTION OF THE EARTH CHEMICAL AND PHYSICAL VERIFICATIONS" In Barto-Kyriakidis, A., ed., "Critical Aspects Of The Plate Tectonics Theory," vol. II, Athens, Theophrastus Publications, p115-140

**Gough**, D.I. 1986, "MANTLE UPFLOW TECTONICS IN THE CANADIAN CORDILLERA", Can. Jour. Geophysics Res., v91, p1909-1919

**Gould**, S.J. 1989, "WONDERFUL LIFE: THE BURGESS SHALE AND THE NATURE OF HISTORY" W.W.Norton & Co., 347pp

**Grauert**, B., Hanny, R., Soptrajanova, G. 1974, "GEOCHRONOLOGY OF POLYMETAMORPHIC AND ANATECTIC GNEISS REGION: THE MOLDANUBICUM OF THE AREA LAM-DEGGENDORF, EASTERN BAVARIA GERMANY" Contrib. Mineral. Petrol., v45, p37-63

**Green**, A.G., Milkereit, B., Mayrand, L., Spencer, C., Kurtz, R., Clowes, R.M. 1987 "LITHOPROBE SEISMIC PROFILING ACROSS VANCOUVER ISLAND: RESULTS FROM REPROCESSING" Geophys. Jour. Roy. Astr. Soc. v89 #1 p85-90

**Griffin**, W.L., Sutherland, F.L., Hollis, J.D. 1987 "CORRELATION OF XENOLITH PETROLOGY AND SEISMIC DATA: AN EXAMPLE FROM EAST CENTRAL QUEENSLAND" USGS Circ. #956, p30-31

**Griffis**, R.A. 1987, "KERN KNOB PLUTON AND OTHER HIGHLY-EVOLVED GRANITOIDS IN EAST-CENTRAL CALIFORNIA" Unpubl. master's thesis. Calif. State Univ.-Northridge, 305pp

**Grigor'yev** & Zolenev - reference unavailable

**Gussow**, W. C., Hunt, C. W. 1959, "AGE OF THE ICE RIVER COMPLEX" Jour. Alberta Soc. Pet. Geol. Bull. v7, #3

**Gutenberg**, B. 1963, "PHYSICS OF THE EARTH" Moscow, IL, 264pp

**Guterman**, V.G. 1977, "THE EVOLUTION OF A MULTI-LAYERED ASTHENOSPHERE" Kiev Naukova, Dumka, 156pp

**Hacquebard**, P.A., Cameron,A.R. 1989, "DISTRIBUTION PATTERNS AND COALIFICATION IN CANADIAN BITUMINOUS AND ANTHRACITE COALS" Int. Jour. Coal Geol., Spec. Issue v13, #1-4, Eds. P.C.Lyons & B.Alpern

**Hague**, J.M., Baum, J.L., Herrmann, L.A., Pickering, R.J. 1956, GEOLOGY AND STRUCTURE OF THE FRANKLIN-STERLING AREA, NEW JERSEY" GSA Bull. v67, p435-474

**Harris**, M., Radtke, A.S. 1976, "STATISTICAL STUDY OF SELECTED TRACE ELEMENTS WITH REFERENCE TO GEOLOGY AND GENESIS OF THE CARLIN GOLD DEPOSIT, NEVADA" USGS Prof. Pap. 960

**Harris**, R.L., Jr., 1977, "DISPLACEMENT OF RELICT ZIRCONS DURING GROWTH OF FELDSPATHIC PORPHYROBLASTS" GSA Bull., v88, p1828-1830

**Hashemi-Nezhad**, S.R., Fremlin, J.H.,Durrani, S.A. 1979, "POLONIUM HALOES IN MICA" Nature, v78, p333-335

**Hauser**, E.C., Gephart, J., Latham, T., Oliver, J., Kaufman, B., Brown, I., Lucchitta, I., 1987 "COCORP ARIZONA TRANSECT: STRONG CRUSTAL REFLECTIONS AND OFFSET MOHO BENEATH THE TRANSITION ZONE" Geology, v15 #12 p1103-1106

**Hedge**, C.E., Noble, D.C. 1971, "UPPER CENOZOIC BASALTS WITH HIGH Sr87/Sr86 AND Sr/Rb RATIOS, SOUTHERN GREAT BASIN, WESTERN UNITED    STATES" GSA Bull., v82, p3503-3510

**Hein**, F.J., Arnott, R.W. 1983, "PRECMABRIAN MIETTE CONGLOMERATES, LOWER CAMBRIAN GOG QUARTZITES AND MODERN BRAIDED OUTWASH DEPOSITS, KICKING HORSE PASS AREA" CSPG Field Guidebook
        - 1987 "TIDAL/LITTORAL OFFSHORE SHELF DEPOSITS  LOWER CAMBRIAN GOG GROUP, SOUTHERN ROCKY MOUNTAINS, CANADA" Sedimentary Geology, v52, p155-182

**Henderson**, G.H. 1939, A QUANTITATIVE STUDY OF PLEOCHROIC HALOS, V. THE GENESIS OF HALOS" Roy. Soc. of London, Proc., Series A, v173, p250-264
        -Sparks, F.W. 1939, "A QUANTITATIVE STUDY OF PLEOCHROIC HALOS, IV, NEW TYPES OF HALOS" Roy. Soc. of London, Proc., Series A, v173, p238-249

**Henshaw**, P.C. 1942, "GEOLOGY AND MINERAL RESOURCES OF THE CARGO MUCHACHO MOUNTAINS, IMPERIAL COUNTY, CALIFORNIA" Calif. J. of Mines & Geol., v38, p147-196

**Herrick**, D.L., Parmentier, E.M. 1991, "INITIATION OF SUBDUCTION ON EARTH AND VENUS BE EPISODIC LARGE-SCALE MANTLE OVERTURN" NASA, LPSC XXII p557-558

**Hewitt**, D.F. 1957, "GEOLOGY OF THE CARDIFF AND FARADAY TOWNSHIPS" Ontario Dept. of Mines, v66, pt 3

**Heyman**, D. 1991, "THE GEOCHEMISTRY OF BUCKMINSTERFULLERENE ($C_{60}$): SOLID SOLUTIONS WITH SULFUR AND OXIDATION WITH PERCHLORIC ACID" NASA, LPSV XXII, p569-570

**Hibbard**, M.J. 1979, "MYRMEKITE AS A MARKER BETWEEN PREAQUEOUS AND POSTAQUEOUS PHASE SATURATION IN GRANITE SYSTEMS" GSA Bull., v90, p1047-1062

**Hietanen**, A. 1947, "ARCHEAN GEOLOGY OF THE TURKU DISTRICT IN SOUTHWESTERN FINLAND" GSA Bull., v58, p1019-1084

**Higgins**, M.W. 1971, "CATACLASTIC ROCKS" USGS Prof. Pap. 687, 98p

**Hildreth**, W. 1979, "THE BISHOP TUFF: EVIDENCE FOR THE ORIGIN OF COMPOSITIONAL ZONATION IN SILICIC MAGMA CHAMBERS" GSA Spec. Pap. 180, p43-75
            -1981, "GRADIENTS IN SILICIC MAGMA CHAMBERS: IMPLICATIONS FOR LITHOSPHERIC MAGMATISM" J. Geophys. Res. v86, p10153-92

**Hobbs**, R.W., Peddy, C. 1987 "IS LOWER CRUSTAL LAYERING RELATED TO EXTENSION?" Geophys. Jour. Roy. Astr. Soc. v89 #1 p239-242

**Hodgson**, G. W., in Fairbridge, R. W. Ed. 1972, "ENCYCLOPEDIA OF EARTH SCIENCES" Dowden, Hutchinson & Ross, Inc., Eight volumes, #IVA, p498

**Hoeve**, J. 1978, "COMPOSITION AND VOLUME CHANGES ACCOMPANYING SODA METASOMATIC ALTERATIONS, VASTERVIK AREA, SE SWEDEN" Geol.Rundsch., v67, p920-942

**Hoffer**, E. 1976, "THE REACTION, SILLIMANITE + BIOTITE + QUARTZ CORDIERITE + K-FELDSPAR + $H_2O$ AND PARTIAL MELTING IN THE SYSTEM $K^2O$ $FeO·MgO·Al_2O_3·SiO_2·H_2O$" Contrib. Mineral. Petrol. 55: 127-130

**Holmes**, David A. 1990, "PACIFIC NORTHWEST ZEOLITE MINERALS UPDATE" In Industrial rocks and minerals of the Pacific Northwest, Oregon Dep't of Geol. & Mineral Indust. Special Pap. #23, p79-88

**Hopson**, R.F. Ramseyer, K. 1990a, "CATHODOLUMINESCENCE MICROSCOPY OF MYRMEKITE" Geology, v18, p336-339. 1990b "Reply" Geology, v18, p1163-1164
        -Cater, F.W., Crowder, D.F., 1970, "EMPLACEMENT OF PLUTONS, CASCADE MOUNTAINS, WASHINGTON" GSA Abstr. with Program, v2, no. 2, p104.

**Hoyle**, F. 1983 "THE INTELLIGENT UNIVERSE" Holt Rinehart & Winston, 256pp

**Hunt**, C.B., Mabey, D.R. 1966, "STRATIGRAPHY AND STRUCTURE OF DEATH VALLEY, CALIFORNIA" USGS Prof. Pap. 494A, 165p

**Hunt**, C.W., 1950 "PRELIMINARY REPORT ON WHITEMUD OILFIELD, ALBERTA, CANADA" AAPG Bull. v34, #9, p1795-1801

**Idiz**, E.F. 1981, "GEOLOGY OF THE MARBLE CANYON PLUTONIC COMPLEX" Master's thesis, Los Angeles, Univ. of California, 189pp

**Ilupin**, I.P., Nagaeva, N.P. 1970, "CHROMIUM AND NICKEL IN ILMENITE FROM KIMBERLITES OF YAKUTIA" In Geol., Petrogr. & Mineralogy of Magmatic Formations of the Northeastern Siberian Platform - Moscow, Nauka, p288-300

**Ishimoto**, 1932 reference in Byerly 1987

**Jahn**, B. 1973, "A PETROGENETIC MODEL FOR THE IGNEOUS COMPLEX IN THE SPANISH PEAKS REGION, COLORADO" Contrib. Min. Petrol., v41, p241-258
        -Sun, S.S., Nesbitt, R.W. 1979, "REE DISTRIBUTION AND PETROGENESIS OF THE SPANISH PEAKS IGNEOUS COMPLEX, COLORADO" Contrib. Min. Petrol., v70, p281-298

**Jahns**, R.H. 1954, "PEGMATITES OF SOUTHERN CALIFORNIA" In "Geology of Southern California." Chapter VII. "Mineralogy and Petrology," No. 5, Calif. Div. Mines & Geol. Bull. 170, p37-50
        - Tuttle, O.F. 1963, "LAYERED PEGMATITE-APLITE INTRUSIVES" Min. Soc. Amer. Spec. Pap. 1, p78-92

**James**, H.L., 1966, "CHEMISTRY OF THE IRON-RICH SEDIMENTARY ROCKS" USGS Prof. Pap. 440, chapt. W, p1-60

**Jeffrey**, E.C. 1913. "NATURE OF THE SUBSTANCE KNOWN AS MOTHER OF COAL AND ITS RELATION TO THE PROCESS OF COAL FORMATION" GSA Bull. v24, p715

**Johns**, R. W. 1970, "THE ATHABASCA SANDSTONE AND URANIUM DEPOSITS" Western Miner, Oct. issue

**Johnson**, R.B. 1961, "PATTERNS AND ORIGIN OF RADIAL DIKE SWARMS ASSOCIATED WITH WEST SPANISH PEAK AND DIKE MOUNTAINS, SOUTH-CENTRAL COLORADO" GSA Bull., v72, p579-589
        -1968, "GEOLOGY OF THE IGNEOUS ROCKS OF THE SPANISH PEAKS REGION, COLORADO" USGS Prof. Pap. 594-G, p1-47

**Johnstone**, S.J. 1954, "MINERALS FOR THE CHEMICAL AND ALLIED INDUSTRIES" John Wiley & Sons Inc.

**Joly**, J. 1917, "THE GENESIS OF PLEOCHROIC HALOS" Phil. Trans. of the Roy. Soc. London, Series A, v217, p51

**Jones**, L.M., Walker, R.L. 1973, "Rb-Sr WHOLE-ROCK AGE OF THE SILOAM GRANITE, GEORGIA: A PERMIAN INTRUSIVE IN THE SOUTHERN APPALACHIANS" GSA Bull., v84, p3653-3658.

**Kamchatka**, 1974 - reference unavailable

**Kaminsky**, F.V. 1984, "DIAMOND-BEARING NON-KIMBERLITE IGNEOUS BODIES" Moscow, Nedra 173pp

**Kenji**, S., Kanamori, H. 1990, "FAULT PARAMETERS AND TSUNAMI EXITATION OF THE MAY 29, 1989 MACQUARRIE RIDGE EARTHQUAKE" Geoph. Res. Lett. v17, #7, p997-1000

**Kerr-Lawson**, D.E. 1927, "PLEOCHROIC HALOS IN BIOTITE FROM NEAR MURRAY BAY" Univ. of Toronto Studies, Geol. Studies, v24, p54-71.

**King** L.C. 1983, "WANDERING CONTINENTS AND SPREADING SEA FLOORS ON AN EXPANDING EARTH" John Wiley & Sons, New York, 232pp.

**Klemperer**, S.L. 1987 "REFLECTIVITY OF THE CRYSTALLINE CRUST: HYPOTHESES AND TESTS" Geophys. Jour. Roy. Astr. Soc. v89 #1 p217-222
        - Matthews, D.H. 1987 "IAPETUS SUTURE LOCATED BENEATH THE NORTH SEA BY BIRPS DEEP SEISMIC PROFILING" Geology v15 #3 p195-198

**Knoll**, A. H. 1991, "END OF THE PROTEROZOIC EON" Sci. Amer., Oct., 1991, p64-73

**Kolesnik**, U.N. 1965, "NEPHRITES OF SIBERIA" Novosibirsk: Nauka 105p.

**Kostopoulos**, D.K. 1991, "MELTING OF THE SHALLOW UPPER MANTLE: A NEW PERSPECTIVE" Jour. Petrol., v32, p671-699

**Kostrovitsky**, S.I., Egorov, K.N. 1982, "A METHOD OF FORMING THE CONDUITS OF KIMBERLITE PIPES" Volcanology and Seismology, #1, p3-12

**Krauskopf**, K.B. 1971, "THE SOURCE OF ORE METALS" Geochim. Cosmochim. Acta, v35, p643-659

**Kravtsov**, A.I. et al. 1981, "THE YAKUTIA DIAMOND PROVINCE" Int. Geol. Rev. v23, p1179-1182

**Kreston**, P. 1988, "GRANITIZATION - FACT OR FICTION?" Geol. Foren. Stockholm Forh., v110, p335-340

**Kudo**, A.M., Aoki, K., Brooking, D.G. 1971, "THE ORIGIN OF PLIOCENE-HOLOCENE BASALTS OF NEW MEXICO IN THE LIGHT OF STRONTIUM ISOTOPIC AND MAJOR ELEMENT ABUNDANCES" Earth Planet. Sci. Let., v13, p200-204

**Kuili**, J., Yong, Q. 1989, "COAL PETROLOGY AND ANOMALOUS COALIFICATION OF MIDDLE AND LATE PLEISTOCENE PEAT AND SOFT BROWN COAL FROM THE TENGSHONG BASIN, WESTERN YUNNAN, PRC" in Int. Jour. Coal Geol., Spec. Issue v13, #1-4, eds. P.C. Lyons & B. Alpern

**Labozetta**, 1916 - reference in Byerly 1987

**Laine**, R., Alonso, D., Svab, M., Eds. 1985, "THE CARSWELL STRUCTURE URANIUM DEPOSITS, SASKATCHEWAN" Geol. Assoc. Can., Spec. Pap. #29

**Lapin**, A.V., Marshintsev, V.K. 1984, "CARBONATITES AND KIMBERLITE CARBONATITES" In Geol. of Ore Deposits, v26, #3, p28-42

**Larin**, V.N., 1980 "HYPOTHESIS OF THE PRIMORDIAL HYDRIDE EARTH" Nedra Publishers, 103633 Moscow K-12, reviewed in AAPG v70/11\p1756-1757
        - 1991 "THE EARTH: ITS COMPOSITION, STRUCTURE, AND EVOLUTION" unpublished manuscript
        - 1992 "RIFT ZONES AS AN INEXHAUSTIBLE SOURCE OF HYDROGEN ON EARTH" unpublished manuscript, 3pp

**Larsen**, J. 1985, "FROM LIGNIN TO COAL IN A YEAR" Nature, v314, p316

**La Tour**, T.E. 1987, "GEOCHEMICAL MODEL FOR THE SYMPLECTIC FORMATION OF MYRMEKITE DURING AMPHIBOLITE GRADE PROGRESSIVE MYLONITIZATION OF GRANITE" GSA Abstr, Programs, v19 (7), p741.
        -Barnett, R.L. 1987, "MINERALOGICAL CHANGES ACCOMPANYING MYLONITIZATION IN THE BITTERROOT DOME OF THE IDAHO BATHOLITH: IMPLICATIONS FOR TIMING OF DEFORMATION" GSA Bull., v98, p356-363
        -Thomson, M.L. 1988, "MYRMEKITE AS A TRANSITIONAL STAGE OF K-FELDSPAR BREAKDOWN" In Kallend, J.S., & Gottstein, G., eds., "Eighth Intern. Conf. on Textures of Materials" (ICOTOM 8): The Metallurgical Society: 817

**Laughlin**, A.W., Brookins, D.G., Carden, J.R. 1972, "VARIATION IN THE INITIAL STRONTIUM RATIOS OF A SINGLE BASALT FLOW" Earth Planet. Sci. Let., v14, p79-82

**Leake**, B.E. 1990, "GRANITE MAGMAS: THEIR SOURCES, INITIATION AND CONSEQUENCES OF EMPLACEMENT" J. Geol. Soc. London, v147, p579-589

**Lebedev**, V.M., Staroseltsev, V.S. 1972, "CONTOURS ON THE TRIASSIC IN THE TUNGUSS SYNECLISE" In Contributions to the Platform Regions of Siberiasc, Novosibirsk, Nauka: p29-32

**Lee**, D.L., Van Loenen, R.E. 1971, "HYBRID GRANITOID ROCKS OF THE SOUTHERN SNAKE RANGE, NEVADA" USGS Prof. Pap. 668, 48p

**Leeman**, W.P. 1970, "THE ISOTOPIC COMPOSITION OF STRONTIUM IN LATE CENOZOIC BASALTS FROM THE BASIN-RANGE PROVINCE, WESTERN UNITED STATES" Geochim. Cosmochim. Acta, v34, p857-872
        -1974, "LATE CENOZOIC ALKALI-RICH BASALT FROM THE WESTERN GRAND CANYON AREA, UTAH AND ARIZONA: ISOTOPIC COMPOSITION OF STRONTIUM" GSA Bull., v85, p1691-1696

**Leung**, I., Giro, W., Friedman, I, Gleason, J. 1990, "NATURAL OCCURRENCE OF SiC IN A DIAMONDIFEROUS KIMBERLITE FROM FUXIAN [CHINA]" Nature, v346, #6282, p352-354, 7/90

**Levinson-Lessing**, F.Yu. 1949, "TRAPPS OF THE TULIN-UDA AND BRATSK REGION OF EAST SIBERIA" Selected works, v1, Moscow-Leningrad, Acad. Sci. USSR, p228-253

**Li**, X, Jeannioz, R. 1991a, "PHASES AND ELECTRICAL CONDUCTIVITY OF A HYDROUS SILICATE ASSEMBLAGE AT LOWER-MANTLE CONDITIONS" Nature, v350, p332-334, also in 1991b J. Geophys. Res., v96, p6113-6120

**Litke**, R., Haven, H.L.T. 1989, "PALEOECOLOGIC TRENDS AND PETROLEUM POTENTIAL OF UPPER CARBONIFEROUS COAL SEAMS OF WEST GERMANY AS REVEALED BY THEIR PETROGRAPHY AND ORGANIC GEOCHEMICAL CHARACTER" Int. Jour. Coal Geol., Spec. Issue v13,#1-4, Eds. P.C.Lyons & B.Alpern

**Liu**, I. 1979, "PHASE TRANSFORMATION AND THE CONSTITUTION OF THE DEEP MANTLE" in McElhinny, M.W., ed., The Earth in Origin, Structure, and Evolution, London, Academic Press, p177-202

**Louie**, J.N., Clayton, R.W. 1987, "CONSTRAINTS ON THE PHYSICAL NATURE OF DEEP CRUSTAL STRUCTURES IN SOUTHEASTERN CALIFORNIA FROM THE CALCRUST SEISMIC REFLECTION EXPERIMENT" USGS Circ. #956 p48-50

**Lowman**, P. D., Jr. 1985 "MECHANICAL OBSTACLES TO THE MOVEMENT OF CONTINENT-BEARING PLATES" Geoph. Res. Lett. v12,#5,p223-225
          -1989, "COMPARATIVE PLANETOLOGY AND THE ORIGIN OF CONTINENTAL CRUST" Precamb. Res., v44, p171-195
          -1990, "ASTROPHYSICS, PLANETOLOGY, AND GEOLOGY: THREE APPROACHES TO THE ORIGIN OF CONTINENTAL CRUST" unpubl. ms. with personal communication, 17pp.

**Lukjanova**, L.I., Smirnov, J.D., Zilberman, A.M., Chernishova, E.N., Mikhailovskaya 1978, "DIAMOND DISCOVERY IN PICRITES OF THE URALS" Rept. of VMO, pt 107, #5, p580-585

**Lund**, C.E., Heikkinen, P. 1987, "REFLECTION MEASUREMENTS ALONG THE EGT POLAR PROFILE, NORTHERN BALTIC SHIELD" Geoph. Jour. Roy. Astr. Soc., v89, #1. p361-364

**Lutz**, B.T. 1965, "REACTIONS OF ECLOGITIZATION IN DEEP-SEATED ROCKS" Geol. of Ore Deposits, #5, p18-30

**Lyons**, P.C. et al. 1989, "CHEMICAL ORIGIN OF MINOR AND TRACE ELEMENTS IN VITRINITE CONCENTRATIONS FROM A RANK SERIES FROM THE EASTERN UNITED STATES, ENGLAND, AND AUSTRALIA" in Int. Jour. Coal Geol., Spec. Issue v13, #1-4, Eds. P.C. Lyons & B. Alpern

**Macdougall**, J.D., ed. 1988, "CONTINENTAL FLOOD BASALTS" Kluver Academic Publ.

**Makhotko**, 1981 - ref. unavailable

**Malkov**, A.B. 1975, "CARBONATITE-KIMBERLITES: A NEW TYPE OF DIAMOND-BEARING ROCKS" USSR Acad. Sci. v221, #5, p580-585

**Markovsky**, B.A., Rothman, V.K. 1981, "GEOLOGY AND PETROLOGY OF ULTRAMAFIC VOLCANISM" Leningrad, Nedra 247pp

**Markovsky**, A. P. 1975, "GEOLOGICAL MAP OF EURASIA" Moscow, NPO Aerogeologia

**Marshintsev**, V.K. 1970, "EXPLOSIVE CARBONATITE BRECCIA OF THE EAST SLOPE OF THE ANABAR ANTICLISE" In Geol., Petrogr., & Mineralogy of Magmatic Formations of the Northeastern Siberian Platform - Moscow, Nauka, p129-169

**Martin,** B.D. 1991 "CONSTRAINTS TO MAJOR RIGHT-LATERAL MOVEMENTS, SAN ANDREAS FAULT SYSTEM, CENTRAL AND NORTHERN CALIFORNIA" in press, in Chatterjee, S., and Hotton, N., Eds. "Surge tectonics, a new hypothesis of global tectonics" Texas Tech. Univ. Press

**Martjanov**, N.E. 1968, "ENERGY OF THE EARTH" Novosibirsk, 84pp

**McGraw Hill** 1982, "ENCYCLOPEDIA OF SCIENCE AND TECHNOLOGY" 5th edition

**Meier**, H., Hecker, W. 1976, "RADIOACTIVE HALOS AS POSSIBLE INDICATORS FOR GEOCHEMICAL PROCESSES IN MAGMATITES" Geochem. Jour., v10, p185-195.

**Meissner**, R., Wever, T.H., Bittner, R. 1987, " RESULTS OF DECORP 2-S AND OTHER REFLECTION PROFILES THROUGH THE VARISCIDES" Geophys. Jour. Roy. Astr. Soc. v89 #1 p319-324

**Meltzer**, A.S., Levander, A.R. 1991, "DEEP SEISMIC PROFILING OFFSHORE SOUTHERN CALIFORNIA" Jour.Geoph. Res. v96,#B4, p6475-6491

**Menyajlov**, A.A. 1963, "POSTMAGMATIC MINERALIZATION AND ZONATION IN DIAMOND DEPOSITS" In Problems in Postmagmatic Mineralization - v1 Prague, p343-347

**Metsger**. R.W., Tennant, C.B., Rodda, J.L. 1958, "GEOCHEMISTRY OF THE STERLING HILL ZINC DEPOSITS, SUSSEX COUNTY, NEW JERSEY" GSA Bull., v69, p775-788

**Meyster**, A.K. 1908, "GEOLOGIC MAP OF ENISEY GOLD DISTRICT" -S-P 147pp

**Michael**, A.J., Ellsworth, W.L., Oppenheimer, D.H. 1990, "COSEISMIC STRESS CHANGES INDUCED BY THE 1989 LOMA PRIETA, CALIFORNIA, EARTHQUAKE" Geoph. Res. Lett., v17, #9, p1441-1444

**Michel-Levy**, A.M. 1874, "STRUCTURE MICROSCOPIQUE DES ROCHES ACIDES ANCIENNES" Soc. Fr. Mineral. Crystallogr. Bull., v3, p201-222

**Milashev**, V.A. 1972, "PHYSICAL CONDITIONS OF KIMBERLITE FORMATION" Leningrad: Nedra 104pp

**Miller**, C.F., and Bradfish, L.J. 1980, "AN INNER CORDILLERAN BELT OF MUSCOVITE-BEARING PLUTONS" Geology, v8, p412-416

**Mishina**, A. V. 1978, "GEOLOGICAL MAP OF THE ZOND AND PHILIPPINE ISLANDS AND ADJOINING TERRITORIES" Moscow, VNIIZarubezgeologia

**Miyashiro**, A., Aki, K, Sengor, A. 1985 "OROGENY" Mir, Moscow, 390pp

**Moazed**, C., SPECTOR, R.M., WARD, R.F. 1973, "POLONIUM RADIOHALOS: AN ALTERNATIVE INTERPRETATION" Science, v180, p1271-1274.

**Moench**, R.H., Harrison, J.E., Sims, P.K. 1962, "PRECAMBRIAN FOLDING IN THE IDAHO SPRINGS-CENTRAL CITY AREA, FRONT RANGE, COLORADO" Geol. Soc. America Bull. 73: 35-58.

Mooney, W.D., Ginzburg, A. 1986 "SEISMIC MEASUREMENTS OF THE INTERNAL PROPERTIES OF FAULT ZONES" In Internal Structure of Fault Zones, Ed. Chi-yuen Wang, Pure & Appl. Geoph., v124, #1/2 Birkhauser Verlag, p141-155

-Brocher, T.N., 1987 "COINCIDENT SEISMIC REFLECTION/REFRACTION STUDIES OF THE CONTINENTAL LITHOSPHERE, A GLOBAL REFIEW" Geophys. Jour. Roy. Astr. Soc. v89, #1, p1-6; also in Rev. Geophys. v25, #4, p723-742

Moore, B. J., Sigler, S. 1985, "ANALYSES OF NATURAL GAS" US Bur of Mines Info. Circ. #9129

Moore, D.E., 1987, "SYNDEFORMATIONAL METAMORPHIC MYRMEKITE IN GRANODIORITE OF THE SIERRA NEVADA, CALIFORNIA" GSA Abstr.Programs, v19 (7), p776

Moore, J.G. 1959, THE QUARTZ DIORITE BOUNDARY LINE IN THE WESTERN UNITED STATES" J. Geol. v67, p198-210

-Grantz, A., Blake, M.C., Jr. 1963, "THE QUARTZ DIORITE LINE IN NORTH-WESTERN NORTH AMERICA" USGS Prof. Pap. 450E, p89-93

Moore, P.D. 1987, "COAL AND COAL-BEARING STRATA: RECENT ADVANCES AND FUTURE PROSPECTS" Blackwell Scientific Publ. Inc.

Moore, R. 1933, "HISTORICAL GEOLOGY" McGraw-Hill, p242

Moorhouse, W.W. 1956, "THE PARAGENESIS OF ACCESSORY MINERALS" Econ. Geol., v51, p248-262

Morin, J.A. Turnock, A.C. 1975, "THE CLOTTY GRANITE AT PERRAULT FALLS, ONTARIO, CANADA" Can. Mineral., v13, p352-357

Mozley, E.C., Goldstein, N.E., Morrison, H.F., 1986 "MAGNETOTELLURIC INVESTIGATIONS AT MOUNT HOOD, OREGON" Jour. Geophys. Res. B91 #11 p11596-11610

Mueller, R.J., Johnston, M.J.S., Langbein, J.O. 1991, "POSSIBLE TECTONOMAGNETIC EFFECT OBSERVED FROM MID-1989 TO MID-1990 IN THE LONG VALLEY CALDERA, CALIFORNIA" Geoph. Res. Lett. v18, #4, p601-604

Munksgaard, N.C., 1988, "SOURCE OF THE COOMA GRANODIORITE, NEW SOUTH WALES A POSSIBLE ROLE OF FLUID-ROCK INTERACTIONS" Australian J. Earth Sci. v35, p363-377

Navon, O., 1991 "HIGH INTERNAL PRESSURES IN DIAMOND FLUID INCLUSIONS DETERMINED BY INFRARED ABSORPTION" Nature v353, p746-748

Nelson, C.A., Sylvester, A.G. 1971, "WALL ROCK DECARBONATION AND FORCIBLE EMPLACEMENT OF BIRCH CREEK PLUTON, SOUTHERN WHITE MOUNTAINS, CALIFORNIA" GSA Bull. v82, p2891-2904

Nemilova, A.V. 1956, "SYNTHESIS OF DIAMONDS" Repts. VMO v85, #2, p202-204

**Nikishov**, K.N., Zolnikov, G.V. Safronov, A.F., Kornilova, V.P., Makhotko, V.F. 1979, "PECULIAR COMPOSITION OF GARNETS, OVIVINES, CHROME SPINELS, AND ILMENITES OF A YAKUTIYA PIPE" In Mineralogy and Geochem. of Kimberlite and Trapp Rocks - Yakutsk, USSR Acad. Sci. p52-61

**Offman**, P.E. 1959, "TECTONICS AND VOLCANIC PIPES OF THE CENTRAL SIBERIAN PLATFORM" In Tectonics of the USSR, v4, Moscow Acad. Sci. p5-344
    -1981, "OIL AND GAS GEOLOGY OF THE SIBERIAN PLATFORM" Moscow, Nedra, 552pp

**Oldham** - reference unavailable

**Omory,** 1905 - reference in Byerly 1987

**Oppenheimer**, D.H. 1990, "AFTERSHOCK SLIP BEHAVIOR OF THE 1989 LOMA PRIETA, CALIFORNIA EARTHQUAKE" Geoph. Res. Lett. v17, #8, p1199-1202

**Parrot**, M. 1990, "ELECTROMAGNETIC DISTURBANCES ASSOCIATED WITH EARTHQUAKES: AN ANALYSIS OF GROUND-BASED AND SATELLITE DATA" Jour. Sci. Exploration, v4, #2, p203-212

**Pennisi**, E. 1991, "BUCKYBALLS STILL CHARM" Science News, v140, p120-123

**Peterman**, Z.E., Carmichael, I.S.E. Smith, A.L., 1970, "STRONTIUM ISOTOPES IN QUATERNARY BASALTS OF SOUTHEASTERN CALIFORNIA" Earth Planet. Sci. Let., v7, p381-384

**Petraschek**, F. 1931, "THE FORERUNNERS OF COAL" In Handbuch der Mineralchemie, Bd.4, T.3, p16

**Phillips**, E.R. 1974, "MYRMEKITE - ONE HUNDRED YEARS LATER" Lithos, v7, p181-194
    -1980, "ON POLYGENETIC MYRMEKITE" Geol. Mag., v117, p29-36
    -RANSOM, D.M., 1970, "MYRMEKITIC AND NON-MYRMEKITIC PLAGIOCLASE COMPOSITIONS IN GNEISSES FROM BROKEN HILL, NEW SOUTH WALES" Min. Mag. v37, p729-732
    -VERNON, R.H. 1972, MYRMEKITE AND MUSCOVITE DEVELOPED BY RETROGRADE METAMORPHISM AT BROKEN HILL, NEW SOUTH WALES" Min. Mag. v38, p570-578

**Pinet**, B., Montadert, L., 1987, Geophys. Jour. Roy. Astr. Soc. v89 #1
    - "DEEP SEISMIC REFLECTION AND REFRACTION PROFILING ALONG THE AQUITAINE SHELF" p305-312
    - "PRE-DRILLING REFLECTION SURVEY OF THE BLACK FOREST, SW GERMANY" p325-332
    - "RESULTS OF DEEP SEISMIC PROFILING IN THE OBERFALZ (BAVARIA)" p353-360

**Pitcher**, W.S. 1953, "THE MIGMATITIC OLDER GRANODIORITE OF THORR DISTRICT, CO. DONEGAL" Quart. J. Geol. Soc. London, v108, p413-446
    - et al. 1959, "THE MAIN DONEGAL GRANITE" Quart. Jour. Geol. Soc. London, v114, p259-305
    - Berger, A.R. 1972, "THE GEOLOGY OF DONEGAL: A STUDY OF GRANITE EMPLACEMENT AND UNROOFING" New York, Wiley Intersci. 435p
    - Hutton, D.H.W. 1982, "DISCUSSION ON A TECTONIC MODEL FOR THE EMPLACEMENT OF THE MAIN DONEGAL GRANITE, N. W. IRELAND" J. Geol. Soc. London, v141, p599-602
    - Hutton, D.H.W., Atherton, M.P., Stephens, E. 1987, "GUIDE TO THE GRANITES OF DONEGAL" from symposium on the 'Origin of Granite', Edinburgh, Scotland, Sept. 14-16, 1987

**Ponomarenko,** A.I., Pankratov, A.A., Poberezjksky, V.A. 1970 "INDICATIONS OF KIMBERLITE MAGMATISM ON THE SOUTH FLANK OF THE ANABAR SHIELD" In Geol., petrogr.,and mineralogy of magmatic formations on the northeast portion of the Siberian shield" Moscow, Nauka, p33-47

**Price,** R.A., Mountjoy, E.W. 1970, "THE GEOLOGICAL STRUCTURE OF THE CNADIAN ROCKY MOUNTAINS BETWEEN THE BOW AND ATHABASCA RIVERS" Geol. Assoc. of Can., Pap. 6:7-25 Special Issue, Feb.

**Pushkar,** P., Condie, K.C. 1973, "ORIGIN OF THE QUATERNARY BASALTS FROM THE BLACK ROCK DESERT REGION, UTAH: STRONTIUM ISOTOPIC    EVIDENCE" GSA Bull., v84, p1053-1058

**Quirke,** T.T. 1927, "KILLARNEY GNEISSES AND MIGMATITES" GSA Bull., v38, p753-770
    -1940, "GRANITIZATION NEAR KILLARNEY, ONTARIO" GSA Bull. v51, p237-254

**Ramaekers,** P. 1990, "GEOLOGY OF THE ATHABASCA GROUP (HELIKIAN) IN NORTHERN SASKATCHEWAN" Sask. energy & mines, report 195

**Rankama,** K. 1963, "PROGRESS IN ISOTOPE GEOLOGY" New York, Interscience Publ., 705p

**Reynolds,** D.L. 1947, "THE ASSOCIATION OF BASIC "FRONTS" WITH GRANITIZATION" Sci. Prog., v35, p205-219

**Rich,** V.I., Chernenko, M.V. 1976, "UNFINISHED HISTORY OF ARTIFICIAL DIAMONDS" Moscow, Nauka 137pp

**Richter,** C. F. 1963, "ELEMENTARY SEISMOLOGY" Moscow, Mir, 670pp

**Rikitake,** T. 1979, "EARTHQUAKE PREDICTION" Moscow, Mir, 370pp

**Ringwood,** A.E. 1975, "COMPOSITION AND PETROLOGY OF THE EARTH'S MANTLE" New York, McGraw-Hill, 677pp
        -1991, "PHASE TRANSFORMATIONS AND THEIR BEARING ON THE CONSTITUTION AND DYNAMICS OF THE MANTLE" Geochem. et Cosmica Acta. v55, p2083-2110

**Robertson,** F.S. 1962, "CRYSTALLIZATION SEQUENCE OF MINERALS LEADING TO FORMATION OF ORE DEPOSITS IN QUARTZ MONZONITE ROCKS IN THE NORTHWESTERN PART OF THE BOULDER BATHOLITH, MONTANA" GSA Bull., v73, p1257-1276

**Rock,** N.M.S., Bowes, D.R., Wright, A.E. 1991, "LAMPROPHYRES" New York, Van Nostrand Reinhold, 285pp

**Roddick,** J.A. 1982, "ON GRANITE LOGIC" in Drescher-Kaden, F.K. & Augustithis, S.S., eds., "Transformists' Petrology," Athens, Theophrastus Publ., p87-104

**Rosen,** O.M. 1972, "ON THE QUESTION OF THE ORIGIN OF ECLOGITES" Acad. Sci. USSR, v203, #3, p674-676

**Rotstein,** Y, Yuval, Z., Trachtman, P. 1987, "DEEP SEISMIC REFLECTION STUDIES IN ISRAEL" Geophys. Jour. Roy. Astr. Soc. v89 #1 p389-394

**Rubey,** W.W. 1951, "THE GEOLOGIC HISTORY OF SEA WATER" GSA Bull., v62, p1111-1147

**Ruff**, L.J., Given, J.W., Sanders, C.O., Sperber, C.M. 1989, "LARGE EARTHQUAKES IN THE MACQUARRIE RIDGE COMPLEX:TRANSITIONAL TECTONICS AND SUBDUCTION INITIATION" Pure & Appl.Geophsics, v29, p71-129

**Ryzhov**, Y.K., Mordovskaya, T.V. 1979, "NEW DATA ON TRAPP MAGMATISM ON THE SOUTH SIBERIAN PLATFORM" Geol. & Geophys. #4, p139-141

**Salle**, C. 1987, "LE PROGRAMME ECORS" Ann. Mines v194, #4-5, p86-102

**Savage**, J.C., Liswoski, M., Prescott, W.H. 1990, "AN APPARENT SHEAR ZONE TRENDING NORTH-NORTHWEST ACROSS THE MOJAVE DESERT INTO OWENS VALLEY, EASTERN CALIFORNIA" Geoph. Res. Lett. v17, #12, p2113-2116

**Scheidegger**, A. E. 1958, "PRINCIPLES OF GEODYNAMICS" Springer-Verlag 280pp

**Scholz**, C.H. 1969 "WORLDWIDE DISTRIBUTION OF EARTHQUAKES" Nature, v221, p165

**Sharov**, 1987 - Reference not available

**Sheppard**, R.A. 1986, "FIELD TRIP GUIDE TO THE DURKEE ZEOLITE DEPOSIT, DURKEE, OREGON" Oregon Geol., v48, #11

**Shestopalov**, M.F. 1938, "ULTRAMAFIC SILLS AND RELATED DEPOSITS OF THE KITOYSKY ALPS, EAST SAYAN" In About Precious Stones, Moscow-Leningrad, p84-99

**Shilo**, N.A., Kaminsky, F.V., Palandzyan, S.A., Tilman, S.N., Tkachenko, L.A., Lavrova, L.D., Shepeleva, K.A. 1978, "THE FIRST DIAMONDS FOUND IN ALPINE-TYPE ULTRAMAFICS IN NORTHEASTERN USSR" Acad. Sci. v241, #4, p933-936
    -Kaminsky, F.V., Lavrova, L.D., Dolmatov, B.K., Pleshakov, A.P., Tkachenko, L.A., Shepeleva, K.A. 1979, "THE FIRST DIAMONDS FOUND IN ULTRAMAFICS IN KAMCHATKA" USSR Acad. Sci. v248, #5, p1211-1214

**Simandl**, G.J., Hancock, K.D. 1991, "GEOLOGY OF THE MOUNT BRUSSILOF MAGNESITE DEPOSIT, SOUTHEASTERN BRITISH COLUMBIA" B.C. Ministry Geol. Field Work, Pap. 1991-1, p269-278
    -Hora, Z.D. 1991, "GEOLOGY OF CARBONATE-HOSTED MAGNESITE DEPOSITS, MOUNT BRUSSILOF AREA, SOUTHEASTERN BRITISH COLUMBIA" Abstr. Forum '91, Industrial Mineral Conference, Banff, Alberta, May 5 - 10, 1991

**Simpson**, C. 1984, "BORREGO SPRINGS-SANTA ROSA MYLONITE ZONE: A LATE CRETACEOUS WEST-DIRECTED THRUST IN SOUTHERN CALIFORNIA" Geology, v12, p8-11
    -1985, "DEFORMATION OF GRANITIC ROCKS ACROSS THE BRITTLE-DUCTILE TRANSITION" J. Struct. Geol., v7, p503-511
    -& Wintsch, R.P. 1989, "EVIDENCE FOR DEFORMATION-INDUCED REPLACEMENT OF K-FELDSPAR BY MYRMEKITE" Jour. Metamorph. Petrol., v7, p261-275

**Sims**, J. D. 1989 "CHRONOLOGY OF DISPLACEWMENT ON THE SAN ANDREAS FAULT..." USGS Open File Rep. 89-571

**Sims**, P.K. 1958, "GEOLOGY AND MAGNETITE DEPOSITS OF DOVER DISTRICT, MORRIS COUNTY, NEW JERSEY" USGS Prof. Pap. 287,162pp
    -and Gable, D.J. 1967, "PETROLOGY AND STRUCTURE OF PRECAMBRIAN ROCKS, CENTRAL CITY QUADRANGLE, COLORADO" U. S. Geol. Surv. Prof. Paper 554-E, 56pp

**Skobelin**, E. A., , Sharapov, I. P., Bugayov, A.F. 1990, DELIBERATIONS OF STATE AND WAYS OF PERESTROIKA IN GEOLOGY (HAS PLATE TECTONICS RESULTED IN A REVOLUTION IN GEOLOGY?)" Critical Aspects of the Plate Tectonics Theory, vol 1, p17-37, Theophrastus Publications, Athens
- 1991, "THEORY OF MAGMATISM AND ENDOGENIC MINERALIZATION" unpubl. manuscript

**Slodkevich**, V.V. date? "PARAMORPHS OF GRAPHITE BY DIAMOND" Repts. of VMO Pt.III, #1, p13-33

**Smedes**, H.W. 1965, "GEOLOGY AND IGNEOUS PETROLOGY OF THE NORTHERN ELKHORN MOUNTAINS, JEFFERSON AND BROADWATER COUNTIES, MONTANA" USGS Prof. Pap. 510, 116pp

**Smith**, D. 1991,"BUT WILL IT WORK ON A HIBACHI?: THE WORK OF PROFESSOR DAVID GOODWIN" Engineering and Science, Winter, 1991, p40-42

**Smith**, R.I.L., Clymo, R.S. 1984,"AN EXTRAORDINARY PEAT-FORMING COMMUNITY ON THE FALKLAND ISLANDS" Nature, v309, p617

**Smith**, R.L. 1979, "ASH-FLOW MAGMATISM" GSA Spec. Pap. 180, p5-27.

**Sobolev**, V.S. 1951,"GEOL. OF DIAMOND DEPOSITS OF AFRICA, AUSTRALIA, BORNEO ISLAND, AND NORTH AMERICA" Moscow Gosgeolizdat 126p

**Sobolev**, N.V., Botkunov, A.I., Lavrentjev, Y.G., Usova, L.V. 1976,"NEW DATA ON THE MINERALOGY ASSOCIATED WITH DIAMONDS IN THE MIR PIPE" Geol. & Geophys. #12,p3-15

**Sologub**, V.B., Chekunov, A.V. 1981,"EARTH'S CRUST IN CENTRAL AND SOUTHEASTERN EUROPE" In geol. and geochem. fundamentals of oil and gas prospecting, Kiev, Naukova Dumka, p5-23

**Solonenko**, V.P. 1958,"ORIGIN AND CLASSIFICATION OF GRAPHITE DEPOSITS" Proc. of Siberian Br. USSR Acad. Sci. #5,p12-18

**Somwill**, 1925 reference in Byerly 1987

**Stanton**, R.L. 1972, "ORE PETROLOGY" New York, McGraw-Hill, 713pp.

**Staroseltsev**, V. S. 1981 "TECTONICS OF BASALTIC PLATEAUS: RELATIVE TO EVALUATING THE HYDROCARBON POTENTIALS OF ANCIENT PLATFORMS" in Neftegazonosost' Sibiri i Dal'nego Vostoka, Ed. V. S. Surkov, 513pp

**Stel**, H., Breddveld, M. 1990, "CRYSTALLOGRAPHIC ORIENTATION PATTERNS OF MYRMEKITIC QUARTZ: A FABRIC MEMORY IN QUARTZ-BEARING GNEISSES" J. Struct. Geol., v12, p19-28

**Subhasissen**, 1983, "RECENT DEVELOPMENTS IN GLOBAL TECTONICS" Metals and Mines Review, v22, #10, p229-231

**Suzuki**, T., Akimoto, S., Yagi, T. 1989, "METAL-SILICATE-WATER REACTION UNDER HIGH PRESSURE.I. FORMATION OF METAL HYDRIDE AND IMPLICATIONS FOR COMPOSITION OF THE CORE AND MANTLE" Physics of the Earth and Planetary Interiors, v56, p377-388

**Swanson**, S.E. 1977, "RELATION OF NUCLEATION AND CRYSTAL-GROWTH RATE TO THE DEVELOPMENT OF GRANITIC TEXTURES" Amer. Min. v62, p966-978

**Sylvester**, A.G. 1978, "PAPOOSE FLAT PLUTON: A GRANITIC BLISTER IN THE INYO MOUNTAINS, CALIFORNIA" GSA Bull. v89, p1205-1219.
    -Miller, C.F., Nelson, C.A. 1978, "MONZONITES OF THE WHITE-INYO RANGE, CALIFORNIA, AND THEIR RELATION TO THE CALC-ALKALIC SIERRA NEVADA BATHOLITH" GSA Bull. v89, p1677-1687.

**Tarling**, D.H. 1981, "MAGMATIC SNAP, CRACKLE AND POP" Nature, v291, p108-109

**Taylor**, B.E., Foord, E.E., Friedrichsen, H. 1979, "STABLE ISOTOPE AND FLUID INCLUSION STUDIES OF GEM-BEARING GRANITIC PEGMATITE-APLITE DIKES, SAN DIEGO CO., CALIFORNIA" Contrib. Miner. Petrol. v19, p1-71

**Taylor**, H.P., Jr. 1968, "THE OXYGEN ISOTOPE GEOCHEMISTRY OF IGNEOUS ROCKS" Contrib. Min. Petrol., v19, p1-71
    -EPSTEIN, S., 1962, "RELATIONSHIP BETWEEN $^{18}O/^{16}O$ RATIOS IN COEXISTING MINERALS OF IGNEOUS AND METAMORPHIC ROCKS" Earth Planet. Sci. Lett., v47, p243-254

**Thompson**, G.A. 1987, "NEW VIEW OF THE CONTINENTAL CRUST FROM SEISMIC REFLECTION PROFILING" USGS Circ. #956, p44-45

**Thompson**, D'Arcy 1917 "ON GROWTH AND FORM" Reprinted, two vol., 1942, Cambridge Univ. Press

**Thompson**, G.A., 1987 "A NEW VIEW OF THE CONTINENTAL CRUST FROM SEISMIC REFLECTION PROFILING" USGS Circ. #956 p44-45

**Tichelar**, B.W., Ruff, L.J. 1990, "RUPTURE PROCESS AND STRESS-DROP OF THE GREAT 1989 MACQUARRIE RIDGE EARTHQUAKE" Geoph Res. Lett. v17, #7, p1001-1004

**Toksoz**, M.N. 1975, "THE SUBDUCTION OF THE LITHOSPHERE" In Volcanos and the Earth's Interior, Publ. Sci. Amer., p6-16

**Trehu**, A.M., Wheeler, W.H., 1987 "POSSIBLE EVIDENCE FOR SUBDUCTED SEDIMENTARY MATERIALS BENEATH CENTRAL CALIFORNIA" Geology, v15 #3 p254-258

**Tripol'skiy**, 1987 - Reference not available

**Trofimov**, G.N. 1939, "DIAMOND LODES IN NON-KIMBERLITE HOST ROCKS" Soviet Geol., #4-5, p40-59
    -1980, "GEOL. OF NATURAL DIAMOND DEPOSITS" Moscow, Nedra 304p
    -1967, "THE PATTERN OF OCCURRENCE AND ORIGIN OF DIAMOND DEPOSITS ON THE OLD PLATFORMS AND GEOSYNCLINAL REGIONS" Moscow, Nedra 300p

**Trusova**, I.F. 1956, "PARAGENETIC ANALYSIS OF ARGILLITES OF LOWER ARCHEAN OF KOKCHETAV MASSIF" Soviet Geol., #5, p45-74

**Turekian**, K. K. & Wedepohl, K. H. 1961 "DISTRIBUTION OF THE ELEMENTS IN SOME MAJOR UNITS OF THE EARTH'S CRUST" GSA Bull v72, p175-191

**Tuttle**, O.F., & BOWEN, N.L., 1958, "ORIGIN OF GRANITE IN THE LIGHT OF EXPERIMENTAL STUDIES IN THE SYSTEM $NaAlSi_3O_8 \cdot KAlSi_3O_8 \cdot SiO_2 \cdot H_2O$" GSA Mem. 74, 153pp.

**Tuyezov**, I.K. 1987, "GEOELECTRIC MODEL ON A PROFILE FROM LAKE BAIKAL TO THE PACIFIC OCEAN" Pacific Geology, #2 p125-127

**Valasek**, P.A., Hawman, R.B., Johnson, R.A., Smithson, S.B. 1987, "THE NATURE OF THE LOWER CRUST AND MOHO IN EASTERN NEVADA FROM WIDE-ANGLE REFLECTION MEASUREMENTS" Geophys. Res. Lett. v14 #11 p1111-1114

**Vanjan**, L.L., Shilovsky, A.P., Okulessky, B.A., Semjonov, V.Ju., Sidelnikova, T.A. 1987, "ELECTRICAL CONDUCTIVITY OF EARTH'S CRUST, EASTERN SIBERIAN PLATFORM" News of Akad. Ski. USSR, in Physics of the Earth, #6, p83-93

**Varshavsky**, A.V. 1981, "ETCHEBILITY OF DIAMONDS*" (*Ed. Note: The translation of this title may be inaccurate.) In Physical Properties and Mineralogy of Diamond, Yakutsk, p5-16

**Vasco**, D.W., Smith, R.B., Taylor, C.L. 1990, "INVERSION FOR SOURCES OF CRUSTAL DEFORMATION AND GRAVITY CHANGE AT THE YELLOWSTONE CALDERA" Jour. Geoph. Res., v95, #B12, p19839-19856

**Vasiljev**, L.A., Belykh, Z.P. 1983, "DIAMONDS: THEIR PROPERTIES AND APPLICATIONS" Moscow, Nedra 102p

**Vernadsky**, V.I. 1959, In Selected Works, v2, Moscow, USSR Acad. Sci. 624p

**Vernon**, R.H. 1969, "THE WILLYAMA COMPLEX, BROKEN HILL AREA" Jour. Geol. Soc. Australia, v18, p267-277

**Vilke**, K.T. 1977, "GROWING CRYSTALS" Leningrad, Nedra, 600p

**Vlasov**, K.A. ed. 1966, "GEOCHEMISTRY AND MINERALOGY OF RARE ELEMENTS AND THE GENETIC TYPES OF THEIR DEPOSITS" Distrib. D. Davey & Co., Inc., NYC
-1986, "K-FELDSPAR MEGACRYSTS IN GRANITES PHENOCRYSTS, NOT PORPHYROBLASTS" Earth-Sci. Rev., v23, p1-63

**Wake-Dyster**, K.D., Sexton, M.J., Johnstone, D.W., Wright, C., Finlayson, D.M. 1987, "A DEEP SEISMIC PROFILE OF 800-KM LENGTH RECORDED IN SOUTHERN QUEENSLAND, AUSTRALIA" Geophys. Jour. Roy. Astr. Soc. v89 #1 p423-430

**Wakefield**, J.R. 1987-88, "GENTRY'S TINY MYSTERY - UNSUPPORTED BY GEOLOGY" Creation/Evolution, v22, p13-33
-1988, "THE GEOLOGY OF 'GENTRY'S TINY MYSTERY'" Jour. Geol. Educ., v36, p161-175

**Wakita**, H., Sano, Y., Urabe, A., Nakamura, Y. 1990, "ORIGIN OF METHANE-RICH NATURAL GAS IN JAPAN: FORMATION OF GAS FIELDS DUE TO LARGE-SCALE SUBMARINE VOLCANISM" Applied Geochem. v5, p263-278
-1983 "$^3He/^4He$ RATIOS IN $CH_4$-RICH NATURAL GASES SUGGEST MAGMATIC ORIGIN" Nature 305, 792-794
-Mizoue, M., 1987 "HIGH $^3He$ EMANATION AND SEISMIC SWARMS OBSERVED IN A NON-VOLCANIC FOREARC REGION" Jour. Geoph. Res., 92, 12539-12546

**Walker**, F., Poldervaart, A. 1950, "DOLERITES OF THE KAROO UNION OF SOUTH AFRICA" In Geol. & Petrogr. of Trapp Formation, Moscow, IL p8-182

**Wang**, C-Y., Rui, F., Zhengsheng, Y., Xignjue, S. 1986, "GRAVITY ANOMALY AND DENSITY STRUCTURE OF THE SAF ZONE" In Internal Structure of Fault Zones, ed. Chi-yuen Wang., Pure & Appl. Geoph., v124, #1/2, Birkhauser Verlag, p127-140

**Watson**, F.B. 1979, "ZIRCON SATURATION IN FELSIC LIQUIDS. EXPERIMENTAL RESULTS AND APPLICATIONS TO TRACE ELEMENTAL GEOCHEMISTRY" Contrib. Mineral. Petrol., v70, p407-419

**Weda**,  1980 Ref. unavailable
  -1989 "   "

**Wedepohl**, K.H. 1978, "HANDBOOK OF GEOCHEMISTRY" Berlin, Springer-Verlag (in five volumes)

**Weigand**, P. W., Thomas, J. M. 1989 "MIDDLE CENOZOIC VOLCANIC FIELDS ADJACENT TO THE SAN ANDREAS FAULT, CENTRAL CALIFORNIA: CORRELATION AND PETROGENESIS" South Coast Geol. Soc. Guidebk #17, v1, p207-222

**Wetherill**, G.S. 1956, "DISCORDANT URANIUM-LEAD AGES" Trans. Amer. Geophys. Union, v37, p320-326

**Whipple**, J.W. 1989, "MIDDLE PROTEROZOIC TECTONICS OF THE BELT BASIN" In IGC Field trip T334 guidebook, p43-46

**White**, A.J.R., Chappell, B.W. 1977, ULTRAMETASMORPHISM AND GRANITOID GENESIS" Tectonophysics, v43, p7-22

**White**, G.V., 1985 "LANG BAY GERMANIUM PROSPECT" B.C. Ministry of Energy, Geological Fieldwork, Pap. 1986-1

**Whitney**, J.A. 1989 "ORIGIN AND EVOLUTION OF SILICIC MAGMAS" in "Ore Deposition Associated with Magmas," eds. J.A. Whitney & Naldrett, A.J., Soc. Ec. Geol. Reviews in Economic Geology.

**Wiebe**, R.A. 1970, "RELATIONS OF GRANITIC AND GABBROIC ROCKS, NORTHERN SANTA LUCIA RANGE, CALIFORNIA" GSA Bull., v81, p105-116.

**Williams**, W.M. 1886, "BITUMINOUS SHALES AND THE ORIGIN OF COAL" Knowledge, v9, #111

**Wiman**, E. 1930, "STUDIES OF SOME ARCHAEAN ROCKS IN THE NEIGHBORHOOD OF UPPSALA, SWEDEN, AND THEIR GEOLOGIC POSITION" Bull. Geologie Inst. Universitet Uppsala, 23

**Winston**, D. 1989 "INTRODUCTION TO THE BELT" In IGC Field trip T334 guidebook, p1-6

**Wright**, W.I. 1938, "THE COMPOSITION AND OCCURRENCE OF GARNETS" Amer. Min., v23, p436-449

**Yegorokin**, et al. 1984 - Reference unavailable

**Yoder**, H. 1973, "THE FORMATION OF BASALT MAGMAS" Mir, 238pp

**Yoos**, T.R., Potter, C.J., Thigpen, J.L., Brown, L.D. 1991, "THE CORDILLERAN FORELAND THRUST BELT IN NORTHWESTERN MONTANA AND NORTHERN IDAHO FROM COCORP AND INDUSTRY SEISMIC RELECTION DATA" AAPG Bull. v75, #6, p1089-1106

**York**, 1979 - Reference unavailable

**Yoshino**, T. 1991, "LOW-FREQUENCY SEIMOGENIC ELECTROMAGNETIC EMISSIONS AS PRECURSORS TO EARTHQUAKES AND VOLCANIC ERUPTIONS IN JAPAN" Jour. of Sci. Explor. v5, #1, p121-144

**Young**, F.G.,1979, "THE LOWERMOST PALEOZOIC MCNAUGHTON FORMATION AND EQUIVALENT CARIBOO GROUP OF EASTERN BRITISH COLUMBIA" GSC Bull. 288

**Zayachkovsky**, 1972 - reference unavailable

**Zorin**, Y. A. 1972 "GEOPHYSICAL ASPECTS OF THE PLANATION OF EPOCHS OF PLANATION OF RELIEF AS ILLUSTRATED IN EASTERN SIBERIA" Geomorfologiya v2, p13-18

**Zysman**, M., 1990 "CATASTROPHISM 2000", Heretic Press, Toronto

# INDEX-ALPHABETICAL

# GLOSSARY

## A

**Alaskite**: a granitic rock containing only a few percent dark minerals.

**Almandine garnet**: $Fe_3A_{12}(SiO_4)_3$.

**Amphibolite**: a rock formed from a group of dark ferromagnesian silicate minerals.

**An$_{xy}$**: footnote, p89.

**Andalusite**: $Al_2SiO_5$.

**Anatexis**: partial melting of rock involving only certain minerals.

**Anorthite**: $CaAl_2Si_2O_8$.

**Anorthosite**: igneous rock composed almost entirely of plagioclase.

**Aplite**: fine-grained granitic rock.

**Arkose**: footnote p102.

**Asthenosphere**: a weak layer below Earth's lithosphere in which seismic waves are attenuated and magmas may be generated.

## B

**Bioherm (-al)**: colonial, mound-like growth of sedentary organisms.

**Biota**: all living organisms of an area considered as a unit.

**Biotite**: a mica, $K(Mg,Fe^{2+})_3(Al,Fe^{3+})Si_3O_{10}(OH)_2$.

**Blastic deformation** : dynamothermal metamorphism resulting in crystal elongation.

**Bond energy**: see p55.

## C

**Cataclasis**: rock deformation with fracture and rotation of mineral grains.

**Carbides**: a mineral formed of carbon and a metal; see p52.

**Charnockite**: a hypersthene-bearing granite.

**Clarain**: bright, sheet-like coal with a silky sheen.

**Clinoptilolite**: see p284;

$(Na_2O)_{0.70}\cdot(CaO)_{0.10}\cdot(K_2O)_{0.15}\cdot(MgO)_{0.05}\cdot Al_2O_3(SiO_2)_{8.5\text{-}10.5}\cdot6\text{-}7H_2O$

**Clinopyroxene**: a pyroxene crystallizing in the monoclinic system, often calcium-rich.

**Cordierite**: described on p100.

**Cymatogen**: a welt in the Crust.

## D

**Desert varnish**: an iron- and manganese-rich veneer that coats rock in desert climates.

**Desmite**: transparent coal residuum, usually high-grade.

**Diagenesis**, low PT changes in sediment after its deposition.

**Diapir**: upward-directed intrusion that pierces cover rock.

**Diorite**: plutonic rock intermediate between felsic and mafic composition.

**Durain**: dull, grey-looking coal.

## E

**Eclogite**: metamorphic rock of very high-T and high-P characteristics.

**Electromagnetic fields**: VLF, ELF see p58.

**Electronegativity**: see p54.

**Endogeny**: the processes resulting from energy of the Earth's interior.

**Epidote**: $Ca_2(Al,Fe)_3Si_3O_{12}(OH)$, footnote p128.

**Exsolution**: the separation of a homogeneous material into two separate crystalline phases without addition or loss from external sources.

## F

**Felsic**: dominantly composed of feldspar and quartz and containing few dark minerals.

**Fullerenes**: see p287.

**Fusain**: mineral charcoal, light and porous unless mineralized, fibrous, very black-looking, with a silky sheen.

## G

**Gabbro**: a mafic intrusive igneous rock mainly plagioclase and pyroxenes.

**Gastrobleme**: a crater resulting from endogeny but lacking evidence of solid volcanic ejecta.

**Gedrite**: footnote. p102.

**Germanide**: see p52.

**GPa**: billions ($10^9$) of pascals pressure.

**Graben**: a downthrown, linear, crustal block bordered by normal faults.

**Granite**: a plutonic rock in which quartz is prominent in the felsic fraction, and alkali feldspar is 65-90 % of total feldspars.

**Granodiorite**: coarse-grained plutonic rock between quartz diorite and quartz monzonite in composition.

**Granophyre (-ic)** an extrusive igneous porphyritic rock (coarser crystals set in fine-granular matrix).

**Granulite**: a high temperature and pressure metamorphic rock; generally equigranular in appearance and lacking hydrous minerals.

**Graywacke** : footnote p102.

**Graphic (micrographic) granite**: pegmatite in which quartz crystals have grown in runic patterns (resembling writing) in feldspar.

**Greenalite**: an accessory mineral often found in iron formation; $(Fe^{2+},Fe^{3+})_{5-6}Si_4O_{10}(OH)_8$.

**Greenstone**: dark green metamorphosed mafic igneous rock.

**Grossularite garnet**: $Ca_3Al_2(SiO_4)_3$.

## H,I

**Hanging wall**, upper block above an inclined fault surface (lower side is known as footwall).

**Heat of formation**: see p55.

**Hedenbergite**: a black mineral of clinopyrxene group, $CaFeSi_2O_6$.

**Hertz**: unit of frequency in radiation, 1 hertz = 1 cycle per second.

## J,K

**K-feldspar**: potassium feldspar, $KAlSi_3O_8$, orthoclase or microcline.

**Kimberlite**: chaotically-emplaced breccia with Mantle-type mineralogy.

**kb**: thousands of atmospheres of pressure.

**kJ**: thousands of Joules (units of energy or work in the SI system).

**KJ, K/J**: Cretaceous-Jurassic time boundary.

**Kyanite**: $Al_2SiO_5$.

## L

**Lapilli**: fine volcanic fragments, usually 1-64 mm diameter.

**Lamprophyre(-ic)**, a rock with phenocrysts of mafic minerals set in a finer-textured groundmass of the same plus felsic minerals.

**Laterite soil**: footnote p125.

**Leucogranite**: footnote p126.

**Listric faulting**: faulting on a curved plane that is concave upward.

# M

**Mafic:** composed dominantly of ferromagnesian silicate minerals, such as biotite, hornblende, pyroxenes, or olivine: see p33 footnote.

**Metacryst:** (porphyroblast), a large crystal that grew *in situ* during cataclastic metamorphism in finer-grained groundmass.

**Metamorphic gneisses:** crystalline, foliated, and banded rock derived from pre-existing rock.

**Metasomatism:** a term meaning replacement, volume-for-volume.

**Microcline:** a light-colored alkali feldspar, $KAlSi_3O_8$.

**Micrographic:** a texture distinguishable only by microscopy.

**Migmatites:** "mixed rocks" consisting of granitic pods and lenses in dark metamorphic gneisses.

**Mol:** weight in grams equal to atomic weight.

**Muscovite:** a mica, see footnote p125,128.

**Myrmekite (-oid):** See Appendix. (-myrmekite-like).

# O,P,Q

**Omphacite:** sodic-clinopyroxene, $2NaAlSi_2O_6 \cdot nCaMgSi_2O_6$; or, $NaCa_2MgFe^{2+}AlSi_6O_{18}$.(Alderman, 1936)

**Orogeny (-esis):** the formation of mountains.

**PDB:** Peedee Belemnite, a standard to which O and C isotopic abundances are referred.

**Pegmatite:** exceptionally coarse-grained igneous rock with interlocking crystal structure, usually granitic in composition.

**Pelite:** footnote p139.

**Peraluminous granite:** see p112.

**Peridotite:** a family of coarse-grained ultramafic igneous rocks made up of olivine and various accessory minerals.

**Perlite:** volcanic glass with rhyolite composition and spherular form.

**Perovskite:** $CaTiO_3$.

**Perthite:** a K-feldspar containing islands of albite plagioclase lamellae.

**Phenocryst:** large crystal in finer matrix.

**Phyllite:** footnote p125,128.

**Pingo:** an ice diapir.

**Pluton (-ic):** igneous rock intruded at depth.

**Pyrope garnet:** $Mg_3A_{12}(SiO_4)_3$.

**Quasar**: quasi-stellar radio source.

**Quartz diorite**: syn. tonalite; plutonic rock with composition of diorite but having > 20 % quartz.

**Quartz monzonite**: granitic rock with composition of 10-50 % quartz and alkali feldspar 35-50 %.

## R,S

**Restite**: non-melting fraction in partial melting leading to migmatite.

**Roof pendant**: a remnant of earlier terrane cradled within the top of an intrusive igneous body.

**Runic**: graphic, like writing.

**Sialic**: granitic, see p33 footnote.

**Silanes**: (mono-, di-, tri-, etc.): see p52.

**Silicides**: a mineral formed of silicon and a metal; see p52.

**Sillimanite**: $Al_2SiO_5$; a needle-like mineral of metamorphic association.

**Spinel**: an oxide mineral of the form $AB_2O_4$, where A may be $Fe^{2+}$, Mg, Zn, or Mn, and B may be Al, $Fe^{3+}$, or Cr.

**Subsolidus**: "below melting temperature."

## T

**Tectonosphere**: the "Crust," where tectonic adjustment takes place.

**Tephra**: volcanic pyroclastics, generally airfall variety only. .

**Terrane**: a group of geologically-related rocks in a given area.

**Tripoli**: finely divided, light-weight, porous, powdery silica.

**Tonalite**: syn. quartz diorite; plutonic rock with composition of diorite but having > 20 % quartz.

**Trapp** (trap): dark-colored, fine-grained igneous rock.

**Trondhjemite**: alkali-deficient granodiorite.

**Turbidite**: a sedimentary deposit of submarine avalanching.

## U,V,W,X,Y,Z

**Ultramafic**: see p33 footnote.

**Vitrain**: very black-looking, vitreous, conchoidal-fracturing coal that breaks with cubic bias.

**Xenolith**: a foreign inclusion in igneous rock.

**Zeolites**: see p284.